'Climate is, in some respects, highly predictable; yet, in other respects, highly unpredictable. But there is no contradiction. The resolution of this seeming paradox in *Predicting Our Climate Future* leads in turn to a vision for how humankind must respond to this most important problem of all time.'

George Akerlof, Nobel Laureate in Economics, 2001

'A profound yet very accessible guide to climate science, highlighting the significant uncertainties without apology. This book explains clearly why doubt creates a greater and more urgent need to act now to build a better future.'

Trevor Maynard, Executive
Director of Systemic Risks, Cambridge Centre for Risk Studies

'The immense complexity of the climate system raises deep questions about what science can usefully say about the future. David Stainforth navigates philosophical and mathematical questions that could hardly be of greater practical importance. He questions what it is reasonable to ask of climate scientists and his conclusions challenge the way in which science should be conducted in the future.'

Jim Hall, Professor of Climate
and Environmental Risk, University of Oxford

'Is the science settled? Are climate models rubbish? Stainforth's book serves up nuanced answers to big questions in climate science, in an easy conversational style.'

Cameron Hepburn, Professor of Environmental Economics, University of Oxford

'A thoughtful exploration of the foundations and limitations of climate prediction that explains how its chaotic and probabilistic nature leads to deep uncertainty when assessing climate risk.'

Ramalingam Saravanan,
Professor of Atmospheric Sciences, Texas A&M University

'*Predicting Our Climate Future* is an erudite and very personal reflection on climate change, the state of climate science, and their implications for the decisions society needs to take. It should be top of the reading list for scientists, practitioners, and anyone who wants to truly comprehend the challenge of climate prediction.'

Simon Dietz, Professor of
Environmental Policy, London School of Economics and Political Science

Predicting Our Climate Future

Predicting Our Climate Future

*What We Know, What We Don't Know,
And What We Can't Know*

David Stainforth

*Professorial Research Fellow,
Grantham Research Institute on Climate Change and the Environment,
London School of Economics and Political Science
Honorary Professor,
Department of Physics,
University of Warwick*

OXFORD
UNIVERSITY PRESS

OXFORD
UNIVERSITY PRESS

Great Clarendon Street, Oxford, OX2 6DP,
United Kingdom

Oxford University Press is a department of the University of Oxford.
It furthers the University's objective of excellence in research, scholarship,
and education by publishing worldwide. Oxford is a registered trade mark of
Oxford University Press in the UK and in certain other countries

Published in the United States of America by Oxford University Press
198 Madison Avenue, New York, NY 10016, United States of America

British Library Cataloguing in Publication Data

Data available

Library of Congress Control Number: 2023934092

ISBN 9780198812937

DOI: 10.1093/oso/9780198812937.001.0001

Printed in the UK by
Bell & Bain Ltd., Glasgow

Contents

PART 1

CHARACTERISTICS

1
The obvious and the obscure

This book is not about climate change.

Okay, that's not true. This book is about climate change. Almost entirely about climate change but only because climate change reveals some of the most fascinating and exciting challenges in scientific research today. This book is about those challenges. Many of them are new; they are challenges which arise from and reflect the times in which we live. Think about issues such as how we gain new understanding about reality using computer simulations, or how we combine diverse sources of knowledge in a world with unprecedented access to data, and unprecedented access to opinions. Almost all these challenges also apply to other areas of research and other fields where science and society meet, but climate change is perhaps unique in the way so many challenging issues come together in one problem. It's also unique in the way it influences politics and affects how we build our futures. So while this book is about a collection of fundamental and conceptual challenges facing researchers in the twenty-first century, the starting point is to think about climate change. That means beginning by laying down, here in part one, the particular combination of characteristics that make it unique. After that, in part two, come the conceptual and practical challenges they lead to: the fun stuff.

You almost certainly have an opinion about climate change. Most people do. Maybe you think it's the most serious issue facing humanity today. Maybe you think it is overhyped and a minor problem at worst. Maybe you're somewhere in between: it's happening and probably very important but only one of a large number of societal problems and personal issues that you care about, and anyway, there's not that much that you think you personally can do about it.

Whose view is right? Well, it's not up to me to say whose view is right or whose perspective is the 'correct' one because the answer depends on what you value. In any case, I wouldn't want to come down on one side or the other on the first page. If I did that, a large fraction of you would stop reading now because climate change is one of those issues, characteristic of modern society, where many people look for information which supports the view they already hold. You may be in a bookshop or at a friend's house and I'd like you to continue reading at least until you decide to purchase a copy or two.

It is a notable feature of humans that we differ in what we value. In this case, some care passionately about the risk of polar bears or Spanish butterflies going extent— I'm thinking of the Sierra Nevada blue[a]—or about receding ice sheets and threatened coral reefs. For others these things are not so important. Even so, you may not be 'green', you may not care about polar bears, but you might still consider it a bad thing if in thirty years' time you won't be able to get sufficient food or water to sustain you and your friends and family. You might think it a bad thing even if the only consequence of climate change were that you won't be able to get the foods you like best; maybe mangoes or pomegranates or turnips. But then again, if you're already ninety years old you might not be so bothered. Or then again, you might be ninety and care passionately about ecosystems and what we leave for the next generation. Or just care about the lives of younger friends and family. What you care about and how much you care about different things depends on your context, your perspective, your temperament, you. Why do we disagree about climate change? Partly it's simply because we're all different.

But some things don't depend on personal values. Some things are just true; or rather, they are true given a small number of very, very reasonable assumptions that practically all of us can agree on. At this point—just six paragraphs in—I perhaps risk irritating the philosophers among you. How do we know what's true? What does it mean to be true? Isn't everything open to question? We'll come back to philosophers later but let's not worry about them for now. I'm willing to gamble that they'll stick with me for a little while longer, if only to see what other wild statements I might make.

The thing is, there are some things I think I can rely on with a confidence that borders on certainty. If I throw a ball up into the air, it's going to come down. If I switch on the kettle in our kitchen after filling it with water, the water will heat up. These are predictions. Rather general predictions I grant you, but still predictions. Of course there are ways in which these predictions might fail. Maybe a swallow will fly by, grab the ball, and fly off with it. Maybe I'm inside a building and the ball is sticky, so when I throw it up it sticks to the ceiling. More likely, perhaps, is that I'm outside, there are trees around, and the ball gets stuck in some branches. What about the kettle? Well maybe it has broken since I last used it. Or maybe there is a power cut. It might be wise to be specific about some of the ways I can see my prediction going wrong; about some of the assumptions underlying it. But honestly, we all know that the basic and most reliable expectations are that notwithstanding some rather unlikely events, the ball will come down and the water will heat up. To consider what else might happen might be a fun academic exercise, it might be a fun discussion in a pub, a park, or a coffee shop but we all know what is by far the most likely outcome. We **know** how some things behave.

Since we value things differently, we are likely to disagree about the relative importance of climate change to us and our societies, and therefore on the degree of action which is justified. Our differing political perspectives are also likely to lead us to disagree on the most appropriate and effective ways of achieving changes in society, even when we agree that a certain type of change is desirable. It would be tremendously helpful, however, if such discussions and debates could take place on a foundation of what we can all agree is well-known and well-understood. That's to say, things that we know at the level of expecting a ball to come down when you throw it up into the air. So what do we know about climate change? Not what does the latest research science tell us. Not what does the latest economic assessment report say. Not what are the latest model predictions. Not what is the latest statement from some august institute or international body. What do we know with confidence bordering on certainty? How do we know it, and what is the basis for any such confidence?

Answering these questions involves aspects of knowledge which are rather uninteresting. It involves science which is not at the forefront of research. Indeed much of it is pretty basic and even rather boring. It is so 'well-known' and uninteresting that it rarely makes it into even the pared-down, simplified discussions of this subject in the media. This robust knowledge is not what this book is about. Well, not directly. This book is about how what we know sits in a sea of what we don't know: the vast realm of what is still to be understood. This book is about the boundaries of knowledge; the questions we don't yet know how to answer.

There is much we know with confidence about climate change but there are also many questions which we simply don't know how to answer. Yet. These are not just questions to which we don't have the answer but questions for which we don't have a method which we think could give us the answer. We're struggling to even know how to approach some of these questions. They are foundational challenges for science, the structure of science, and for academic study more broadly.

Scientists and other academics disagree strongly and fervently about our level of knowledge on many issues relating to climate change. There is confidence and consensus about some things but disagreement and confusion about others. The climate challenge is about getting a handle on how these fit together and how to push back the boundaries of knowledge on those aspects that are most urgent to understand. Many difficult questions in the study of climate change are as conceptual as those found in other great scientific challenges such as the search for dark matter in the cosmos, the development of a grand unified theory of particles and forces, and an explanation for the origins of consciousness. The core issues in climate change are fascinating in the same way—they go to the core of what we know and what we can know.

This book is about these deep challenges but also about how we identify and use the information we have today. For that reason I am, later on, going to address the basic, really well-understood stuff. In any case, we need to be clear why we are confident about some things in order to grasp why other things are so uncertain and represent such fundamental problems. The opposite is also true—grasping why some things are so uncertain can help us identify other things which we are more confident about.

A serious problem for much climate change discourse is that climate science is not terribly good at separating the solidly known from the seriously questionable. The latest results tend to be presented as if they have the same pedigree as well-tested foundational physics; there is little acknowledgement of the really interesting, limits-of-knowledge questions in the subject. This is a shame because we need the best young minds to be working on them but if they don't hear about them, why should they be interested in studying them? Failure to acknowledge the boundaries of our knowledge is also bad for society because it muddies the waters of policy discussions by allowing what is known confidently to be mixed up with what is still very much open to debate. Discussions that should be based on well-founded, agreed-upon understanding are derailed by confused interpretations of the science and misinformed arguments. In practice, what we know well is quite enough to support well-founded, constructive debates about our response to climate change at the level of communities, nations, or humanity as a whole. But if we aren't clear about the separation between what we know and what we don't know, then such debates—which should be founded on our differing values—get muddled and confused and are unlikely to achieve actions that reflect the values we share.

This lack of clarity in the reliability of different types of climate change information also means that scarce research funds are not well directed because the big research questions are simply not in the limelight.

Of course, if the subject of climate change were solely about pushing back the limits of human understanding or developing exciting new technology such as holographic phones or space tourism, then this wouldn't really matter. Getting it wrong for a few decades or centuries would be no big deal. After all, scientific research has rarely been well-optimized; it rambles along, taking detours and going down blind alleys before eventually finding a way forward. Strategic planning and oversight of the process of fundamental research might be desirable but in the historic scheme of things, it's not that important and is rarely successful. Climate change, however, is time critical. If academic understanding is to be useful it is needed as soon as possible, so optimizing research investments is extremely important.

How though can we tell the difference between what we know with huge confidence, and the latest research results which hindsight may (or may not) reveal as examples of the many blind alleys in the history of human understanding? What is confidently understood, and what is open to serious questioning? Differentiating between the two is the key to both revolutionizing climate research and responding effectively as a global society to the problems of climate change. I hope that by the end of this book you, and I, will be clearer on our own opinions of how to do this.

First things first, though. To begin with, we need to consider what it means to study climate change. The subject is complex and involves many realms of knowledge. It requires us to bring together a wide range of elements which must all be considered together, at the same time. There are, nevertheless, a number of characteristics which frame its study. Part one introduces these characteristics: what is climate change all about and what makes climate prediction peculiarly difficult? Later, in part two, the implications of these characteristics will come to the fore, along with the fascinating, fundamental, and sometimes conceptual challenges they lead to.

2
A problem of prediction

The first characteristic of climate change: Climate predictions

Climate change is intrinsically about predicting the future behaviour of a complex, multi-component system which includes everything from clouds to oceans to ice sheets to trees to soils to ecosystems to economic growth, stock markets, industry, and governance.

'Det er vanskeligt at spaa, især naar det gælder Fremtiden'.
'It's difficult to make predictions, especially about the future.'
Danish Parliament, 1937–1938[a]

The first, and arguably most fundamental, characteristic of climate change is that it is all about prediction. Understanding climate change is not about understanding how our climate system works but about predicting its future behaviour. The two are related, of course, but they are fundamentally quite different tasks and they involve very different challenges. Approaching an understanding of how the climate system works can be achieved a bit at a time but predicting its future behaviour has to be done while considering everything at once.

The phrase 'climate change', as it is commonly used, refers to human-induced climate change and the associated threats to our environments and our societies. The justification to take action on climate change is founded on the basis that we have an understanding of what the future will be like if we don't take action and also what it will be like if we do. It's this understanding that provides us with a choice over which future we prefer. Studying human-induced, sometimes called anthropogenic, climate change is therefore unavoidably related to questions of prediction and the reliability of predictions.

But hold on, I hear you cry (well, I imagine I hear you cry)—there's nothing special about prediction. All scientific research is about prediction. It is intrinsic to the scientific method. Scientists observe some aspect of how the world behaves, come up with a theory to explain it, use the theory to predict the outcome of an experiment, and then do the experiment to gain support for the theory. Prediction is at the very heart of scientific research, of the scientific method. There is nothing special about prediction in science.

This, of course, is true but a prediction of this sort is a prediction in a very different context; it is a far more restricted type of prediction than that associated with climate change. To see why, it is useful to reflect, briefly, on some aspects of the scientific method. Science often doesn't progress in the simple way outlined above, but where it does is in research domains which have the flexibility to choose or even design experiments. In such situations, an experiment is constructed to test a theory or hypothesis. The experiment is designed to focus on the particular aspect of interest and to minimize the influence of extraneous factors. Furthermore—and this is perhaps the most important bit—the experiment should be, if at all possible, repeatable.

Consider, for instance, Boyle's law. In physics, Boyle's law tells us how the pressure of some fixed amount of gas is related to its volume at a constant temperature: it says the pressure times the volume is constant. It is precisely accurate for an idealized approximation of the way the molecules of a gas interact—something called an 'ideal gas'—but most of the time it also does a very good job at describing real gases as well. Back in the seventeenth century, Robert Boyle tested it in the laboratory multiple times and found it to be the case. He was building on the work of Richard Towneley and Henry Power. Many others followed who also demonstrated the law's efficacy and now secondary school children can do the experiment[b] and they too demonstrate its reliability—well they do if they carry out the experiment carefully with good-enough equipment. The experiment is repeatable, it has been repeated, and furthermore it is now well-understood how the law comes about. So the law is demonstrably accurate and is also physically understood. Furthermore, it is also understood under what conditions it breaks down and becomes a poor guide to the behaviour of a real gas, for instance close to the point that the gas condenses to a liquid or gets sufficiently hot that it breaks down into a plasma. This is very much trustworthy science.

Take another area of physics—particle physics. The Large Hadron Collider in Switzerland was built to test predictions first made in the 1960s of the existence of the Higgs boson, an elementary particle associated with explanations of why matter has mass. It seems to have done so. In 2012 a new particle was detected that was described as being 'consistent with' the Higgs boson. Over the following years the experiment was repeated and more details were gathered such that the new particle is now simply described as the Higgs boson.

How about in medicine? In drug development one can test the behaviour of anti-cancer drugs by multiple trials with patients suffering from the relevant cancer. Again this is repeatable.

These are all examples of how building up understanding and faith in scientific conclusions involves repeating repeatable experiments. The concept of 'repeatable experiments' is useful because some experiments are not repeatable, some are repeatable and have been repeated, and some are repeatable in theory but may not have been repeated in practice. The difference between these situations can be important for separating what we know, from what we could know, from what we cannot know.

The problems of undertaking science without repeatability will crop up as soon as the next two chapters, but it is not just repeatability that is important. The experiments just described are also designed to maximize the study of the subject or facet of interest. That's to say the cancer drug is not given randomly to members of the population but to those suffering from cancer. The Large Hadron Collider collisions are designed to collide protons with sufficient energy to potentially generate the Higgs boson given the proposed theory. There would have been little point in building the Large Hadron Collider a bit smaller, thereby saving a lot of money, because it wouldn't have been able to achieve the necessary energy in the colliding particles to even potentially find the Higgs boson. It wouldn't have been able to test the theory. These experiments are designed to focus on the issue of interest and this too is a valuable characteristic.

Sometimes experimentation—and in particular repeatable, focused experiments of the sort just described—are either not possible or are unethical. This is often the case in astrophysics, cosmology, solar physics, and some aspects of evolutionary studies and psychology, for instance. A hypothesis for how stars behave is not something we can create versions of stars to test. We can design experiments to observe particular aspects of the information coming from the stars and we can choose to direct our observing equipment towards particular parts of space, but we are at the whim of nature regarding what is going on in the universe and we are constrained in what we can observe.

The study of the climate system, that is 'the climate system' not 'climate change', is somewhat like this. We can't design, run, and observe repeatable experiments on climate but it is possible to focus our attention on the parts of the system which we're interested in. In practice this can be fairly easy because the interesting features often come around again and again: think hurricanes in the Carribbean, heavy rainfall events in central Europe, the way weather systems track across the North Atlantic, or the behaviour of the Indian summer monsoon, for example. These can be observed, analysed, and theories developed to explain their behaviour. Furthermore, one can use those theories to predict how subsequent instances of these events will behave and new observations can be taken to test those theories. This provides an element of repeatability. It's not as good as a laboratory experiment where one can focus specifically on the aspect of interest and to a large extent exclude extraneous influences. It also doesn't provide the possibility to make however many repeated observations we like under almost exactly the same conditions. Nevertheless, it is a good source of information for understanding these aspects of the climate system.

To test a theory more thoroughly though, one might want experiments which look for consequences which have not previously been observed: new phenomena predicted by the theory. If a theory makes such predictions and experiments subsequently conform with the theory, then it gives us greater confidence that the theory is not just describing what we've seen but capturing some fundamental underlying behaviour. Such experiments are perhaps easiest to achieve in a laboratory setting but they are not impossible outside the laboratory. There are many examples in the history of science where theory has predicted previously unseen behaviour which has later been looked for and observed. The Eddington experiment of 1919, for instance, utilized a solar eclipse to measure the deflection of starlight by the sun, thus testing and confirming predictions made earlier by Einstein's general theory of relativity.

Observations of this nature would presumably please the philosopher of science Karl Popper (1902–1994), who argued that scientific theories should be falsifiable. If a theory predicts behaviour which has not previously been observed but which can be observed, then it is falsifiable because subsequent observations could show it to be false. If subsequent observations of this behaviour actually conform with the theory, then our trust in the theory grows.

This is the wider scientific context for the study of the climate system and the climate system is studiable using these conventional scientific protocols. It has been studied this way for centuries and as a result you can now find fascinating explanations of everything from cloud types to tornado formation to the behaviour of weather fronts and storms.

But climate change, specifically climate change in the twentieth and twenty-first centuries, is a very different kettle of fish (Figure 2.1). The goal is not to understand isolated parts of the climate system or to test our theories. It's not, at least for the most part, about prediction as part of the scientific process of building better understanding. No, prediction within the study of climate change is principally about predicting the future behaviour of the whole thing—the whole, huge, complex, interacting muddle of physical components from clouds to ocean ecosystems, from ice sheets to rain forests, from hurricanes to the water and carbon bound up in soils. It's about predicting how the whole physical climate system will behave in the future. And that's before we start to consider non-physical aspects: the role of humans, questions of policy, and the interactions with economics.

Now that's quite a task. Even in the context of the whole history of science it sounds tricky. In science we usually try to focus on some particular aspect, while excluding all other factors as much as possible. Here we are throwing everything, or almost everything, into the fish kettle.

Our goal is to describe the future. We want to know what might actually happen. Even if we're only interested in one part of the climate system, such as heatwaves in California or water flows in the Ganges, we want to know how that aspect will change in the context of all the other changes: we can't treat it in isolation. We

Figure 2.1 A fish kettle.

want to predict the future dependent on our actions because climate change is not just of academic interest, it provides the information we are using to construct our futures. Climate change research is used to guide policy choices and support preparations for anticipated changes and threats. Ultimately, therefore, climate change—not just the science part but also climate change policy, economics, social impacts, and opinions—comes down to climate prediction. The really big questions are all associated with either trying to make climate predictions, or trying to use climate predictions.

At this point it's worth reflecting on the old adage: 'It is difficult to make predictions, especially about the future'. This is a phrase ascribed to Danish origins and sometimes attributed to the physicist Niels Bohr, although there doesn't seem to be any evidence that he coined it. It's a snappy phrase and exudes a healthy sense of scepticism, but like so many soundbites it hides as much as it illuminates. It is frequently used to imply that predictions of the future are all fundamentally untrustworthy. The problem is that it lumps all types of predictions together whereas in practice some predictions are much easier than others, even among those made about the future. We can often predict some things but not others. For instance, today[c] I can, with reasonably high confidence, predict that there will be a US presidential election in 2028 but I have absolutely no idea who will win or even who will run. The point is that some things are confidently predictable while others are not, and the central issue in understanding climate change is separating the two. This is essentially a rephrasing of the message of the first chapter that we need to be clearer about the separation between what is confidently understood and what is open to serious questioning.

The first characteristic of climate change then, is simply that it is all about prediction; prediction of a complex system with many interacting components from the atmosphere and oceans to ecosystems and social systems. Consideration of such predictions leads us to the next three characteristics. The first of these is that we're entering new territory; we're going beyond anything we've seen or experienced before. The second is that there'll only ever be one twenty-first century and there is inherent, unavoidable uncertainty about what it will look like. The third is that the climate prediction task involves going beyond the implications of 'the butterfly effect' to ask ourselves what a computer model can and cannot tell us about reality. This third characteristic may sound a little mysterious and esoteric but don't worry, its meaning and importance will become clear in a couple of chapters' time.

3
Going beyond what we've seen

The second characteristic of climate change: Extrapolation

Predictions of climate change in the twenty-first century are a matter of extrapolation, of predicting behaviour under conditions we've never seen before.

Later I'll come to the difficulties in defining quite what we mean by 'climate'—an interesting exercise in itself—but for the moment, consider the second half of the phrase 'climate change', the bit about change. Climate change is intrinsically about change. It's inherent in the concept. The reason the subject exists is founded on the idea that the present is not like the past and the future will not be like the present.

It may seem odd to even raise such a trivial point. Surely all prediction is about predicting change—otherwise it would be fairly boring. Nobody would give me a prize for predicting that my sitting room walls will be green tomorrow. They've been green every day for the last seven years at least. It's not a very interesting prediction. Predicting that something will stay the same is often rather boring. But not always. A prediction that it will rain tomorrow might be interesting even if it is raining today because I know that the weather can change from day to day. Similarly predicting that a stock market will rise tomorrow is something you might be interested in whether or not it rose today. Nevertheless, aren't predictions only interesting if things are, or at least might reasonably be expected to be, changing?

The answer is yes, but what's important here is that things can change while also staying the same. This concept is essential to grasping the fundamental difference between weather prediction and climate prediction. The thing is that the probabilities of different outcomes can stay the same even while the specific outcomes vary. A familiar example of this is the throw of a dice. If I throw a standard six-sided dice, then the number I get is uncertain but the probability of each outcome is fixed and unchanging. If it is a fair, unloaded dice then there is a one in six chance of getting a one each time I throw it. The same is true for a two, a three, a four, a five, and a six. This is a situation where there is randomness, variability, and change within a setup which is unchanging because the probabilities stay the same.

It's similar with weather. If the climate isn't changing, then you might expect the probability of snow in, say, Oxford, England, on a day in March, to not change from one year to the next. You might deduce this probability by looking back over many years and calculating the fraction of March days that have seen snow in Oxford.

A possible approach to weather forecasting would then be to simply give this probability as the forecast for every March day in Oxford. A more comprehensive forecast would report the probabilities for different types and intensities of snow, and for many different aspects of weather—rain, hail, wind, heat, humidity, etc. These probabilities are the 'background' probabilities, the default expectations; you might call them the March climate of Oxford. They can be valuable in many ways, although they are far less easy to measure than the probability of different dice roles because many types of weather don't come up very often. Even getting the dice probabilities can take a large number of throws (see Figure 3.1). We would have to have very many years of data to be able to confidently deduce the probabilities of all the different possible weather conditions in a particular place.

Let's not worry about that for now though: let's set aside the practical difficulties of finding these background probabilities because typically we expect a weather forecast to do better than this anyway. We want more than just the historic probabilities and nowadays, most of the time, weather forecasts are indeed much better than this. They are better because knowing the state of the weather today gives us more knowledge, more information which can be used to refine the probabilities for what will happen tomorrow. Knowing that it snowed today means that the chance of snow tomorrow is greater because snowy days tend to come in runs. Furthermore, knowing

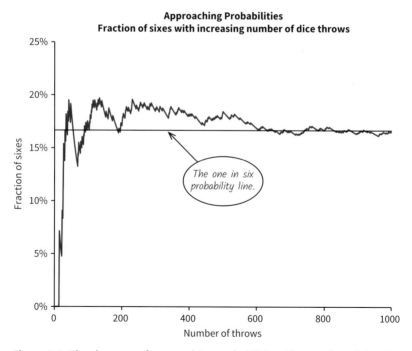

Figure 3.1 The slowness of approaching probabilities. The number of sixes in a particular sequence of (digital) dice throws only gradually converges to the one in six probability we expect.

the conditions in the atmosphere today across a wide area—from the North Atlantic to Scandinavia if we're interested in Oxford, England—can help refine the probabilities of different outcomes for tomorrow's weather still further. In a similar way, once the dice has been thrown and is bouncing on the table, we could potentially have more knowledge about the outcome than the simple one in six probability. This turns out to be extremely difficult in practice but in principle it would be possible. In both cases the underlying processes which govern what happens are unchanging. As a result the probabilities for the next dice roll stay the same and in an unchanging climate this is also somewhat true for the probability of snow on a March day in Oxford from one year to the next. Put another way, the chances of snow in Oxford in March next year would remain largely unchanged even if this year happens to be particularly cold and snowy, just like the chance of rolling a six on your next go is unaffected by you rolling a six this go. What we have here are predictions of particular outcomes within an overall unchanging setup. This state of affairs is called **a stationary system**. There can be lots of change from day to day, year to year, even decade to decade, or longer, but in one very important sense nothing is changing at all.

Weather forecasting is intrinsically about improving on the background probabilities based on what is happening now, to give better information about tomorrow and the next few days. It's like going to meet someone you've never met before and wondering how tall they might be. Maybe it's a blind date or a job interview. I don't know why you would be interested in their height. Maybe you particularly like tall or short or average-sized people. Whatever the reason, your initial expectation might be based on the distribution of heights across the national population (an average of about 169 cm in England[a]). Then you learn that the individual you're meeting is called Brian and you jump to the conclusion this person is most likely to be male which leads you to refine your expectation upwards (an average of 175 cm in England). Then somebody tells you that this person plays basketball as a hobby and you refine your expectation upwards again; maybe he is closer to 200 cm. The overall distribution of heights across the national population hasn't changed but your expectation about the particular example from that distribution has changed a lot.

The crucial difference with climate **change** predictions is that the background probabilities themselves vary. It's not a stationary system. The probabilities of different weather events are themselves changing. The probability of rain or snow[1] around Oxford on a day in March in the observational record of the last 71[2] years is 39%[b] but if climate is changing there is no reason to think that this will be the probability this year,[c] let alone in the future.

Imagine you have a dice and have thrown it thousands of times before. Maybe you noted the outcome of each throw. Why? Well maybe you're keen on checking your dice is fair or maybe you just find it relaxing. Whatever the reason, it turns out that up to now the dice has been fair. There is a one in six chance of each number from one to six. But today you throw it and the probabilities have changed—some numbers are

---

[1] Data for rain and snow combined is easier for me to get hold of than for snow alone.
[2] The particular data set I used here runs from 1950 to 2020 inclusive: seventy-one years.

more likely to come up than others. Or maybe there has been a more radical change. Maybe the numbers on each side have changed; they are now one, three, four, six, seven, and eight. Maybe even the number of sides has changed; today it is a ten-sided dice or twenty-sided dice. This is climate change. You're not just unsure what specific number will come up on a particular throw; you're also unsure what the probabilities of different outcomes will be. They may have changed from what you've seen in the past, perhaps dramatically. There may even be possibilities which you've never seen before.

Of course with climate change we're not starting from scratch today; we're not replacing one dice with another one overnight. No, with climate change, changes happen over time—although to say they are *gradual* gives a misleading impression. As a consequence the probabilities of different types of weather observed over the last seventy-one years don't represent the probabilities now because they have been changing over that time.

Nevertheless, observations can help us paint a picture of changes in climate over the past century and these are one source of information about how it **might** change in the future. We also have information from much further back: there are all sorts of brilliant techniques—some involving drilling fantastically long tubes of ice out of ice sheets in Greenland and Antarctica—which provide information on paleo-climate going back hundreds of thousands of years. These provide background information regarding just how much the climate has varied in the past, and how such variations might relate to atmospheric carbon dioxide concentrations. We do therefore have useful observations but they are observations of the climate system in a different state to that of today or that of the next 100 years. We know this because the planet has not seen the concentrations of greenhouse gases in the atmosphere that we have today at any point in the last million years. A fundamental aspect of climate prediction on timescales of decades to centuries, is therefore that the probabilities for various types of climate behaviour in the future will be different from any we have seen in the past. The way they change—in response to future and present-day levels of atmospheric greenhouse gas concentrations—is unobserved. We have never seen similar circumstances before. There have never been similar circumstances before. The world has never been in the same state it is in today, let alone with greenhouse gases changing in the same way. Climate change is taking us into a never-before-experienced situation. Forget about repeatability—we don't even have one observation of the behaviour we're trying to predict in the future. Climate prediction is therefore a task of extrapolation. It's about predicting how the probabilities of different types of weather will change in circumstances never before experienced. How we go about that task, what assumptions we make and how to test whether our methods work, are questions which challenge the approaches of modern science to their extremes.

The second intrinsic characteristic of climate change is therefore simply that climate prediction is about predicting what has never been seen before. It's not about predicting behaviour within the context of previously observed events as you would be doing if forecasting the weather, or the path of a hurricane, or the outcome of a dice roll, or the time of train arrivals at London Euston, or the efficacy of a new medical drug which has already been through medical trials. It's not about predicting the outcome within a known set of probabilities. It's not even about predicting how the probabilities will change within a setup or experiment for which we have some, if limited, data regarding the change we're interested in. It's about extrapolating to something entirely new.

This characteristic is important for many reasons. It raises questions about how we use observations: to what extent does what we have seen in the past inform us about how things will change in the future? How do we use fundamental understanding of the physical world? How do we know what is reliable?

Think again about throwing a ball in the air. I may expect the ball to come down after I throw it up but that prediction could be based simply on experience, on past observations. That's fine and in this case it is reliable but what about predicting what would happen if I did the same thing on Mars or on Halley's comet or on the International Space Station? I have never been there so I don't have experience; I don't have observations. Nobody has ever done it on Mars or Halley's comet so it is a fundamental fact that no observations exist. If your perspective on the world is heavily based on what has been experienced before, your instinct might tell you that we can't make such a prediction but in this case, fundamental physical principles come to the rescue. We do have sufficient knowledge—reliable knowledge based on a huge body of evidence—to make such predictions. Indeed I'm confident we can make good predictions of its behaviour including details of how it will accelerate and decelerate and of the conditions under which I should expect it to come down again rather than continuing up into space. Such predictions are predictions of behaviour which is different to any I've experienced or observed on earth, but they are nonetheless of incredible reliability because they are based on robust understanding of the physical world. Reliable, extrapolatory predictions are by no means impossible—sometimes, they are quite easy. The key question though, is how we judge when and how we can make reliable predictions of situations we've never observed. What constitutes reliable evidence? What sort of predictions can we trust and when should we be sceptical?

More on this later but for the moment, imagine that we did indeed have reliable predictions of future climate, of the probabilities of different types of weather in the future. In that case, how would we use them?

4
The one-shot bet

The third characteristic of climate change: The one-shot bet

In responding to climate change we only get to choose what type of dice we're going to throw, not what the outcome of the roll is. And we only get one throw. Put another way, climate predictions can't tell us what **will** happen if we do or we don't take certain actions; they can only, at best, give us the probabilities for different outcomes. And in reality, there will only be one outcome. Climate change is a one-shot bet.

I'm not a betting person. The prospect of losing something I value makes me stressed. I don't find it enjoyable. As a consequence I have a tendency to be risk averse. Only a tendency mind you—my attitude to risk varies massively from one situation to the next: I get frustrated when dealing with banks or financial advisors who want to categorize me as simply seeking low, medium, or high-risk products. I see their categorizations as often misrepresentative of me and sometimes of their savings and investments options.

On a similar point but a trivial level, I know that I should be more relaxed about getting to the train station in time to catch my train but the hassle and cost of missing it leads me to always leave lots of spare time. I don't recall ever missing a long-distance train, I've never missed a flight, and I've only once missed a boat. Some might say this is not the optimal way of running my life, that I should actively plan my journeys to miss some fraction, maybe one in twenty, because overall I would gain by having more time to do other things that are useful or enjoyable.[a] I wouldn't waste so much time sitting in stations even though occasionally I would miss my train. The idea behind such a perspective is to balance the cost (not just financial) of missing my train, with the cost (perhaps entirely non-financial) of reducing the chances of doing so. This concept of balancing the chance of some negative consequence against some lesser but perhaps more certain inconvenience, is not unrelated to approaches in the economics of climate change. But sticking, for the moment, with public transport, I personally prefer to reduce the risk of missing my train to a pretty low level and instead use the time at the station to read a book, write, or listen to some music. In this situation I'm risk averse.

On the other hand, I love games which have an element of chance, of randomness. A mixture of control and luck in a game is something I find enjoyable. Card games often provide this. Cribbage and bridge are fun but poker illustrates it best. The only problem with poker is that I don't enjoy the possibility of losing significant amounts of money. Or indeed, much money at all. I'm quite happy to play for Monopoly money or for nominal chips. I enjoy the game itself, not the thrill of potentially losing or winning something of value to me. I love playing games of risk but there are many situations in which I prefer to avoid real-world risks. That's just me. And in any case, it is a generalization that only applies to some aspects of my life. People are complicated.

Climate change is, in many ways, a matter of risk management. It has elements of control (the choices we make about greenhouse gas emissions and how we develop our societies) and of chance (we only have limited knowledge about the consequences of our choices due to our limited understanding of climate combined with a degree of unavoidable randomness). However, climate change involves real-world damages, benefits, and investments so it is not just a game but also a gamble. We are playing the game whether we like it or not and we are playing with real money—real-world risks—whether we like it or not. Understanding how all the controllable and chance elements interact is the basis for building informed, coherent opinions about how we should respond to the issue.

In making decisions about the things we can potentially control (e.g. emissions of greenhouse gases), we want to know the consequences of our choices. That means we want climate predictions for each of the different choices we might make. The predictions are therefore **conditional predictions**: predictions that are dependent on an assumed set of actions and future developments. The assumptions might be about the success of a particular international agreement to reduce greenhouse gas emissions, or about the development of commercially viable fusion power, technology to extract carbon dioxide from the atmosphere, or agricultural practices that obviate the need for nitrogen fertilizers that lead to the release of nitrous oxide,[b] a greenhouse gas much stronger than carbon dioxide.

The predictions are conditional but we still need to consider what a useful and reliable one would look like. For a start it might include details of many different aspects of climate, from global average temperature to local seasonal rainfall and changes to ecosystems, from agricultural production to flood risks. It might also include changes in economic activity and evaluations of the social consequences, the impact on jobs, the price of rice or concrete, or the excess lives lost due to heat waves. But let's not get distracted by all this complexity. Just consider one aspect. It could be any one. How about the twenty-year average summer temperature in Chicago? That is, the temperature in Chicago averaged over all the summer months (June, July, and August) over two decades. This may seem like a peculiar choice but this is the type of information that tends to be provided by the scientific community. It is close to the science, if not necessarily close to what most of us actually care about.

Imagine that we are predicting the average summer Chicago temperature for the period 2041 to 2060, and that the prediction is based on some assumed future greenhouse gas emissions. It is a conditional prediction of what would happen **if** mankind takes some course of action regarding its use of fossil fuels. The assumptions might imply that we do nothing to curb emissions or that we do a lot, but once we've made those assumptions the prediction itself is not uncertain because of uncertainty in the future actions of mankind. There is no uncertainty there. Nevertheless, the climate prediction for average Chicago temperature, like one for any other quantity, is still uncertain. This uncertainty has various sources that we'll come to later. What's important here is that the best possible climate prediction would encapsulate today's best knowledge of all the remaining uncertainties beyond the uncertainties in the actions of mankind, and it would provide probabilities for a collection of different possible outcomes. That's to say, the prediction would be a probabilistic prediction. It would be presented as a **distribution of probabilities** for different outcomes in the same way as one might make a prediction of the outcome of the roll of a dice or the height of a randomly chosen male over the age of eighteen from Australia (Figure 4.1).

How one produces climate prediction probabilities need not concern us for now—it is the subject of later chapters—but it is worth noting here one important aspect of these probability distributions: the ones available from the scientific community today[d] tend to give non-negligible probabilities to quite a wide range of outcomes. That's to say, a wide variety of outcomes are considered plausible. The 2021 report of the Intergovernmental Panel on Climate Change (IPCC) provided conditional predictions (they call them projections) for global average temperature change based on assumptions for future atmospheric greenhouse gas concentrations. For all but the very lowest increases in concentrations, the predictions have a **range** of at least 0.9°C for the mid-twenty-first century and at least 1.1°C for the end of the century.[e] For instance, based on an assumption of substantial reductions in human

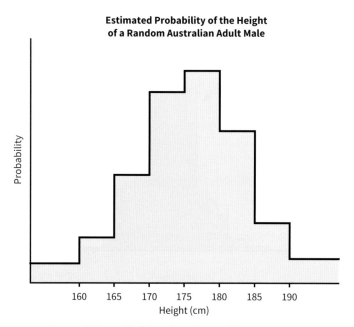

**Estimated Probability of the Height
of a Random Australian Adult Male**

Figure 4.1 An estimate of the probability distribution for the height of Australian adult males.[c]

greenhouse gas emissions they suggest a range of warming of between 1.3°C and 2.2°C for mid-century and between 1.3°C and 2.4°C for the last part of the century, each by comparison with the late nineteenth century. With little or no effort to reduce emissions the numbers go up to 1.9°C to 3.0°C for mid-century and 3.3°C to 5.7°C for the end of the century. These ranges represent the IPCC's assessment of where more than 90%, nine-tenths, of the probability is. They allow for up to a one in ten chance, a 10% probability, that reality could be outside these ranges for each scenario.

Predictions of local and seasonal temperature changes are typically much more uncertain than those for global average temperature, so a 1.1°C uncertainty at the global scale is likely to translate into a much greater uncertainty for summer time in Chicago.

Now maybe a degree or two doesn't sound like much to you but the last time the global average temperature was as high as it is today, which is about 1.1°C higher than the late nineteenth century, was more than 110,000 years ago—before the last ice age.[f] At that time sea levels were 6 m or more higher than today,[g] forests extended further north into regions that are now tundra, and there were hippos in England and Germany. A few degrees difference in the global average temperature could mean a very different world, so we should consider **an uncertainty** of a degree or more to be very significant indeed. At local scales a very basic calculation (Figure 4.2) suggests that a 1.1°C increase in summer temperatures in Chicago could be associated with an 18% increase in the number of days above 28°C (from 54% to 64% of days),

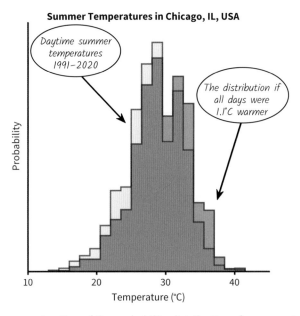

Figure 4.2 An approximation of the probability distribution of summer daytime temperatures in Chicago (based on data from 1991 to 2020) along with the same distribution shifted by 1.1°C.[i]

Summer is defined as June, July, and August.

and a 140% increase in the number of days above $35°C^h$ (from 2.5% to 6% of days). Again, the message is that a degree or so of uncertainty matters. In practice, we expect uncertainties at local scales to be much greater than at global scales so the relevant uncertainties for Chicago associated with a 1.1°C uncertainty in global temperature, is likely to be much bigger than these numbers.

The point here is simply that at all scales—global, regional, and local—the **range** of outcomes with non-negligible probabilities is not tiny. That's to say, the uncertainty in our Chicago prediction will not be irrelevantly or uninterestingly small but rather will include outcomes that would impact Chicago in very different ways. This is important. It tells us that trying to tie down the probabilities is not just an academic exercise. The uncertainties they encapsulate represent very different outcomes that could really matter for things we care about, and could impact how we behave and prepare for future changes. This is why I said that climate change is a matter of how we respond to risk.

Our best possible climate predictions (well, our best **potentially** possible climate predictions) will therefore inevitably consist of distributions of probabilities. This doesn't make these predictions strange and unusual beasts though. Predictions in terms of probability distributions are found all over the place and are often very useful. If, rather than Chicago temperatures in the 2050s, we were predicting the number of hurricanes which will make landfall during the next hurricane season, then we could also provide it in terms of probabilities. Such probabilities would be useful for disaster preparedness and relief agencies, as well as for insurance and reinsurance companies who could potentially use the predictions to set their premiums to reflect the payouts they might have to make. If we were predicting who will be next year's winners at Wimbledon, then probabilities for different players could be useful to bookies in setting the odds they're willing to give or to gamblers in laying bets. If we were predicting the number of people going into central London for New Year's Eve celebrations, then the probabilities might be useful for putting in place sufficient transport and in ensuring emergency cover for accidents and security, and so on. If we were predicting whether a particular stock or stock market will rise or fall in the next year, hour, or millisecond then they could be useful to an investor. If I were predicting the time it takes me to cycle to the station, then the probabilities might be useful to me in optimizing how long I spend waiting at the station for my train. All these predictions could be made probabilistically and there are ways in which the probabilities of different outcomes could be used to optimize the use of resources and investments.

However, these examples of using probability predictions have one aspect in common which is not found in climate change: they all involve a repeated situation. Insurance companies set new premiums each year, gamblers tend to make multiple bets, public celebrations in central London come around regularly, investors make repeated investments, and I cycle to the station several times a week. If the prediction—the probabilities of different outcomes—is accurate, then we could calculate the optimal insurance premium, the best bet, the best use of resources, the optimal investment strategy, and the best time for me to leave my house. Any particular outcome, however, will not be like the prediction: it will not be a distribution. The outcome on any particular occasion is a fixed number of hurricane landfalls, a particular player at Wimbledon, a specific gain or loss on an investment, and a certain number of minutes on my bike. So the insurer, investor, and bookie will make or lose money, and I will catch or miss my train. The outcome is a fixed result not a probability distribution but that's okay because the situation is repeated. By using the predictions repeatedly, we can achieve the optimal outcome **on average**. Sometimes we win, sometimes we lose, but if we use the prediction often enough then a good probability prediction tells us the best way of making our choices given our preferences. If the prediction has the probabilities correct, then over time it tells us the best way to use our resources, whether that be money, time, equipment, or whatever.

This works pretty well for my journey to the station and potentially so in the other situations, although practical, organizational, and value issues often confound the belief that this approach provides a simple solution in many societal situations: understanding the maths isn't necessarily the only issue in solving the problem, particularly when humans are involved. With climate change, however, this approach collapses entirely. That's because climate change is like having a probabilistic prediction but acting on it only once; it's like a one-shot bet. Like one bet on one person to win one Wimbledon. Like one investment over a fixed time period. Like making only one trip to the station.

If we had climate predictions that instead of probabilities gave a single precise picture of what the future would look like under each of a number of options for future greenhouse gas emissions, then making the choice between them would be simpler; it would 'simply' be a matter of personal values and priorities. Balancing these values across people and nations would still be difficult, but hey—let's deal with one thing at a time. It would be the same 'simple' situation if the predictions were probability distributions that were so narrow that it made no practical difference—but as we've seen earlier, the uncertainties are not small, so this isn't the case.

There would also be a clear route to the choice of the best policy option, even with large uncertainties, if we had lots of copies of the earth and were looking to choose one set of policies to apply to all of them. With probabilistic predictions and many copies of the earth in its current state, we could use the predictions to choose the best policy option **on average**. That would be complicated because it would involve putting numbers on how good or bad we consider the different outcomes to be. This would take us into the realms of economics and ethics. Nevertheless, conceptually it could be done. Any other than the best option would be worse because, while even in the best option the outcome in some worlds would be better than in others, overall, any other policy option would represent a balance where there was more 'badness' or less 'goodness' across all the copies of the earth. This can be tricky to get your head around. It sounds obscure but it's just like the examples above. If I make a thousand investments in a year, I might lose money on some, I might lose a lot of money on a few, but if I make enough on the others then overall I'm happy. If I'm managing 1000 earths then, setting aside some very serious hypothetical ethical issues, I might feel the same way. With the best policy, a few earths might suffer badly, but most would do okay and some would do really well. In that case I know how to use the probabilistic climate predictions to choose the best policy options.

But we don't have a thousand earths. We have one. If we consider the potential consequences of increasing atmospheric greenhouse gases to be important, then we have only one chance to take action. We don't get to run the twenty-first century many times. The actions we have taken in recent decades and those we take in the next one or two will define the probability distribution of outcomes that describe how the climate could change this century, as well as constraining what happens beyond. Reality, however, will turn out to be just one of those possibilities—like one roll of the dice. Other possibilities would be real possibilities; they wouldn't necessarily be wrong or bad predictions, but there will only be one outcome. Our choice, given good predictions of the probability distributions for different options, is which distribution we want the one outcome to come from. Whether we choose to roll a standard six-sided dice or a loaded ten-sided dice—that's the kind of choice we have. What numbers are on each type of dice, and how they are loaded, is what we look to climate science to tell us, but that's the best information we can hope for. There will inevitably still be uncertainty and that means it is fundamentally important what attitudes we have to planetary and societal risk. How much do we value our societies and our current way of life? What level of risk are we willing to take with them? It's a question of risk, but not in a familiar way because we only get one go. This is a fundamentally different type of risk-based decision to the examples above or to what we would do with 1000 earths.

Climate policy gives us the power to shift the probabilities and maybe rule out some particularly good or particularly bad outcomes. But we can't choose the outcome we want, only (at best) what distribution it comes from. Consequently, deciding what actions we will take involves considering the risks we are willing to accept. Of course, even talking about 'deciding what actions we will take' assumes we have the potential to act as a global society towards some common goal. Let's just accept that for now. It may not be the case but first things first. If we don't understand the type of decision we're trying to make, and the type of information we have, then we have no good basis for understanding what we individually, or as communities, think is best. In that situation the idea of a common goal is itself nebulous. Accepting that we are choosing not outcomes but what dice to roll is therefore important for climate policy even if the practical global implementation of policy faces many additional challenges. Understanding what science can tell us about what is on each dice and how they might be loaded is therefore central to the climate change debate and is why we need better climate science. Also central to the climate change debate is how we respond to one-shot-bet risks of this nature—something that is little studied.

5
From chaos to pandemonium

> ## The fourth characteristic of climate change: Nonlinearity
>
> You may not be familiar with the concept of 'nonlinearity' but don't worry, I'll explain its key features in the next few pages. It is, however, tremendously important in the climate change story because it provides the context for how we might go about predicting future climate. Indeed it raises deep challenges and fundamental conceptual questions regarding what predictions are possible and what the role of computer models should be. It pulls the rug out from under the most obvious methods and requires us to think much more deeply about exactly what we are trying to do.

How scientific institutions generate the probability distributions that are nowadays taken to represent climate predictions is dealt with in the next chapter. There is, though, one more generic feature to consider first: one which gives rise to many of the biggest challenges in making and using such predictions. The previous two characteristics—extrapolation and the one-shot bet—relate to the kind of predictions we're making. This one is different. It is a generic feature of meteorology, of climate, and indeed of a very wide variety of mathematical and physical objects and collections of objects. In the business we refer to such objects and collections as 'systems': mathematical and physical systems such as 'the climate system', 'the economic system', 'an ecosystem', or a collection of equations designed to represent them—even just a collection of equations conjured up by mathematicians to be interesting. The feature we're interested in is widely studied. There are entire conferences and multiple journals devoted to it, but nevertheless the task of understanding its implications for climate predictions is very much in its infancy. The feature is the broad mathematical concept of nonlinearity and its close cousin, chaos.

The key aspect of nonlinear systems—well the one we're initially interested in here—is the potential for outcomes to be extremely sensitive to where you start from, the conditions from which you are making a prediction, the so-called initial conditions. The forecast of next week's weather, for example, can be very sensitive to the details you assume about the weather today. This sensitivity is commonly referred to as the butterfly effect.

The butterfly effect is the epitome of uncertainty. The first butterfly reference in this context appeared in a talk given by Edward Lorenz at the 1972 meeting of the American Association for the Advancement of Science. The talk was entitled

'Does the flap of a butterfly's wings in Brazil set off a tornado in Texas'. That the flap of a butterfly's wings in Brazil could even potentially influence when and where there is a tornado at some future point in time pretty much captures the concept. The phrase encapsulates the idea that even if we understand perfectly what the rules are that govern how something behaves, it might still be impossible to predict what it will be doing in the future—at least beyond a certain period of time ahead. The finest details of the state of a system now affect how it will behave in the future. Uncertainty in its current state, however small that uncertainty may be, can eventually have a large impact on what happens. That part about 'however small the uncertainty may be' is 'like, duh, important'.[1]

Contrast this with throwing a ball in the air and catching it. The time it takes for the ball to return to your hand depends on how hard you threw the ball: how fast it was going when it left your hand—the initial conditions. But it depends on this in a very predictable way. If you throw it a bit harder, it will take a bit longer. If you throw it a bit harder still, it'll take a bit longer still. A little extra effort doesn't suddenly make a huge difference. Rather, the time it takes to return to your hand is proportional to how hard you threw it. (Well, almost.) Such a situation is called linear (well, almost linear), and a great characteristic of linear systems is that they are predictable. Even if you don't know the starting conditions exactly, from a good estimate of them you can make a good estimate of what will happen. Furthermore, from an estimate of how wrong your knowledge of the initial state might be, it's quite easy to make a reliable estimate of how wrong your prediction might be. The prediction error is proportional to the error in your initial conditions. Knowing roughly how hard you threw the ball is enough to know roughly when it will return to your hand. Here, everything is clear and we can make confident predictions. We don't need to worry about extrapolation if we think that what we are predicting is linear and going to continue to be linear because in this situation, the past provides a good constraint on the future. Throwing the ball gently tells us about throwing the ball hard. If climate change were linear, then climate change in the twentieth century would give us good information about climate change in the twenty-first century, even though the greenhouse gas concentrations will be much higher.

Something which is nonlinear simply means that it isn't like this. Most systems are nonlinear: the mathematician and physicist Stanislaw Ulam described the term 'non-linear science' as being like 'calling the bulk of zoology the study of non-elephants'.[a] Nevertheless, making the assumption that the situation of interest is linear makes it easier to study. Results can be obtained, and published, much more quickly. The important question is whether that is a reasonable thing to do: to what extent can we learn about mice and otters by assuming they are like small elephants? When do we need to consider all the potential consequences of nonlinearity from the start of our research—from the point we design our experiments rather than when we already have some results? This question is important for climate predictions because in this subject, nonlinearity raises its head in many new and unfamiliar ways. But before

---

[1] Quote: Dedee Truitt in the film *The Opposite of Sex*.

getting to that, it would be useful to have a bit more of a handle on what nonlinearity actually is, rather than what it is not.

As I've already said, many nonlinear systems behave very differently at some point in the future if you make small changes to the details of how they are now. Dice rolling is one example: very small changes can result in the outcome being a six rather than a three. A further small and similar change could make it a five. The butterfly tornado concept is the same idea. It's not that the butterfly creates the tornado but that the tiniest shift in the conditions at one point in time can lead to huge differences later. It could be that if there were no butterfly, there would be no tornado in Texas; if there were a butterfly there would be a tornado, but if the butterfly were in a very slightly different place, then again there would be no tornado. It's about how and whether one thing leads to another, to another, to another . . . One can be confident that there will be tornados in Texas next year, but not when and where. The when and where are affected by the butterfly. The situation is one of substantial confidence in some aspects and substantial uncertainty in others. You might be detecting a common theme.

The mathematical description of the physics involved in rolling a dice is pretty complicated. Trying to predict the outcome of any one throw faces practical problems related to gathering data on the way the dice bounces on the particular table in question, as well as challenges in solving the mathematical description of how it bounces and rotates and how the rotation changes on each bounce,[b] etc. It's fairly obvious that it's a difficult problem. The tornado situation is also very obviously complicated and difficult to solve. Sensitivity to the details of the starting conditions is not, however, a consequence of the complexity of the situation. It can happen in really very simple, if not necessarily familiar, situations.

An example of a simple nonlinear system with high sensitivity to the starting conditions can be found in something almost, but not quite, as simple as a pendulum. A pendulum swings to and fro in a regular and predictable way. It is so predictable that for centuries it was used in clocks as the basis for measuring time. It still is; we have a pendulum clock in our house—he's called Fred (Figure 5.1). The description of the physics of a pendulum is pretty simple. It consists of the forces on the pendulum (described by Newton's law of gravity and some basic geometry) and the resulting movement (described by Newton's second law of motion). The physics is simple and it is well understood. For small angles of swing the motion is almost linear, so if you set two pendulums going side by side from approximately the same starting position, they will behave in a very similar way; they will swing in time. Even as they begin to behave differently, going out of sync due to (mainly) slight differences in the friction they experience, it happens gradually. The difference between one and the other changes only very slowly, something that is a consequence of the (almost) linearity. At this point, you might want to go out and spend some time on the swings in a nearby park, ideally with an excitable child. Set two swings going at the same time with a child or adult on each, and watch how they gradually go out of sync. It may illustrate the point but mostly it's just fun.

The physics of a pendulum is understood and its behaviour is predictable. Now take two and attach one to the end of the other. The physics is still well understood.

Figure 5.1 Fred.

The mathematical description of what is going on is a little more complicated because there are two rather than one, but the maths arises from the same physics of balancing forces and Newton's second law of motion. It doesn't seem to be horrendously complicated.

We understand how it works, there is really no mystery as to what is going on, but the two together are nonlinear and their behaviour is quite different to one alone. Now it is very difficult to predict where each one will be at some point in the future, or how they will be behaving, because the answer changes a lot as a consequence of how you set them going. If you have two identical double pendulums side by side and you set them going as close as you can in the same way from the same starting position, before long they will be doing quite different things (Figure 5.2). Even for an idealized mathematical version where there is no friction and no air resistance, where we know precisely what all the forces are and there is no messy real-worldyness to the problem—even then, the smallest difference in the starting conditions rapidly leads to different behaviour. That's the butterfly effect. That's the consequence of nonlinearity, and it sounds a bell of caution that while sometimes a good understanding of how something works is enough to make good predictions, sometimes it is not. We need to be careful—particularly with climate predictions, because the extrapolation and one-shot-bet characteristics mean we have no chance of repeating the 'experiment' and seeing how the results might actually vary. We have no chance of testing our conclusions. There's only one twenty-first century and we don't have the results from that yet.

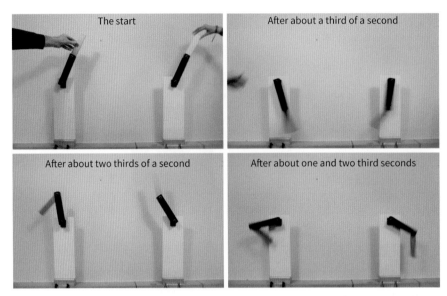

Figure 5.2 This shows two double-pendulums set off from close to the same position but subsequently behaving quite differently.

So far, so tricky. Maybe you find these ideas novel and exciting. I do, even though I first read about them thirty years ago. Maybe you find them old hat (Figure 5.3). That's not unreasonable. Books, ideas, and even films riffing off the butterfly effect are not new. Indeed the concept of mathematical chaos, for that is broadly speaking what I've been describing, has been widely studied in recent decades. It is the reason weather forecasting is difficult. It crops up all over the place in science and in popular culture so there is a risk that rather than being excited at this point, you're yawning and thinking you already know what's going on. You may have jumped to the conclusion that the difficulty in making climate predictions is all about the inherent uncertainty due to the butterfly effect. Nothing new to see here—move along.

If we were talking about weather forecasting, then this conclusion might be a reasonable summary of the situation but the implications of nonlinearity for climate prediction are not the same as for weather forecasting. They raise new, intriguing and unresolved issues and there is little agreement among experts as to what the implications are and how they should be handled. The role of the butterfly effect in climate predictions is more complicated, more nuanced, and much less understood. Furthermore, the butterfly effect is only one manifestation of sensitivity to the finest details of the situation. There are more wings flapping in the woods.

To see why the implications of the butterfly effect are different for climate predictions it's useful first to say a few words about the terms 'nonlinearity' and 'chaos'. A system which has certain types of nonlinearity can show high sensitivity to its starting conditions. A chaotic system always shows this sensitivity. However, a chaotic system also has an extra feature. It is usually taken to include an aspect of constraint: its behaviour has to stay within some limits, some region. Sometimes it may be possible

Figure 5.3 An old hat.

to know quite a lot about the restrictions—there may be absolute physical limits. In other situations, it can be harder. Tying down how to express this constraining feature is difficult. In his book *A Very Short Introduction to Chaos*, Leonard Smith says only that 'we require the system to come back to the vicinity of its current state at some point in the future, and to do so again and again.'ᶜ This expresses the concept of constraint in a very general way which is useful for a wider discussion of chaos, but for our purposes we can get away with thinking about it as simply some physical limits within which even a chaotic system must stay. For the climate system these might be, for instance, maximum and minimum temperatures or maximum and minimum river flows. The system can't do just anything. It can't go anywhere it pleases. It's just difficult to know, even approximately, where it will be **within its limits** at any particular point in the future. Nevertheless we might know what those limits are.

The double pendulum, for instance, is constrained in the positions it can take. It is fixed at the top, of course, but also its motion is constrained by the length of each part of the pendulum and by whether you just let it go at the start or gave it a hefty push which means it swings with more energy. Tornados in Texas are somewhat limited in their potential strengths in a similar way, and with respect to climate there are rough boundaries to how hot and how cold it can get on a summer's day in Sydney, for instance.

That's chaos for you, but in all these situations the system is 'stationary'—the probability distributions are not changing (see Chapter 3). The climate system shows such chaotic characteristics but with climate change the situation is more complicated. With climate change, the constraints on the climate system are themselves changing as energy is being trapped in the lower atmosphere at an increasing rate due to increasing atmospheric greenhouse gases. This means that the domain of possible, and of likely, behaviour is itself changing. Under climate change, previously impossible behaviour may become possible and things that have previously been possible, or even commonplace, may become impossible or very, very unlikely. Whatever aspect of climate we're interested in might never return to something close to its current state. We're therefore talking about something related to but more complicated than chaos, and as a result, many of the tools used to study chaotic systems simply no longer work; they don't apply. Yet the sensitivity to initial conditions is still there. This situation is perhaps better described as pandemonium.[d] It requires our study of chaos and nonlinear systems to move to a new level.

In weather forecasting, there is sensitivity to the starting conditions—that pesky butterfly makes things difficult—but we might nevertheless have a good estimate, from observations, of the probabilities for different types of weather. In that case, we are just trying to predict where tomorrow will be within the underlying probability distribution of possible weather conditions: the constrained range of possibilities. That's a very hard task but one that the weather forecasting centres are getting increasingly good at. It is a great success story of modern computer-based science but it's not what we're interested in when we talk about climate change.

With climate prediction the underlying distribution of possible weather events is itself changing. We are not trying to predict where we will be within these changing probability distributions decades ahead: we are not trying to predict the particular details of future weather that might happen in April 2054, for instance. The butterfly makes that impossible beyond a few weeks ahead regardless of climate change. We are, however, trying to predict how the probability distributions will change. Unfortunately that too is a nonlinear problem. That's because everything in the climate system is connected to everything else, and the connections are themselves often nonlinear.[2]

Consider, for instance, a situation where we know the climatic distribution of summer temperatures in Madrid for the current climate, and we're interested in what will happen if the world warms by 2°C. How do we relate an increase in global average temperature to a change in the distribution of summer temperatures in Madrid? Our first instinct might be that what is most important to understand is the regional behaviour of central Spain; surely that is where we should focus our research efforts? In a complex nonlinear system, however, the answer might be profoundly affected by the details of the circulation patterns in the pacific ocean, or the behaviour of sea ice in the Arctic, or the changes in forests in sub-Saharan Africa, or any number of other large and small details. Predicting future climate can never be achieved just by focusing on the bit we're interested in because it is all nonlinearly connected. The curse of nonlinearity is that predicting any one thing may well involve predicting everything, or almost everything. Get some particular detail wrong and the whole thing, the whole prediction, might fall apart.

So far, so problematic, but what does this all mean for the study of climate change? The combination of climate change and nonlinearity, when taken together, has two important implications. The first relates to how we study and interpret the sensitivity to starting conditions—the butterfly effect. The second relates to how we use computer models, and what conclusions we can draw from them.

---

[2] The connections are also complicated but as illustrated by the double pendulum the bizarre results of nonlinearity do not rely on the system being complicated. That they can arise in relatively simple systems raises the prospect of studying and understanding the problems of climate prediction without having to study something as complicated as the climate system itself. Much more on this later in the book.

Sensitivity to initial conditions is widely studied in maths and physics departments but analyses are often carried out on systems which are themselves unchanging, like the double pendulum, or almost unchanging, like the weather. Such studies are not much use in helping us understand climate change. With climate change the question is how the probability distributions themselves will change in the future but this too depends on the starting conditions. The distribution of temperatures which will actually be experienced in Madrid or Chicago during the 2050s, under some scenario for greenhouse gas emissions, also depends on the finest details of the state of the climate system today[e]. The future distributions are themselves influenced by the butterfly; it doesn't just affect where you happen to end up within the distribution on any particular day. The distributions are also influenced in an even more substantial way by large-scale, not fine-detail, aspects of the present-day climate, such as the particular state of ocean circulation patterns. These are new and different types of sensitivities from those usually studied by experts in chaos and nonlinear systems. They mean that nonlinearity affects the types of information we can plausibly get from climate predictions and also how we should be going about trying to make them with climate models.

Yet there is one, even more conceptual consequence of nonlinearity to consider. How good does a computer model have to be to make reliable predictions? If 'predicting any one thing may well involve predicting everything, or almost everything', as I said earlier, then what are the consequences for using computer models to help us with climate predictions? An inaccurate representation of arctic sea ice could lead to misleading predictions of Madrid summers. An inaccurate representation of the Indian summer monsoon could lead to misleading predictions of Chicago winters. In a nonlinear system it is difficult to rule out anything as not mattering. Computer-based climate predictions could be extremely sensitive to which processes are included and which are excluded. They could also be sensitive to the finest details of the way the processes are represented within the computer model. This is an effect akin to the butterfly effect but associated with sensitivity to how a model represents reality rather than to the starting conditions. Dr Erica Thompson from the London School of Economics has named it 'the Hawkmoth effect'. It has been demonstrated in a simple chaotic system but not yet studied in climate models. It raises questions over how similar to reality a climate model has to be to give useful predictions. Can a good but not perfect computer model give a good but not perfect prediction? Or is that not to be expected? If it can, what are the requirements for it being able to do so?

Even if a good but not perfect model can potentially give useful information, the existence of nonlinearity within and between the different components of a climate model raises questions regarding how we should interpret the results from collections of models. This is important too, because interpreting multiple models is a significant activity in climate change research, particularly in relation to the results presented by the Intergovernmental Panel on Climate Change (IPCC). One climate model might represent clouds better while another might represent ocean circulation better, but their simulations of the future are always a messy mixture of everything they include. How then, should we interpret multiple models in a way that provides useful guidance to society? Or even just useful guidance for scientific research? What does it mean to have a 'better' model? Realistic-looking simulations of historic observations could potentially be achieved more easily with a bad model than a good one, but the good one would provide better predictions. The consequences of nonlinearity are that this is a plausible situation.

These issues are all about how nonlinearity affects where and how we can use computer models to tell us about reality in a much broader sense than just climate change. It's the stuff of philosophy as well as mathematics, physics, computer science, and the like. It affects many areas of modern scientific endeavour, from cosmology to brain research.

Climate prediction is hard, very hard. It's both harder and different to weather forecasting. Nonlinearity raises new conceptual forecasting challenges which scientific research is only beginning to take a few tentative footsteps towards addressing. Chaos makes weather forecasting difficult, but with climate change we face pandemonium—all the same problems and a host more because the constraints on the system are changing. In weather forecasting, sensitivity to initial conditions makes it difficult to predict the weather in a few days' time within the distribution of possibilities; in climate forecasting, the way the distributions are changing is itself sensitive to the starting conditions. In weather forecasting we can evaluate our predictions by taking new observations; in climate forecasting we can't do this so we rely on the ability of our models to represent reality well. But nonlinearity raises questions over how much of reality the models need to represent. And how well. And what it means to make conclusions from a collection of models which all omit certain elements. How do we tackle these issues? How do we make predictions in this context? How do we use computers wisely?

These issues require new research, new thinking, and more effort on how to address the problems. They are of importance for the way we pursue a wide range of twenty-first-century science, way beyond climate science. They are a framing characteristic for climate change because they substantially affect how we should use science to understand and guide our responses to the issue.

6
The curse of bigger and better computers

The fifth characteristic of climate change: The lure of powerful computers

Massively increasing computing capacity provides exciting new opportunities for understanding potential futures, but it also distracts from careful experimental design and reflection over when and where computer models can provide useful information.

I'm part of the Spectrum generation. The Sinclair ZX Spectrum was a computer, a home computer, launched in 1982. In a world where computers are now everywhere, the term 'home computer' has somewhat lost its meaning but in 1982 almost no one had a computer at home. Computing was a nascent subject mostly found in very particular research institutions but it was beginning to enter the realm of the enthusiast. And from there, the whole of society.

Into this void were launched a number of products. For me, in Britain, the ZX Spectrum was the big hit. It was a good machine. There were other good machines coming to the market but where the ZX Spectrum won hands down was on price. Its main UK competitor, the BBC Micro, was perhaps a better machine but its price placed it out of bounds for me.

I was in there at the start, ordering one just as they came on the market. It was the first generation and more than that, it was the deluxe model with not just sixteen kilobytes of memory but forty-eight kilobytes! Wow. I was chuffed. Things progress of course, and I'm writing this now on a personal computer with more than sixteen million kilobytes of memory; even my phone has more than two million kilobytes—more than forty thousand times more than my old Spectrum.

At that time computing seemed to hold so many possibilities. For someone like me, with an inclination towards mathematics and physics, their stupidity—the way they only do exactly what you tell them to do—was attractive. The result of a computer program is precise. It might be precisely wrong because you made a mistake in writing the program or because you didn't think about whether the program you were trying to write would actually answer the question you wanted to ask, but you can't blame that on the computer—that's your fault. The computer would do exactly what you told it to do.

With my new Spectrum, I played games, mostly *The Hobbit* (there wasn't much choice), and I began to learn programming. The games at that time had very basic graphics, if they had graphics at all, so one of my first goals was to get control over the visual output. I set to learning the programming language BASIC. I thought I would start with something really, really easy, like getting a dot to move across the screen. From such a starting point, I was sure I could rapidly build up to wonderful colourful animations. How naïve I was. I wrote the programme, it wasn't hard, and sure enough the dot appeared and moved across the screen. Yawn—it took perhaps ten seconds to get from one side to the other. BASIC was just too slow. So I gave up on that and got a book on assembly language—a programming language which works in terms of what the computer 'understands' rather than what we understand. This was a lot harder, but after a while I had a new version of my programme. I ran it. Nothing happened. I checked it and ran it again. Nothing happened. Eventually, I worked out that while BASIC was too slow, this was too fast. The programme worked, but the dot moved across the screen faster than the refresh rate of the screen, which made it too fast to see. So I inserted a piece of code telling the computer to sit and wait for a while each time it had moved the dot by one pixel. Finally, I had my dot moving across the screen. Hurray! With a small sense of triumph, I went back to playing *The Hobbit*.

From this I learned that computers are tremendously powerful and can be very fast but getting them to do exactly what you want is hard work, takes a lot of time, and can be very, very tedious. So I gave up. It was eight years later, in 1990, that I returned to programming. Computers were much faster by then and because I was now working in a university, I had access to some of the fastest ones available. Now, however, I wasn't just playing around; I was trying to represent scientific understanding. I was attempting to simulate some approximation to reality. Furthermore, I wasn't starting from scratch but developing and expanding on programmes built by others. My job was to improve an already existing computer model designed to simulate the circulation patterns of an ocean. This was my introduction to climate modelling.

The 1980s and 1990s saw a revolution in computing and the way computers are used. In 1982 very few people had any interaction with computers at all but by 2000 they had become familiar items, commonplace in many societies. Nowadays, in many nations few people have no interaction with computers, if only through their mobile phones, and worldwide it is estimated that over half of the planet's population has access to the internet.[a]

Even though computers were far from commonplace in the 1980s, their use in research establishments had been growing for a couple of decades. Their use to simulate aspects of climate, in particular, was well underway. Computer models of the atmosphere and of the oceans were becoming key tools for research. At that time while their outputs were considered a source of potential insight, they weren't seen as being in any sense directly representative of reality. Indeed one of the winners of the 2021 Nobel prize for physics, Syukuro Manabe, said in a paper in 1975: 'Because of the various simplifications of the model [. . .] it is not advisable to take too seriously the quantitative aspect of the results obtained in this study. Nevertheless, it is hoped that this study [. . .] emphasizes some of the important mechanisms which control the response of the climate to the change of carbon dioxide'.[b]

Things are quite different today. Today computer models of the climate system have become the foundational tools for both studying climate change and providing quantitative predictions of future climate. The idea that we can use a computer model to represent reality and then study the representation as if they were the same thing now pervades climate research. And not just climate research. It is increasingly the case in disciplines as diverse as hydrology and economics. It has become so commonplace, so quickly that it is easy to miss that this is a profound change in the way research is carried out. It is a change that raises deep questions regarding what qualifies as scientific understanding—what qualifies as scientific evidence.

The characteristic of this chapter—the lure of powerful computers—is about how the massive and expanding availability of computer power is fundamentally changing what it means to do climate science. Computer models of the climate system are presented as the laboratories of climate science.[c] Studying a model is, though, profoundly different to studying reality. Apart from higher resolution and better graphics, one might reasonably argue that nothing has fundamentally changed, with respect to model interpretation, since Syukuro Manabe expressed his words of caution in 1975. Like me with my ZX Spectrum in 1982, much research today is driven by what the latest machines can do rather than an assessment of what we want to do or what we think needs to be done. What the latest machines can do is so exciting and so time-consuming that it distracts us from questioning whether it is actually what we are interested in. Are they actually answering the questions we want answering? Are we going about the problem the right way? Have we even identified what the problem is?

Let's take a step back and reflect on how this change sits in the history of climate science. Most of the science of climate change is really quite young. For sure, some bits stretch back a hundred years or so but it doesn't have the history and pedigree of Newton, Galileo, and Aristotle. At the same time though, neither was it developed yesterday. Aspects of climate prediction do date back to at least 1896 when Svante Arrhenius, a Nobel prize-winning Swedish scientist, evaluated how the temperature of the planet would change if carbon dioxide levels in the atmosphere were increased. At that time he concluded that a doubling of atmospheric carbon dioxide would lead to a warming of between five and six degrees, depending on latitude.[d] He later seems to have revised these estimates down to about 4°C on average, which is pretty close to what academic papers today present as the most likely value. It's not clear how he made the revisions but the method behind the original estimate is described in his paper. It is well documented. It was done by hand and brain; no computers were used in the making of that prediction.

Climate science done in this way has continued through the twentieth century and continues today. Mathematical concepts and ideas are developed and evidence for them is studied in observations. It is the same type of science as was done by Arrhenius. Nowadays, computers are often used to crunch observational data and to solve the equations which are proposed as a hypothesis for how an aspect of the climate system behaves, but fundamentally this approach sits in the regime of conventional science. It leads to much knowledge which sits in, or at least very close to, the basket of things we can say we know well; it's often knowledge we can trust, like expecting a ball to come down when you throw it in the air. Here computers simply make it easier to pursue science in the way it has been done in the past. The computer is just a handy and powerful tool for crunching data or solving a set of equations which represent a hypothesis, work that would otherwise have to be done by hand. It's like a washing machine: it makes life easier but it doesn't fundamentally change what we are doing.

If you decide to take up a career in climate change science tomorrow, however, you're unlikely to spend much of your time in that regime. Indeed you'll find it difficult not to get heavily, perhaps overwhelmingly, involved in the study of very complicated computer representations of the whole climate system. These representations are called global climate models (GCMs). They are quite different to a concept or an idea. Rather they represent a host of interconnected elements of the climate system. Their developers aspire, it seems, to represent the whole of the climate system as comprehensively as possible. Their models are reductionist in nature: they build up a representation of the whole from representations of the many different parts. They try to create replica earths on a computer. They start from three-dimensional descriptions of how air and water move, and they add in descriptions of clouds, sea ice, land cover (so as to differentiate between, for instance, forests and savannah), how light and heat interact with various gases, clouds, and ice, and even descriptions of soils and their water content. Some have been developed out of weather forecasting models by adding processes which are important for climate prediction but relatively unimportant for weather forecasting. The most significant example of this is the ocean. The state of the water in the ocean doesn't change much on the timescale of a weather forecast but for climate prediction, how the oceans will change is one of the most important considerations. The ocean model I worked on in 1990 became the basis for the ocean component of one of these GCMs.

The models themselves are used in many ways but the most influential experiments are simulations of twentieth- and twenty-first-century climate change. These simulations incorporate greenhouse gas emissions to reflect different possibilities for how humans will respond to concerns about climate change. One scenario might represent the rapid reduction to zero of all human emissions of greenhouse gases, while another might reflect significant human emissions continuing unabated throughout the twenty-first century. These simulations are important because they provide a picture of potential future climates (well, potential future climates within the model). They are studied directly in climate science research but they are also used to guide national and international policy, as well as being presented as sources of information to guide climate-sensitive decisions across society. Furthermore, the model futures that they generate are used in a plethora of other academic disciplines: disciplines which study the consequences of climate change for floods, river flows and water availability, crop yields and agricultural planning, vector-borne diseases and the health impacts of heatwaves, and many more. Consequently, these models play a very significant role across a wide range of scientific and social science research and are central to the connection between science and policy in climate change. They are treated as though the physical behaviour of all relevant climatic processes is well understood and can be represented on a computer. They are often perceived as good representations of reality and specifically of the processes important for future changes in climate. This may or may not be how they are seen by those who build them but this is definitely the role they play in the wider study of climate change.

These GCMs are not in stasis, of course. They are continually changing and improving. Resolutions are regularly increased to represent finer details, and components are added to represent additional aspects of, for instance, the carbon cycle, chemical interactions between gases in the atmosphere, and the behaviour of the stratosphere.[1] The availability of rapidly increasing computational power influences, indeed drives, climate change research by continually providing the prospect of increasing the level of detail possible within these models. As a result much time, money, and effort is spent increasing their complexity. To a certain degree, this is a good thing. The development of new, more complicated models enables the study of interactions between physical processes which may previously have been unstudiable. There are, however, some fundamental questions which often get overlooked in this race for the biggest, most complicated model.

Some of the most challenging questions relate to how 'good' a model needs to be to answer a particular question or to make a 'reliable' climate prediction. The scare quotes are important. What do I mean by 'good'? Well, 'good' in an assessment of a GCM refers to how accurate a representation of reality it is. And 'reliable'? 'Reliable' relates to having confidence that the prediction is sufficiently accurate to guide or support whatever the planned use is. The use might be building better scientific understanding of how climate change will affect the circulation patterns of the Antarctic oceans, or it might be to assess whether a particular flood defence system will be sufficient to protect a particular Bavarian town for the next fifty years. A model might be reliable for one purpose but not for another.

GCMs are not exact representations of the real world, of course. There are significant aspects that are often missing, aspects which we might expect to be important for how the climate responds to increasing levels of atmospheric greenhouse gases; consider for instance ice sheet dynamics, ocean acidification, ocean ecosystems, aspects of atmospheric chemistry, the stratosphere, and so on. It's not, however, just the presence or absence of these big things that matter. As raised in the last chapter, there is a fundamental question of how different from reality they can be and still provide reliable guidance about the future. The answer, of course, depends on what they are being used for, so you might expect this to be a question that is asked again and again in the academic literature. Unfortunately, that is not the case. But then climate

---

[1] The stratosphere is the atmosphere at heights above about 10–15 km and up to about 50 km. It has historically been largely excluded from most climate change simulations of the twenty-first century although it has nevertheless been widely studied with these models using different types of simulations.

change science and particularly climate change modelling are young endeavours. Like adults with a new toy, we're still excited by discovering the thrilling things our computer models can do. So excited that we don't spend much time asking ourselves whether they can do what we want them to do, or to consider how we might design our modelling experiments to be most informative about the real world.

Weather forecasting models are of course also imperfect representations of reality but they are nevertheless extremely valuable forecasting tools. We now have much more reliable weather forecasts than we had 30 years ago,[e] largely as a result of these models. So why aren't the same model issues of concern for weather forecasting? The answer is threefold. First, we can test whether a model makes good weather forecasts because we have a frequent cycle of forecasts and outcomes. Weather forecasters are able to evaluate the reliability of their forecasts because they make them every few hours, and within a day or so, they know what actually happened.[2] It doesn't matter therefore if the model gets the right results for the wrong reasons because we can know how good the forecasts are regardless of the degree of realism of the model.[3] If it makes good forecasts, then we know that any inadequacies in the model don't matter for what we are trying to predict. This is not possible in climate forecasting because of the one-shot-bet and extrapolation characteristics.

The second benefit of the short timescales of weather forecasting is that the model doesn't even have to attempt to get the whole system right—only those bits that can influence the weather in the place we're interested in over the next few days. There are still connections between all the different aspects of the world's climate but some can't influence others very quickly, so for a weather forecast it doesn't matter if you get them wrong or don't include them. To forecast the next week in Europe, the southern hemisphere and most of the circulation of the world's oceans are unimportant, so the weather forecasting models can get away without them.

And thirdly, weather forecasting is about predicting variability within a system that is largely unchanging, like a dice, so past observations do largely tell us about current behaviour—about how the various physical processes behave and interact under current conditions.

Climate prediction is much more difficult. With increasing atmospheric greenhouse gases the constraints on the climate system itself are changing, so we don't know to what degree past observations tell us what we want to know about the future behaviour of processes such as clouds, ocean circulations, and storms—and how they interact. We don't have a collection of forecasts and observations of what actually happens, so we can't measure how good our prediction system is, which means we rely on the processes being correctly represented in the model, rather than it simply

---

[2] The large number of forecasts and outcomes means they can assess the reliability of the probabilities they provide.

[3] An additional benefit of this constant cycle of forecasts and outcomes is that weather forecasters can identify the conditions in which they can make good forecasts, along with those where the forecast is likely to be less reliable. Sometimes they may be confident about the forecast four days ahead, sometimes only one, but they can often know which is the case when they make the forecast because they have experience with forecasts and outcomes. This information is itself extremely valuable and it comes largely from the models.

being a demonstrably reliable method. Most challenging of all, though, is that the timescales mean that huge numbers of additional processes might well be critically important. Consider, for instance, the stability of ice sheets, the variability of the El Niño oscillation in the equatorial Pacific, the response of rain forests and of deciduous forests, the ability of the oceans to transfer heat and carbon dioxide to the depths, changes in the ability of ecosystems to absorb atmospheric carbon dioxide, and many more. All these matter for climate predictions but not weather forecasts.

How good, then, does a model have to be to make reliable climate predictions? For advancing scientific understanding even simple models which only include a small set of processes are sometimes sufficient. Such models can be useful in clarifying what's going on and they can even provide guidance about the plausible consequences of climate change. However, in terms of forecasting the details of future climate, such as the probability distribution of summer temperatures in Chicago or Madrid in the 2070s, the interconnectedness of the climate system really matters. We might reasonably expect this type of prediction to be influenced by everything from changes in ocean currents in the Arctic to changes in rainforests in Indonesia. Our models need to include everything that we think could possibly be important.

The problem is how do we know what's important and how accurate does our representation need to be of what we include? What has to be included and what can be left out? These are difficult questions. Some of the processes which these models simulate (e.g. the large-scale movement of air and water) are well-founded in physical understanding. For these, it's true to say that the models are founded in physics. There are many others, however, that are not well understood in terms of their basic underlying behaviour or that cannot be represented at the relevant scale: for instance, clouds, forests, soil respiration, and even the consequences of small-scale movements and waves in air and water. These are included in the models as flexible elements which can be adjusted to ensure that the model as a whole simulates observed behaviour reasonably well. This may sound like a good way to proceed but remember, the aim of these models is to simulate **future climate change**. In order to have faith that they represent the future, we rely on the models getting the processes of the climate system right, not on them simulating the past in a reasonable way. It's much easier to build a model which can generate some set of observations than to build one which represents the real-world processes which led to those observations. In any case, our yardstick for assessing the model should be its ability to represent our best understanding of how these processes may change in the future, not how they have behaved in the relatively stable climate of the past. We don't have observations of the future, so we can't assess if these processes are represented sufficiently well to capture the changes we're most interested in.

My description of GCMs has so far taken a physical science perspective which focuses on the processes we believe to be important. This perspective assumes we can judge what details matter most and therefore where we should put the bulk of our effort. The mathematics of nonlinearity, however, provides a very different perspective. It is plausible that because of nonlinearity, climate predictions could be extremely sensitive to the finest details of how a model is constructed—this is the

Hawkmoth effect. Getting some innocuous detail wrong could potentially have a big impact on the details of the simulated changes of the future; details such as the probability of winter snowfall in Aspen, Banff, or Grindelwald in the 2060s, for instance. From a nonlinearity perspective, we should not expect that we can easily guess what the most important processes are; physical intuition or experience may not be reliable. It may even be that sensitivity of predictions to the finest details of the model is a desirable property because it reflects the way the physical universe would behave. That would represent a substantial challenge to creating reliable model-based climate predictions. Alternatively, though, it may be that for something as complicated as climate, the complexity itself limits this type of sensitivity; different processes could interact with each other in such a way as to make some details of the model unimportant. Or it could be that this is not a problem for some things (say continental temperatures) but is for others (say Californian heatwaves). The essential question though is—**how would we know**?

Climate predictions are important way beyond the realm of academia, so this is not just an academic question. We need to seek confidence. Building a model and hoping for the best is a very long way from being good enough. Yet today's science largely avoids doing the research we need to understand these aspects of models—aspects which influence how we should build and use computer models of not just climate but many other systems such as brains, galaxies, river basins, and hearts.

These are deep, fundamental, conceptual questions but even if we set them to one side, we know we have major challenges to address because we know that today's climate models are very significantly different from reality in ways that should affect their predictions at regional scales. It's certainly not the case that our models are 'almost perfect'.

All this means that efforts in climate modelling today need to shift from building and running models to posing and answering questions about their role in climate science: how should we use these complicated and powerful computational tools to tell us about future climate; how should we explore and quantify the consequences of errors in our models; how should we relate today's computer models of climate to the real-world climate system; how can we know when we have a model, or a collection of models, or a method for interpreting multiple models, which is reliable for guiding particular decisions in society?

In fact, these are just climate versions of questions we urgently need to be asking regarding the role of computer simulations in research more widely: when and how can computer simulations reveal insight into the behaviour of real-world systems; how do they fit into the scientific method; to what extent is a model simulation evidence and how do we design modelling experiments to get robust conclusions; when are they good enough to answer the question we are asking, and how would we know?

Computing is a difficult activity which often requires a large group effort, as illustrated by how little I achieved with my ZX Spectrum. The challenge of getting a computer programme to work at all, or to get a climate simulation even somewhat reminiscent of reality, often obscures the challenge of understanding what is useful to **try** to get it to do and how we would know when we were doing it right.

The continuing production of bigger and better computers provides a drive to build bigger and better simulation models without necessarily addressing what it is we require of them or the extent to which we can interpret their results in terms of reality. This is what makes the computational context an important characteristic for understanding what we know about climate change. It leads to an urgent need to rework the relationship between climate modelling, climate science, and society.

7

Talking at cross-purposes

The sixth characteristic of climate change: Multidisciplinarity

---

Climate prediction is a multidisciplinary subject. Few aspects of it can be studied robustly without being guided by a diverse range of disciplines.

Making climate predictions is tricky. That much is evident from previous chapters. Trickiness, however, is not in itself unusual. Fundamental research is often tricky. Working out how something behaves in a new situation, or how to do something that has never been done before, is rarely a walk across the moors. By definition there are no guidebooks. The trickiness of climate prediction does, however, have a characteristic that makes it unusual, even in the playground of fundamental research. Making climate predictions requires us to bring together understanding from a much wider range of sources and disciplines than perhaps any other research topic. This diversity characteristic arises partly from its links to society and social science, and partly from its particular brand of nonlinear complexity.

Progress in scientific research, and in academic research more broadly, is about discovering something that was not previously known—discovering something new. Typically this is done by building on current understanding: refining and improving what is already understood. It might involve a leap forward but if so, it is a leap from an already existing platform. As expressed by Isaac Newton in the seventeenth century, and by Bernard of Chartres in the twelfth, we see further by standing on the shoulders of giants.[a]

Of course, we already understand an awful lot of things so it takes a lot of work to climb onto those shoulders: research requires a depth of knowledge, a deep and somewhat comprehensive understanding of a particular field. It's not very easy to discover radical new insights unless you already understand a subject pretty well. The consequence of this, an inevitable consequence, is that researchers specialize. They specialize a lot. They have to if they are to keep their jobs by publishing new results. As a result, academia is full of individuals with tremendously deep and specialized knowledge. Yet climate change, and climate prediction in particular, requires understanding to be brought together from many disciplines. To do it well requires acknowledging that it is intrinsically a multidisciplinary endeavour. It demands deep understanding, not just in one subject but in many subjects, and that is something that makes it peculiarly and unusually difficult. It also makes it unusually time-consuming, at least if it is to be done effectively and in a way that is useful to society.

A consequence of multidisciplinarity being difficult and time-consuming is that not many people do it. In climate change, researchers usually dodge the issue and focus on one particular aspect. This doesn't mean they ignore other aspects or other disciplines altogether. Far from it. Substantial efforts are made to encourage researchers in different disciplines to connect with one another. And they do. The problem is that these connections are not central to their research. They are usually fringe activities. They are rarely about building deep and long-lasting collaborations in which expertise and perspectives are shared. Rather, cross-disciplinary collaboration often comes down to passing data across disciplinary boundaries: one discipline providing model output or observations or the results of some analysis to another. The focus is not on trying to understand, still less question, each other's assumptions or goals from a different perspective. These cross-discipline connections can nevertheless involve long and detailed discussions, and sometimes weeks or months of processing data, but in the long game of developing scientific understanding this is more often than not small talk. The multidisciplinary nature of climate change requires much more than this. It requires connections between disciplines: connections that question what research is needed, what is useful, what is achievable, what constitutes evidence, and how to interpret experiments. It's about the process of generating knowledge, not just about passing data, or even passing knowledge, around.

Stepping back from this big picture, it's worth reflecting on which disciplines are important for climate prediction. So far I've talked about physics in relation to fundamental processes, maths in relation to nonlinearity and chaos, computational science in relation to model construction, and the philosophy of science in relation to what represents reliable evidence. Coming up in chapter 9 I'll discuss the role and perspective of statistics, a discipline which is important in the analysis of both observations and model output. That's five subject areas for a start.

Yet these disciplines are just the tip of the iceberg. There are many related fields of physical science which are fundamental to the problem and must be integrated when trying to study changes in the climate system. These include hydrology, glaciology, oceanography, meteorology, land surface dynamics, atmospheric chemistry, geomorphology, and terrestrial and ocean ecology. Many of these include aspects that might be referred to as branches of physics or physical geography, but there are also aspects from completely different overarching disciplines, including chemistry and biology. Even for those that might sit in a physics department the domains of specialization vary substantially. Why does this matter? It matters because all these aspects interact in the climate system, so climate prediction requires a depth of understanding in each one and in how they combine. The traditional approach of specialization and compartmentalization of knowledge risks letting us down badly. It risks missing both important interactions between components of the system and opportunities to solve problems and improve understanding by being confronted with different methods and perspectives. It risks information being lost or misunderstood as it crosses the disciplinary boundary lines.

Even now though we haven't neared the end of the list of relevant disciplines. There are more subject areas: subjects still in the physical sciences but coming closer to society and societal impacts. These include hydrology (again) for consideration of floods and droughts; agricultural science for evaluation of the potential impacts on food production; various types of health and medical research related to everything from vector-borne diseases (e.g. malaria) to the consequences of heatwaves; engineering and infrastructure design; and many more. And last, but not least, there are the social science disciplines including economics, governance, policy studies, finance, conflict studies, actuarial science, etc. Each of these is involved in assessing the impacts of climate change and informing us about ways of coping with or responding to what we learn from the physical sciences.

These closer-to-society disciplines are of obvious importance for the study of climate change, but one might think they are less important for, or at least separated from the task of, climate prediction. The conventional approach to climate prediction is to use the physical sciences to predict the future evolution of the physical climate system and then feed that information into the impact and social science disciplines. It's a one-way flow of information: physics to physical impacts to societal consequences to societal actions. If robust, reliable, and detailed predictions could be made of the complete physical climate system (under some assumed human behaviour regarding greenhouse gas emissions), then this might be a sensible approach, but in light of the previous chapters (and the rest of this book!) such reliable predictions are in serious doubt.

If reliable predictions-of-everything are not plausible, then the task of climate prediction shifts into one which focuses on the need to inform particular issues and answer specific questions regarding impacts and societal decisions. It's about distilling what we know or could know about particular aspects of the future. It's not about knowing everything but about identifying what we know about some very specific things. In that case, it is critical to know what questions matter most and what aspects of the physical system are the ones we really want to predict, or at least understand. Climate prediction therefore requires bringing very diverse expertise and perspectives together, and that includes the social sciences.

Researchers are not dissimilar to kids in a playground. Researchers, like kids, cluster. They gather in groups who are all interested in the same topics. There is of course some transfer of information between clusters, mediated by kids who span multiple groups, but mostly they stick with those they know well and with whom they can communicate easily. With climate change, though, we want the whole playground to mix up and share perspectives. Some will subsequently go back into their original groups—not everyone has to work on the integrated problem—but everyone will be more aware of how what they do fits into the bigger picture. That knowledge has the potential to change research foci and enable better communication of what is robust and what is useful, as opposed to what is just interesting to each particular clique.

Most multidisciplinary research arises from collaborations between individuals in different fields—emissaries from the swings, the climbing frame, the football pitch, and the bike shed. The thing is that that's difficult—even when the participants are willing. It's difficult because different disciplines effectively speak different languages and have different perspectives regarding how to study a problem and what represents reliable evidence. Some are quite close—atmospheric physics and oceanography are perhaps like Spanish and Italian; they are clearly different, but a grounding in one provides a very good basis for learning the other. Some are a little further apart—atmospheric physics and hydrology are perhaps like French and German; both are Indo-European languages with certain similarities of structure but also many differences. Some require accepting a completely different approach—climate physics and economics are perhaps like Cantonese and English.

The different vocabularies and grammar are a barrier but the different perspectives are more insidious because they embed assumptions and values as to what represents salient information. Modellers often start from the assumption that their model is informative about reality and the future without questioning what is needed of a model to make that the case. Certain types of economist and statistician take a perspective that good information about future changes is necessarily contained within observations of past changes without regard for whether the size or context of the potential future changes undermines this assumption.[1] Their conversations with certain types of physicists and mathematicians have a tendency to focus on the details of the methods they are using and simply miss key differences of perspective; underlying assumptions are not discussed. Climate physicists and modellers may present results as probability distributions but then go on to focus on the most likely outcomes, the average and central band of probability. By contrast, an economic analysis using such physical climate assessments might find the small probabilities associated with very unlikely outcomes are the most critical information. The physicists may think they are being helpful to economists when working to better tie down the central band of uncertainty, but in fact they may be focused on the least useful part of the problem for the economists. In the integrated issue of climate change, much is a problem of communication and an assessment of what matters. Different perspectives are a source of confusion and can lead to experts talking at cross-purposes. With climate change, we all need to think outside the box of our own perspectives.

There are issues of practical consideration here. Even when disciplines stick to their own core foundations, they will often use information and data from other disciplines. Modellers build models by combining physical understanding and theories with techniques developed in computer science and mathematics. Regional modellers use information from global models. Hydrologists use inputs from regional models. Economists use information from statisticians, econometricians, and computer modellers. Policy analysts use input from all these disciplines and more. The question is how should they be linked? How should information pass from,

---

[1] The future is usually different from the past but with climate change the processes which control the future may also be different, and this can undermine the informativeness of observations.

for example, mathematicians, to physicists, to computer modellers, to more computer modellers, to agricultural scientists, hydrologists, and economists, and on to policy analysts and decision theorists? Each discipline incorporates its own assumptions, dependencies, and conditionalities. The danger is that assumptions and uncertainties known at the start of the chain are lost by the end, leading to over-confidence and badly constructed guidance to politicians, planners, and society. How do we keep track of the assumptions? This is a big challenge and not one that present-day academia is well set up to address.

Today's research infrastructures work against tackling these issues. It's hard to find researchers interested and able to address the integrated questions of climate change because it requires individuals with diverse disciplinary interests in a sector— universities and research institutes—which encourages and rewards specialization. It's hard for them to keep their jobs when working to bring together diverse disciplines because their annual appraisal will want to see evidence of research and publications in the field in which their institute or department specializes. Multidisciplinarity might be accepted, even encouraged, as a relatively eccentric sideline but only so long as it doesn't take too much time or distract from their core activities. It is also hard to get funds to support multidisciplinary research for all sorts of institutional reasons. Research funders may throw their hands up in horror at this claim. I sympathize with their frustration. I frequently see opportunities from research funders to apply for interdisciplinary projects but the scale of funding is tiny in relation to single-discipline projects; the activities are seen as perturbations, small additions, to the main, the real, research. There is little recognition by those that fund research that certain types of fundamental insights might require large-scale, in-depth, long-term, blue-sky, risky, multidisciplinary activities. Nor recognition that institutional and organizational barriers substantially restrict the ability of researchers to pursue it.

Why is all this important? Why am I going off on a potentially tedious whinge about the inadequacies of the academic research sector? It really doesn't seem to be as interesting as the earlier stuff about physics and maths and the practical challenges of climate prediction. Why does this matter? It matters because, like the earlier stuff, it fundamentally influences our ability to separate what is confidently understood from what is open to serious questioning. It implies we need to fundamentally change the way we do research in this field if it is to support practical societal planning in the light of climate change. And of course, how we should go about this restructuring is itself a question for policy studies and the social sciences.

My suggestion that these issues aren't as interesting as the earlier stuff about physics, maths, etc. is perhaps an example of my own biases and my own interests, but it's not quite what I believe. The bureaucratic issues of managing research departments are tedious to me, but the questions of how to design effective infrastructure to achieve particular goals can be fascinating. Considering such issues brings in the potential for additional fields to contribute to improving climate predictions and understanding climate change: decision theory and operations research, even psychology and sociology. These domains of expertise are of course of value when considering practical responses to climate change, but they could also valuably contribute to the development of research structures that can better understand and predict climate change and its consequences. There is certainly a need for a profound examination of how research structures can be reworked to encourage multidisciplinary insights.

I am not, of course, arguing here that every researcher should be a specialist in every relevant discipline. Still less am I arguing that some researchers should be specialists in no discipline with only a little knowledge of many; that would be disastrous. There is, however, a need to give researchers who may be specialists in one (or better still, a few) disciplines time to learn about and build on the connections with others. It's about learning the different disciplinary languages and differing inherent assumptions. Like all foreign languages, some people find certain ones easy, others find many easy, and some struggle to learn even one. The issue here is the need to give those interested and able the space to try working in multiple languages.

In the movie *The Blues Brothers*, Elwood Blues asks in a bar, 'What kind of music do you usually have here?' and gets the answer: 'Oh, we got both kinds. We got country **and** western'. In a similar way, there exists multidisciplinary climate change research. Teams specialize in climate modelling **and** climate statistics, or climate economics **and** climate policy, or climate impacts **and** regional modelling, or nonlinear time-series **and** stochastic processes,[2] etc. This is all fine, but no pair of sub-disciplines gets close to encompassing the diversity of the subject. If you only listen to country and western music, you can't claim expertise in music as a whole. For that, you need a bar where they got country **and** western **and** pop **and** rock **and** hip hop **and** jazz **and** classical **and** rap **and** folk …

In 2008, Her Majesty Queen Elizabeth II of the United Kingdom asked a group of economists at the London School of Economics how come the global financial crisis had not been predicted: 'If these things were so large, how come everyone missed them?' In 2009, the British Academy 'Global Financial Crisis Forum' responded to her saying it 'was principally a failure of the collective imagination of many bright people […] to understand the risks to the system as a whole'.[b] There are many many bright people working on climate change but accurately representing the risks to the system as a whole requires the imagination to make connections across all the traditional silos of knowledge.

---

[2] Mathematical systems that are partly random.

8
Not just of academic interest

> ## The seventh characteristic of climate change: Urgency
>
> Climate change research is not just an academic problem. Understanding climate change matters for informing a plethora of societal decisions. It is a matter of urgency both to be clear about what we know and to expand the scope of relevant knowledge.

Climate predictions matter. They impact us. They influence how we manage our societies, govern our nations, and run our businesses, as well as, for some, influencing lifestyle choices. Knowing which aspects are robust is important for ensuring we use them well and the timescales of climate change create an urgency to developing that knowledge. Yet scientific research often takes a long time to achieve a good understanding of a problem, let alone a solution. With climate change, the societal relevance makes that a problem in itself. If it takes decades to find out which methods we can trust, then the understanding may come too late to be useful.

Scientific understanding of climate change presents us with knowledge of a threat to our societies, to global stability, and to the natural world. That knowledge can be used to reduce the threat and to ameliorate the consequences. Consider the Titanic. If the ship had had satellite information systems, radar, and access to the International Ice Patrol, they could have 'seen' the iceberg sooner—could have had better information about where exactly the icebergs were—and altered course. We have those systems now, and collisions with icebergs are far less common—although they still occur.[a] Predictions of future climate provide a similar sort of information—a warning of a damaging, potentially catastrophic, event ahead. There is a crucial difference, however. The Titanic had a Captain, Captain Edward John Smith, and a rudder. It's pretty clear how they could have used better information about an iceberg ahead if it had been available before they saw it. The Captain could have given the order to change course and/or to slow down. Furthermore, this would not have affected much else. The passengers, crew, owners, and so on would have been inconvenienced little, if at all.

By contrast, there is no one in charge of global society or the globalized economic system. We have no Captain to give the order to change direction and we have no simple rudder to effect that change. Changing direction involves a myriad of changes to the way we run our societies and live our lives. Furthermore, these changes do affect, and arguably inconvenience, us. It wouldn't be surprising therefore if across society there were many different views regarding which responses we like and which we hate: which we find desirable irrespective of climate change, which we find tolerable at a pinch, and which we find abominable and the antithesis of progress.

The difficulty in responding to the warning is one difference from shipping and icebergs, but there is another. The warnings about climate change arise not from some obviously trustworthy, maybe even familiar, device but from a diffuse combination of scientific and social science research and understanding. It's not some tool that has been tried and tested, as is the case with radar and satellite monitoring systems. Climate change is an example of how, or whether, we can use scientific and academic understanding in its basic form to help plan the future of our societies. Can we use what we know to design our social systems to provide and protect what we value? Can we use academic knowledge to prioritize and optimize our future as a species, as nations, as communities, and as individuals? If we want to do this, then for a start we need to be very clear on the difference between trustworthy, reliable information, and information open to serious debate and questioning.

Utilizing academic understanding to guide society is quite different from using science to address a known issue (e.g radar and satellites to detect icebergs) or to create new opportunities to make life easier or more enjoyable (e.g. the development of television, mobile phones, contact lenses, and the internet). Predictions of climate change provide the context for a whole host of societal decisions, and they raise the prospect of a threat. This contrasts starkly with the more traditional relationship between science and society in which science creates opportunities and solves acknowledged problems in specific parts of our lives. Climate predictions have very wide-ranging effects on our societies and if we choose not to use them, or more accurately, if we choose not to use their reliable aspects, we will end up much worse off than we were before.

Furthermore, the multi-disciplinary nature of climate change knowledge makes it hard to use. How do we draw all the strands together and ensure they are consistent? Simply gathering multiple views from a variety of specialists won't provide the interconnected picture we want for a coherent, well-designed response because it fails to grasp the interactions between specialisms and the assumptions that get lost as information passes from one to another. Furthermore, the overlap with politics and governance requires that the dependencies and reliability of the information are all communicated to people whose interests and perspectives are quite different to those carrying out the research. Politicians govern and electorates choose them, so accurate and effective communication of scientific knowledge is crucial if it is to be used to redesign our societies. What then is relevant and valuable for people such as politicians and the wider public to be aware of? What would it be useful for them to know?

Early in my career working on climate predictions, colleagues would sometimes argue that politicians and the public needed to understand climate science. I was concerned by such a perspective. For sure it is good to provide information to those who are interested and in my view there is much that is fascinating in climate change science. Yet the idea that the public **needs** to understand climate science is worrying. The implication is that there **needs** to be widespread understanding of the science in order to respond to the threat. If that's actually the case, then I'm very pessimistic—why should most people be interested in attaining such understanding? Populations are diverse and have diverse interests and abilities. It's not plausible to expect most people to want to understand climate science. Nevertheless, it would be useful if individuals, whether politicians or not, had access to information about climate change that was relevant to them. Society would benefit if they were able to make judgements that reflect their values within the context of reliable climate information—judgements about lifestyle choices and what government actions they support and are willing to vote for.

Neither the public nor politicians need to understand climate science, but they do need to understand what climate science and climate social science can and can't tell us. In both physical and social science, some things are clear while others are obscure. Acknowledging this balance between confidence and uncertainty, the robust and the questionable, is critical to having effective social discourse on this issue. As the US Senator Daniel Moynihan said in 1983: 'Everyone is entitled to [their] own opinion, but not [their] own facts'.[1b] One of the most important aspects of climate prediction is to be clear about, and to communicate, what the facts are. Or at least to be clear what comes with high confidence and what with low confidence.

The claim is sometimes made that 'the science is clear', now we must act.[c] It's a fair point but it would be helpful if such statements acknowledged that while indeed some of the science is clear, other aspects are novel, speculative, and open to question. Furthermore, it is not just the science that matters and has uncertainties. So do the social sciences: our understanding of impacts on economies, industries, and the way we will be able to live our lives, as well as how we value the impacts on the natural world.

When we don't have clarity over which aspects of the implications of climate change are robust, our debates about how we respond are liable to become confused. The ability to balance our differing values and priorities, or to carry populations along with plans to change society's course, is made much more difficult by not having a shared understanding of what is open to debate and what is not. We end up arguing past one another rather than with one another. Personal opinions get mixed up with robust knowledge, making it difficult to tell the differences between them. Without clarity over the differences between information that is essentially beyond doubt, information that is subject to reasonable debate, and information that is really just an expression of opinion, we could easily continue going around in circles in our efforts to respond to climate change: of fiddling while Rome, and many a forest, burns.

These concerns are, of course, significant for those passionately concerned about climate change because they undermine our ability to act—or to act effectively. They are, however, equally significant for those who consider anthropogenic climate change less serious because policies are being enacted based on this information; policies which might otherwise be unnecessary and which some may consider

---

[1] The American financier Bernard Baruch made a similar comment earlier, in 1946.

undesirable. Where do I sit on this spectrum of concern? Up to this point I've somewhat dodged the issue because the subjects I'm discussing and the scientific challenges which follow are of interest without relation to any cause. They should not be coloured by the baggage that is assumed to come with one position or another. Nevertheless, the question of my perspective can be avoided no longer. Put simply, I am at the passionately concerned end of the spectrum. This should not be taken as implying any preferred behavioural choices for individuals, businesses, or nations, nor automatically any relative level of seriousness by comparison with other societal issues. I do have opinions on my preferred choices for individuals, businesses, and nations and also on the relative level of seriousness by comparison with other societal issues but those opinions aren't automatically defined by my assessment of the seriousness of this one. You are likely to have different priorities and different opinions even if we agree on the consequences of climate change.

The basis for my concern arises from some very well-understood science combined with some common sense arguments about the consequences for society. There is a chain of robust understanding leading to the conclusion that climate change is a serious threat to the functioning of our societies, and to many aspects of our environment which individuals and communities value highly. The chain is not particularly complex. It's not new. It's not radical. A description of both its science and social science links is left until later in the book—Chapters 22 to 25—but to avoid confusion amidst all the challenges and uncertainties, it is perhaps useful to have a quick summary of them here. So without attempting a full justification of the arguments, here is an eleven-paragraph explanation of why the reality and seriousness of climate change are beyond reasonable doubt. A few ifs and buts have to be glossed over—it is, after all, only eleven paragraphs—but it covers all the main points.

1: To begin, we need just a few aspects of tremendously well-understood, high-school-level physics. Here goes. All things radiate energy in the form of light, well, actually in the form of electromagnetic radiation, of which visible light is one component. At temperatures typical of those found on earth, they radiate a range of infrared radiation usually referred to as 'longwave radiation'. At temperatures typical of the surface of the sun, they radiate visible light along with some infrared and some ultraviolet, a combination known as 'shortwave radiation'. It is a well-tested property of greenhouse gases such as carbon dioxide that they are transparent to shortwave radiation but they absorb longwave radiation. We also know that the atmosphere contains greenhouse gases. Put all this together and our basic expectation is that energy from the sun mostly passes straight through the atmosphere and much of it is absorbed by the surface of the earth. The surface of the earth, however, radiates longwave radiation. This radiation doesn't pass back through but rather gets absorbed by the greenhouse gases in the atmosphere. This energy is then re-emitted by the gases— remember, everything radiates energy. Some of it makes it out to space but some comes back down to the surface or to other parts of the atmosphere and is reabsorbed. This is the greenhouse effect (Figure 8.1).

2: The next basic physics point is 'conservation of energy', a physical law which is also well-understood, high-school physics, and not open to serious debate. It says

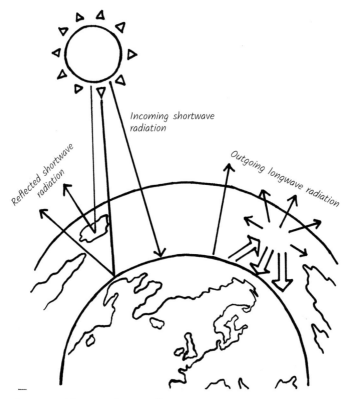

Figure 8.1 The greenhouse effect.

energy cannot be created or destroyed.[2] Alongside energy conservation, we need the concept of 'temperature', and to understand that it is a measure of the amount of a certain type of energy that an object possesses. For the surface of the planet to be at a stable temperature, therefore, it can't be gaining or losing energy overall—it must lose energy at the same rate that it gains it. If it were gaining energy faster than it were losing it, then it would be warming up; its temperature would be increasing.[3] So in a period when the climate is stable, the incoming and outgoing energy must balance each other. We know how much energy is coming into the climate system from the sun, so we can calculate the expected, average temperature of the earth which is necessary to produce the same amount of outgoing energy, from the earth to space, to keep the temperature stable. It's around −18°C. But that number isn't what we see at the surface. Why? Because of the greenhouse effect. When we include in the calculation the absorption and re-emission of energy by the greenhouse gases in the atmosphere, we get an answer of about 14°C. That number pretty

---

[2] Don't get distracted by pedants raising questions of mass/energy equivalence, Einstein, and relativity. They are irrelevant here.

[3] Remember this is a quick summary. One could raise some physicsy details concerning melting ice and evaporating water but they too are unimportant here—in general the expectation would be that in most circumstances the surface temperature of the planet would be rising.

much matches observations and is more than 30°C warmer than without the greenhouse effect. That's what theory says. That's what observations show. Everything fits together. We understand what's going on. So far, so good. The greenhouse effect is keeping the planet more than 30°C warmer than it would otherwise be.

3: So what are these greenhouse gases that are keeping the earth's surface more than 30°C warmer? Well, carbon dioxide accounts for about 20% of the effect, with water vapour, and the liquid water and ice in clouds, representing most of the rest. Put this together with the last two paragraphs and it seems pretty clear that if carbon dioxide concentrations in the atmosphere increase, we should expect the surface temperature to increase because more longwave radiation will be absorbed and re-radiated back to the surface. It's almost like expecting a ball to come down when you throw it in the air. Furthermore, if the temperature of the atmosphere increases, there's good understanding that tells us to expect it to contain more water vapour, evaporated from the oceans. That means that if carbon dioxide is causing some warming, we should expect this to be amplified by an increase in water vapour which further strengthens the greenhouse effect.

4: Are carbon dioxide concentrations increasing? Yes. Observing stations around the world show they are. And not by a little bit, but by a lot. Carbon dioxide concentrations have increased from about 280 ppm (parts per million) in about 1750 to about 410 ppm in 2019.[d] When we account for other human-emitted greenhouse gases (such as methane and nitrous oxide), we get an equivalent carbon dioxide concentration of about 460 ppm[e] in 2019. That's more than a 60% increase since pre-industrial times. It's not a trivial amount. It's not a minor tweak. We know that mankind's activities are the source of the increase because we can add up how much we emit; at one level it's as simple as that. We also know that we're going to continue emitting these gases for at least a few decades so these atmospheric concentrations will rise further. Put it all together and our basic expectation is that the planet should have warmed and that it will warm further. Observations show that it has indeed warmed, by about 1.1°C[4] since pre-industrial times.[5] This warming hasn't been steady, however. Indeed we wouldn't expect it to be. Energy is constantly flowing between the surface and the sub-surface oceans, so the temperature at the surface varies quite a lot from one year to the next, and even from one decade to the next. As a result, there could be multiple years when the average surface temperature doesn't increase at all or may even fall. Nevertheless, over multiple decades we expect to see warming and observations support this expectation. It's also worth mentioning that the trapping of extra heat by greenhouse gases leads to increasing heat in the oceans which leads to water expansion and thus sea level rise—another basic expectation that observations support.

5: The next important part of the story is inertia, thermal inertia: greenhouse gas emissions have led to warming but there's a delay in the system. It's like when you

---

[4] The IPCC sixth assessment report gives a likely range of 0.8 to 1.3 with a best estimate of 1.07.
[5] This warming is less than you might expect by simply scaling the carbon dioxide contribution to the 30°C, for reasons dealt with in Chapter 23. Remember this is only a quick summary.

turn on the heat under a saucepan of tomato soup. The soup doesn't immediately become hot—it takes a while before it is ready to eat. Increasing greenhouse gases is like gradually increasing the heat on the stove from 0 (off) to 6 (max) over a minute or two. If you stop increasing the heat at level 4, the soup nevertheless continues to get hotter. The ultimate behaviour—a slow simmer or a raging boil—is controlled by the level you stop at but whatever the ultimate outcome, it will almost certainly keep warming after you stop turning it up.[6] The **rate** at which it warms after you stop turning it up is controlled by the level you stop at and also by how fast you were increasing the heat before you stopped. Similarly, if greenhouse gas concentrations in the atmosphere remain fixed at today's values, the surface temperature will nevertheless continue to increase for several decades. If, however, the concentrations continue to increase, then the temperatures will increase faster than this and ultimately reach a higher value. Unlike the stove, unfortunately, we can't easily go the other way and turn the heat down because taking greenhouse gases out of the atmosphere is, at best, problematic.

6: Put this all together and mix in a little common sense about the global economy—it's not credible to think we're going to stop greenhouse gas emissions overnight—and we can say with confidence that the planet has both warmed since pre-industrial times and will continue to do so over the next few decades at least. We nevertheless have the opportunity to reduce the pace of warming and limit its ultimate degree.

7: So far, so good—but what does this mean for me? Or for you? This is where it starts to get difficult. We know lots of things which form part of the answer to this question but there's also a lot we don't know. We don't know how much the planet will warm as a result of any assumed future emissions although we do know that under any credible scenario it is a level of change that is significant and takes us beyond anything experienced before by human civilisations.[f] A mixture of theory and observations tells us this. We don't know quite what the warming and other changes in climate will look like in any one place but we do know features such as that land warms faster than the oceans. We know therefore that a 2°C warmer world in the twenty-first century will involve more than 2°C warming in most land areas and much more in high latitude land areas such as northern Europe, Canada, and the Arctic. We understand the theory that goes behind such conclusions and it is consistent with observed changes so it really is not in doubt. We know that a warmer atmosphere can hold more water and contains more energy so we can confidently expect that there will be more rainfall and more energetic extreme events when we look at the planet as a whole. This, however, is entirely consistent with some places getting drier even as others get wetter. Flows of air and water in the atmosphere and oceans already lead to uneven distributions of rainfall across the surface of the planet; with climate change, changes in those flows will lead to **changes** in rainfall that are also unevenly distributed. We know therefore that for most places the climate of the future will be different from that of today, and that the climate of today is different to that

---

[6] Unless you do it incredibly slowly.

of two centuries or even half a century ago. Such knowledge arises as a simple consequence of human activities increasing atmospheric greenhouse gas concentrations and trapping more energy at the surface of the planet and in the lower atmosphere and oceans.

8: These arguments provide information beyond global averages but they don't really answer the question, what does it mean for me? So let me try again. It means change. There are all sorts of specifics that we don't know very well—more on that in the rest of the book—but without doubt it means changes to the physical world we inhabit. There are lots of reasons to expect most of the changes to be for the worse—more heat waves, floods, extreme storms, extreme temperatures—even though for a few places there might be some benefits such as the ability to grow more or different agricultural produce. These physical changes are very likely to affect you directly but let's set these direct effects to one side for the moment. Step back and look at your society as a whole. The mere existence of widespread changes across the planet means that the social structures and physical infrastructures on which we rely today will not be well suited for the future. You might cynically say they are not well suited for today and that may be true, but it is hard to imagine any way in which climate change will not make most of them substantially less well suited for the coming decades.

9: At a very basic level, the geographical distribution of water availability and potential agricultural production will change. That's pretty much inevitable because of changes in temperature and precipitation. Some regions will inevitably see increased water stress, some may see decreased water stress, while some will suffer from increased flood frequencies, some more frequent heatwaves, and some perhaps both. We live in a relatively stable world of nation states but if there is a change in the suitability of places to grow food or access water, or simply be able to live without the risk of regular flooding, then this will inevitably exacerbate current stresses between nations and regions. It will surely also create new ones. The pressures on people to migrate will surely increase which can only further increase stresses on our globalized world. In addition to this, the infrastructure we have built in the past for housing, workspaces, transport, water provision, energy provision, and so on are all likely to become less suitable or effective in one way or another. They will become less optimized for the new state of the physical climate. For instance, a village or a housing or industrial estate might already be at risk of flooding but if the risk goes up from a probability of once every two hundred years to, say, once every ten years, then the viability of living and working there utterly collapses. That's disastrous for the people who live there but it also has wider financial impacts. It requires investments to be written off and it disrupts supply chains. Certain types of investments held by individuals, governments, pension funds, etc. will inevitably fall in value or become worthless. Of course, there's usually an engineering solution for any specific individual threat to infrastrucure but if these issues are arising all over your nation, then the cost of such responses is likely to become massive and/or the solutions unmanageable. If they're happening all over the world then it implies a restructuring of the global economic and trade system towards a battle to maintain facilities which currently we take for granted.

10: You might argue that if some places might benefit while others lose out then climate change is just a matter of reorganization. Unfortunately, that argument isn't a source of consolation for two reasons. First, studies of the consequences of climate change are dominated by findings of negative impacts on industries, regions, and ecosystems.[g] It seems there are vastly more negative consequences than positive ones: they don't balance out. This is not a surprise. It's easy to think of negative potential consequences of climate change (e.g. flooding, heatwaves, inadequate infrastructure) but much harder to think up positive ones (e.g. increased agricultural production in very particular regions). This makes sense because most aspects of our societies, and of the natural world, are in some way optimized for current or past climate and will therefore be less so for a future changed climate. This is the case for ecosystems (e.g. the impact of sea ice loss on polar bears, or temperature increases on corals or the gender distribution of green turtles[h]), infrastructure (e.g. the impacts of sea level rise on coastal cities) and social systems (e.g. the impact of heat waves on tourism and health). The mere fact that climate change involves widespread and substantial **change** leads one to expect that the localized impacts will be mostly bad for most societies around the globe and therefore overall the consequences are negative. There is, however, a second reason not to take comfort from the view that it's just a matter of reorganization. Consider the situation where some places experience negative impacts while for others climate change creates new opportunities and thus positive impacts. Even if on average across the globe these impacts were indeed neutral[7] (e.g. some places get less water, while others get more, some places have less ability to grow food, others more, and so on), the social effects would still be substantially negative because we do not live in a world which allows people to move and resettle wherever they like. The state of the world at the beginning of the twenty-first century is not one common global community. Our ability to re-optimize social structures globally on the timescales of anticipated changes (decades) is extremely limited. The friction between nations and communities would be huge. It is difficult to imagine it not leading to conflict. And even if we did live in a world without borders or barriers, our cultures and our roots are often located in particular geographical locations and often associated with the climate of those locations; shifting populations would therefore still represent vast upheaval and a loss to what many of us value highly. And, of course, the rate of change also works against the potential for ecosystems to adapt, inevitably leading to environmental damage and further losses to what—taking an entirely human perspective—many humans value highly.

11: Still it does not feel like I have quite answered the question, what does it mean for me and you? I have one more paragraph to have a go. It means that we're living in a world in which we are more vulnerable to natural climate hazards than we have been before. Without vast changes to the structures of our societies, we will become even more so. It means that whatever difficult political choices we have in our nations regarding jobs, housing, healthcare, education, poverty, employment, equality, etc.,

---

[7] Which they are not in any way expected to be.

will be made harder by the increasing inadequacy of our infrastructure to cope with the new climate norms: over-heating buildings, health consequences of heat waves, flooding on coasts and rivers, challenges to maintain water supplies, flooded roads and railways, heat-damaged roads and railways, changes to agricultural production, and so on. It means we'll have fewer resources to spend on what we want because we will need to spend much more on simply keeping our systems running. It may be impossible for some, perhaps many, regions and sectors of society to keep up. It means we should expect to be living in an increasingly dangerous world as a result of extreme climatic events and more challenging international relations. It means that the ecosystems of our planet are being put under increasing stress; it is difficult to imagine that the changes won't drive significant numbers of species to extinction. It will affect almost every aspect of the way we run our lives, source our food, trade our goods, use our leisure time, and so on. It will affect everything related to the globalized nature of the modern world and the social and physical infrastructures we use.

So there are my eleven paragraphs on why the reality and seriousness of climate change are beyond reasonable doubt. It's not about the details but about bringing all the issues together. The arguments are deliberately unquantified. This is partly due to the difficulties in climate prediction dealt with in later chapters, but it is useful in any case to remain unspecific about the details when we're just laying out the reasons for concern. The concerns about climate change are not in the details—many details could be responded to individually—but in the interconnected impacts on our global physical, natural, social, and economic systems. Climate change influences pretty much everything. The concerns are based on core scientific understanding combined with simple arguments about pretty much inevitable consequences for our societies and for the natural world. These are enough to place me at the very concerned end of the spectrum regarding the seriousness of climate change. Remember, however, that 'very concerned' does not indicate support for any particular approach to how we should respond or what level of response is possible. Agreeing on the reasons for concern is only a starting point for a much bigger debate on how to respond; on what type of future we want to build, or whether we are just going to allow one to be forced upon us by inaction or inadequate action. Climate change forces us to consider what we aspire to as communities and nations.

In 2015 the members of the United Nations signed up to the 2030 Agenda for Sustainable Development and seventeen associated Sustainable Development Goals. The agenda represents a 'blueprint for peace and prosperity for people and the planet, now and into the future'. The goals include ending poverty and hunger, ensuring healthy lives and well-being, availability of water and sanitation, and promoting sustained, inclusive, and sustainable economic growth. They also include many measures directly related to climate change. The stresses caused by climate change will inevitably make achieving all these goals much harder. Ending poverty and hunger, and ensuring availability of water and healthy lives will be much harder if climate patterns are changing. Local knowledge and experience regarding agriculture and water management will be less applicable or perhaps totally inapplicable. Some of what we have learned from the past about how to achieve these types of goals for a better world will be undermined.

Our goals for a better world are therefore inextricably linked to addressing climate change. Making plans for a better world and getting buy-in for those plans from nations and populations would benefit substantially from a better understanding of what we can and can't say about what the future holds. We don't need detail to know whether to be concerned but details would certainly help in designing the best course of actions. The problems of climate prediction are, therefore, far from being only an academic problem, and addressing them is an urgent issue.

The message of the first part of this book is that climate change is the epitome of a complex problem and has seven core characteristics:

(i)   it's about climate prediction,

(ii)   climate prediction under climate change is about extrapolation,

(iii)   the choices we make represent a one-shot-bet,

(iv)   both the climate system and our computer models involve diverse forms of nonlinearity,

(v)   the expanding availability of computer power tempts us to overlook many fundamental conceptual barriers,

(vi)   understanding the issues requires expertise from multiple disciplines, and

(vii)   tackling the issues is urgent.

This combination of characteristics makes the challenges of climate change research substantially different to those found in other realms of science and academia. Many other fields face a selection of these framing characteristics but not all of them together. If we were able to ignore one or two then the challenges facing climate prediction would be much more easily solved, but we can't do that. With climate change, it is important to always be aware of how all these characteristics combine to influence the reliability and value of research efforts. Armed with knowledge of these framing characteristics, we can however begin to explore both the conceptual and practical challenges we face in providing the detailed climate predictions that we want to support society's debates and decisions.

The climate challenge is the challenge of understanding human-induced climate change. It's a grand challenge. It's about getting a handle on how we separate what we know from what we don't know, how we tie down and respond to uncertainty, and how we push back the boundaries of knowledge on those things that are most urgent for us to understand.

The rest of this book is divided into a series of sub-challenges that contribute to these goals. It begins by focusing on core questions regarding what predictions, and specifically climate predictions, are. After that, it moves through questions of how we can use, and misuse, computer models. It finishes by looking at the consequences for economics and policy, and the difficulties we face in revitalizing climate research and making it more relevant and useful to society.

If your interest is solely on the specific issue of climate change, then the next few chapters might seem a bit remote but trust me, they are central to understanding what we can and cannot know about the issue. Beyond them, it's all about coming closer and closer to the use of climate science by society and about practical knowledge about the real world. Here goes. Good luck. I hope you make it to the end—some of the best bits are there.

PART 2
CHALLENGES

Challenge 1: How to balance justified arrogance with essential humility?

There are many aspects of the future under climate change which we know with substantial, well-justified confidence. There are many other aspects which we don't know well at all and about which we haven't yet understood, or even fully considered, the nature of the prediction task. A core challenge for climate science is recognizing, understanding, and communicating the difference between the two.

9

Stepping up to the task of prediction

9.1 What makes a reliable prediction?

'All hail, Macbeth! hail to thee, thane of Glamis!'
'All hail, Macbeth, hail to thee, thane of Cawdor!'
'All hail, Macbeth, thou shalt be king hereafter!'

[Macbeth, Act 1, Scene 3]

Consider an everyday situation with which I think we can all relate. You're out on a heath in the middle of a thunderstorm and you meet three old women who mention, in passing, that you're going to be king. It's a prediction. Do you trust it? If so, on what grounds? How does the prediction change your actions?

How about my horoscope? Today, Thursday 9th January 2020, mine says: 'Change is in the air in a big way, and you have known this for a while. It may have reached the stage where you feel nothing can stop it. You might feel nervous but knowing this might also be something of a relief'.[1] It's a prediction of change. Not very specific, I grant you, but still, is it trustworthy?

I can make my own predictions. Today[a] I predict that I'll be at the Noel Coward Theatre in London on the evening of 19th February 2020 watching *Dear Evan Hansen*. We have the tickets. We're looking forward to the event. Unlike my horoscope, this is a very specific prediction indeed but again we can ask, is it trustworthy?

Last November, November 2019, the Bank of England predicted that the Gross Domestic Product of the United Kingdom would increase by 1.6% in 2020.[b] They went further and provided a range of possibilities. They said there was a nine out of ten likelihood that the increase would be between about −1% and about +4.5% with 1.6% being the headline value, the 'median', the value with 50% chance of being both an overestimate and an underestimate. They also provided lots of information regarding the assumptions on which their prediction was based: much more information than provided by Shakespeare's witches, today's horoscope, or my prediction of my location in February. It seems great, you might even say scientific, but one might still ask if the assumptions are reasonable, if they are suited to the situation at the time and if the methods they use are appropriate. The question is still, how trustworthy is it?

Predictions are ubiquitous but they are also slippery. They are rarely as clear cut as the witches' 'you will be king'. Part of the reason is the need to balance precision and reliability. If I predict that I will arrive at a meeting in Leeds University at 11:37am next Thursday, given that I'll be coming by train from Oxford on the same day, I am

---

[1] According to the UK Metro.

fairly likely to be wrong. On the other hand if I predict I will arrive there between 10am and 6pm I am very likely to be right—but to be so at the cost of it not being much use to the colleagues I plan to meet. A prediction of between 11:30am and midday is more likely to turn out to be right than the 11:37am prediction because trains are often a few minutes late and how long it takes me to walk from the station to the university depends on how busy it is and how I'm feeling that day. A half hour range is likely to encompass these common factors. It's more likely to be wrong, however, than the 10am to 6pm prediction because train delays of more than half an hour are far from unheard of and sometimes trains are cancelled, and I have to catch the next one. When I make my prediction, I have to choose how to balance precision and reliability. Sometimes it can be better to have a less reliable but more precise prediction, but only up to a point.

This balance is also illustrated by my horoscope. 'Change is in the air' isn't very specific and as a result is unlikely to be wrong. This forecast could be fulfilled by something big like me receiving a job offer on the other side of the world or by something small like a storm coming across Britain and producing a change in the weather. It is unlikely though that with hindsight I wouldn't be able to find something in my life that would feel as though it had changed in the days or weeks following the prediction. The forecast is unspecific, it has very low precision and as a consequence, it has high reliability.

My prediction of where I'll be on the evening of the 19th February is, by contrast, extremely precise. Indeed I could narrow it down to the row I expect to be sitting in—the Grand Circle row E. However, despite its precision I believe this forecast to have high reliability. The reason I believe it has high reliability is that the ways it could fail are all fairly unlikely: severe disruption to the UK rail and road network, a cancellation of the show because the roof of the theatre collapses,[2] a terrorist alert, etc. I'm not 100% confident that it will be accurate but my confidence is pretty high.[3]

When originally writing these paragraphs I failed, of course, to consider the possibility of a global pandemic disrupting the performance, which it would have done only five weeks later. Indeed with hindsight the pandemic is also a rather good interpretation of my horoscope's prediction of change.

The Bank of England's forecasts are more specific than my horoscope but still they are not very precise: they include a wide range of uncertainty. Here, however, the forecasts are of a different character. They don't predict what will happen but rather the probability that the outcome will be in a particular range. My predicted arrival in Leeds between 11:30am and midday can only turn out to be right or wrong, a successful prediction or an unsuccessful one. The predicted −1% to +4.5% range for the change in GDP is, by contrast, accompanied by a likelihood. They say there is a 90% probability of it occurring. They are not precise about what the outcome will actually be but they are exceedingly precise about the probability that the outcome

---

[2] On a previous visit to the theatre—*Death of a Salesman* at the Piccadilly Theatre—the ceiling had collapsed three days previously and the performance was moved to the Young Vic.

[3] Hindsight: It turned out my confidence was not misplaced. I was there.

will be in a certain range. They could have chosen to give a more precise range of outcomes—say 0.5% to 3.0% growth instead of −1.0% to 4.5%—but the more precise the prediction, the lower the associated probability that the forecast will turn out to be true. They might perhaps have had to associate only a 50% probability with the 0.5% to 3.0% range.

One can imagine attaching probabilities to all sorts of forecasts. Instead of just saying that I'll arrive between 11:30am and midday, I could attach a probability of my arrival time being in that range—say 80%, a four out of five chance. But if I do that and I only make one forecast (remember the one-shot-bet characteristic), then you could never say that my forecast was wrong. If I arrive at 2pm it could simply be put down to being in the 20% probability, the one in five chance, that I'd arrive at a different time. You can never show that a one-off probability forecast is successful or unsuccessful.[4]

It turns out that the reliability of the Bank of England forecasts is generally pretty good: they are 'probabilistically reliable'. In nine out of ten similar forecasts, the subsequent change in GDP is indeed in the range provided by the forecast.[c] On this occasion, however, reality was outside the 90% range: UK GDP growth in 2020 turned out to be −9.8%[d] as a result of the pandemic. Of course, that doesn't make the forecast wrong: they didn't rule out such as possibility.

I have a friend who regularly accompanies his predictions with probabilities in almost all aspects of life: arrival times at meetings, success of proposals, management decisions, availability of a particular beer at a particular pub, and so on. I have yet to evaluate the reliability of his probabilities but since he makes them regularly it is something I could do. The existence of lots of forecasts and associated outcomes makes it potentially possible to evaluate the reliability of his probabilities, just as one can the Bank of England's. I would however have to note down each prediction he made and then subsequently each outcome, which would be tedious and rather annoying.

With predictions one can be more or less precise about the outcome, and more or less precise about the probabilities associated with the outcome, but if a prediction involves probabilities then its reliability has to be assessed in terms of those probabilities.

Even a precise and reliable forecast can, however, still be misleading. Let's go back to the heath and the witches. Perhaps on a later occasion you meet them again and you understand them to say that no one can harm you. Put yourself in Macbeth's shoes.

> "Be bloody, bold, and resolute. Laugh to scorn the power of man, for none of woman born shall harm Macbeth." [Macbeth, Act 4, Scene 1]

It's a prediction of sorts. As you understand it, it predicts that you are immune to harm from another person. Perhaps it has turned out that the thing about becoming king has already come true but should that give you confidence in this new information?

---

[4] Unless it rules out certain possibilities entirely and one of those outcomes comes to pass.

Maybe. The witches seem to have demonstrated form in this field. Yet you might want to reflect carefully on the details of the prediction, on the way the information is provided? Do the specifics—the wording—represent some sort of disclaimer which means the information appears to be more useful or interesting or relevant than it actually is? For Macbeth, the answer was a big yes. The prediction was precise and, with hindsight, arguably reliable if we take 'of woman born' to refer exclusively to a natural birth, which is not an unreasonable assumption for Macbeth in the eleventh century. Nevertheless, it was misleading. Macbeth of course was killed by Macduff in the play,[5] who 'was from his mother's womb untimely ripped'. Reflecting on the specifics and on how the prediction might not cover everything it appeared to, might have led Macbeth to be a bit more cautious. There is a lesson here for the interpretation of climate predictions.

For the moment though, let's set aside the potential for miscommunication and misinterpretation because irrespective of such concerns, if we are to act on predictions we want an idea of their reliability. For predictions without probabilities, how likely is it that the predicted outcomes will come to pass? For predictions with probabilities, how reliable are the probabilities associated with the different outcomes? Has the prediction considered all the main ways the future could evolve and the likelihoods of each one coming about? To get a handle on how to answer these questions, we need to consider the different ways that we go about making predictions.

9.2 The foundations of a prediction

I chose to study physics at university. This wasn't a surprising decision; physics was my favourite subject at school. I now work across many different disciplines but I still consider myself essentially a physicist. It's a matter of perspective. For reasons unknown, my inclination has always been towards seeing the world through the perspective of that discipline. I don't look at everything that way, of course; my enjoyment of music, art, drama, history, hiking, and the natural environment all call on very different ways of engaging with the world. I also increasingly work with colleagues who have perspectives from a wide range of other academic disciplines—economists, social scientists, geographers, statisticians, philosophers, etc. It's enjoyable to experience the radically different starting points and assumptions inherent in different fields of research, and it's interesting to reflect on the different ways people interpret the world around them. Nevertheless, when a problem can be viewed from the perspective of physics, I'm usually inclined to start by taking that view.

Different people have different experiences and different perspectives and this colours the way they think about climate change. This is pretty obviously true for the broad social issues of climate policy, personal responsibility, and the like. It is also, but less obviously, true for scientific questions related to the reality and the scale of the potential threat, to what represents reliable and salient evidence, and to how

---

[5] Sorry for the spoiler if you haven't seen or read it.

we approach climate predictions. If we are to understand what represents a reliable climate prediction we need to consider the baggage of our own perspectives. The next few chapters are about the different approaches to prediction taken by different people and different disciplines, along with the assumptions and potential flaws in each. Fortunately, they can be boiled down to just two approaches. I'll call them the **'look-and-see' perspective** (based on observations) and the **'how-it-functions' perspective** (based on understanding the processes and interactions taking place). The former is closely related to a statistical view and the latter to a physics view. A philosopher might relate the former to what is known as inductive reasoning and the latter to deductive reasoning.

For someone with a look-and-see perspective, the evidence that anthropogenic climate change is a reality comes from observations. For someone with a how-it-functions perspective, it comes from understanding how greenhouse gases trap energy which leads to heating and all sorts of related changes.

A how-it-functions perspective starts from a basis that information about the future is best obtained by evaluating the consequences of our understanding of the physical world. This understanding resides in the whole collection of physical and social sciences. By contrast, a look-and-see perspective of climate prediction has its basis firmly in observations of the whole complex thing. You may only be interested in particular bits but those bits are studied only within the context of the whole complex thing. With a look-and-see perspective, how any changes have come about is of secondary importance:[6] the 'how' is not what's important, what's important is the 'is'.

Consider catching a bus. Imagine that there is a bus stop outside my house and every morning (8am to 11am) for a week (Monday to Friday) I make a note of how long it is between one bus and the next. In most cases I find that it's about ten minutes; sometimes a little more, sometimes a little less, with very occasional extreme behaviour when two come along together or there isn't one for twenty minutes. With this information I perhaps feel I understand how the buses work. I start going in to town every weekday at a variety of times through the day—sometimes to meet friends for lunch, sometimes early to go to work, sometimes later to pick up my children from school, or go to a dentist's appointment. It sounds like a pretty relaxed and unrealistic lifestyle since most of us go to work at much the same time every day and kids need taking and collecting from school at regular times, so this 'variety of times' is not very realistic but let's not worry about that. When I want to go into town, I see a bus go by and I predict that the next one will most likely arrive in 10 minutes' time. I give a little leeway and leave the house 8 minutes later. For the next few weeks I use this method and it works fine. This is the look-and-see approach.

The alternative would be to use the published timetable. This approach is based on the times that the bus company plans for the buses to stop outside my house. They presumably have some control over when their buses depart from the bus station, and the timetable contains an understanding of those plans and of the distances travelled,

---

[6] Although one might try to deduce it from correlations in the observations.

speed limits, and likely traffic conditions. It says that from 7am to 6pm, a bus is due to stop outside my house at two minutes past the hour and at 12, 22, 32, 42, and 52 minutes past. So I use the timetable and predict that a bus will arrive at these times. This is a simple version of the how-it-functions perspective, and this works pretty well too. So predictions based on a statistical assessment of the observations (a bus has been seen to go by every 10 minutes) and an understanding of bus company plans (a bus is scheduled to go by every 10 minutes) both work well.

It's worth noting that in this situation the look-and-see approach has some advantages because it doesn't just tell me how frequent the buses are but also how much variability there is around some regular frequency. That's to say it can tell me how likely it is that the bus is actually 1 or 2 or 3 minutes later or earlier than anticipated. On the other hand, the how-it-functions approach is likely to be more reliable if I only have observations on weekday mornings but I want to know how frequent the buses are on Wednesday evenings or Sunday afternoons when the timetable might be very different.

The look-and-see approach assumes that past performance is a good guide to future behaviour and in the world of climate—as distinct from climate change—there are many situations in which past performance might well be expected to be an extremely good predictor of future behaviour. One of these is the seasonal cycle. Every year of my life it has been colder in February (as a monthly average) in Oxford than it has been in the following July. I first wrote a draft of this section in the very cold February of 2017; cold in Oxford that is. Daytime temperatures had been sub-zero (Celsius) every day for the previous five days; not of much note for Edmonton, Chicago, or Moscow but chilly for Oxford, England. I predicted at the time, predicted with great confidence according to my notes, that July 2017 would on average be warmer and that there would be no days with sub-zero temperatures. What was the basis for such a prediction? It could have been my own personal experience (observations), it could have been historical records of temperatures in Oxford (a more systematic set of observations), or it could have been the physical understanding of how it comes about: that in summer the sun is more intense. Again, a look-and-see or a how-it-functions perspective would suffice. The prediction turned out to be correct.

I feel confident in predicting that this characteristic will continue and that February in 2050 will be colder in Oxford than July in 2050. However, my confidence in this prediction comes not from the fact that it has always been this way but from understanding the cause. The effect comes about because the earth's axis is tilted such that in summer in Oxford, the Northern Hemisphere is angled towards the sun with an angle of up to 23.5°; an angle called the earth's obliquity. Consequently, days are longer and the sun rises higher in the sky so each square metre of the earth's surface in the Northern Hemisphere receives more solar energy each day than it does in winter. There is no reason to expect this to change in the next fifty years. Indeed it is pretty difficult to imagine how it could. It would take a pretty big whack from an asteroid or comet or some similar object to substantially change the angle at which the earth rotates in such a short period of time. So I am confident in the prediction, but I am confident not because it has always been that way but because I understand why. Physical understanding (the how-it-functions approach) tells me that the message from historic observations (the look-and-see approach) about July being warmer than February in Oxford is likely to be informative about the future. The two approaches work hand in hand. It is a highly reliable prediction more than thirty years in advance. Reliable long-term predictions are definitely possible.

Such a prediction is, though, dodging the issue. What we really want to know is how February in 2050 will compare to February in, say, 2020; or perhaps how Februarys in the 2050s will compare to Februarys in the 2010s. That is a very different question. Does the sequence of historical values for February temperatures in Oxford provide a good source of information for predicting the 2050 values or must we rely on physical understanding? Is look-and-see as good as how-it-functions for predicting this aspect of climate? Can I use how it has been observed to be changing in the past to tell me how it's going to change in the future (Figure 9.1)? If so, it is an important upgrade to the look-and-see approach because it is not saying the past tells us about the future because the future is like the past (as with the weekday daytime buses), it is saying the past tells us about the future because we expect how things **have been changing** in the past to tell us about how they **will change** in the future. You might call it look-and-see 2.0.

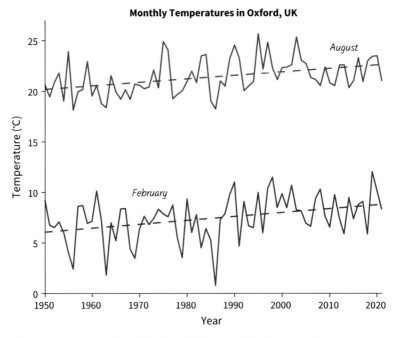

Figure 9.1 Average August (red) and February (blue) temperatures in Oxford, UK since 1950. The straight lines show the trend over this period. The question is what this trend tells us about the future.[e]

To get a grip on the extent to which we can make climate predictions, we need to understand these two perspectives better. We need to understand what assumptions they make and the basis on which they might claim reliability. Why should we choose one over the other and what represents a solid foundation for knowledge?

10

The times they are a-changin'

10.1 Induction, a problem?

Nearly 800 years ago, probably in the 1230s, Roger Bacon began his studies at Oxford University. He didn't study physics as we would understand it today, although his subsequent work was important in laying the foundations of that subject. He seems to have been strongly influenced by Robert Grosseteste—the first chancellor of Oxford University—who was there probably shortly before Roger arrived; it's not clear whether they ever met. Roger Bacon was an early advocate of the scientific method. He wasn't the first, though. Two hundred years earlier in Cairo, Ibn al-Haytham had made similar arguments. All three, Roger Bacon, Robert Grosseteste, and Ibn al-Haytham, were interested in light, vision, optics, and the foundations of observations. The importance of optics for them was not though, as we might consider it today, about the practicality of how one makes observations but about something rather more fundamental. It was about a view, a conceptual approach, that knowledge about the physical world is something one gains by observation, by so-called empirical methods. This went hand in hand with the concept described by al-Haytham that vision was something that occurred by light bouncing off objects and into the eye, something we now consider to be true. It's easy in the current era to overlook quite how fundamental this is. It provides a mechanism for connecting the observed with the observer, the individual with the world around them. It's the foundation of the scientific approach to understanding the world we live in.

This perspective—that knowledge comes from observations—is worth keeping in mind when considering the differences between the look-and-see and the how-it-functions approaches to climate predictions. Both are based on this concept—they are both fundamentally empirical methods—but their application of it differs a lot. The look-and-see approach relies on the idea that what we want to know about climate can be extracted directly from observing climate. The how-it-functions approach relies on understanding drawn from observations taken in a much wider range of settings over hundreds or even thousands of years. It's founded on our understanding of the physical world which arises from observations of physical objects well beyond climate and in circumstances which often allow for directed, repeatable, and repeated experiments—characteristics not easily found in the study of climate itself.

The look-and-see approach to climate prediction relies on an assumption that behaviour seen in climate observations in the past will continue to apply in the future. This assumption is an example of the 'problem of induction'—a problem beloved of many philosophers. The issue is whether such an assumption is reasonable: does behaviour seen in the past provide a reasonable guide to the future? It is nicely illustrated by the phrase 'a black swan event'. From the time of the Romans through to the seventeenth century, the term 'black swan' was used in Europe to indicate something that was impossible.[a] No European had ever seen a black swan and they assumed that they did not exist. In the late seventeenth century, however, Europeans came across black swans in Australia and so the meaning of the phrase has changed and now refers to an event that was previously unexpected or thought to be impossible or extremely unlikely. It captures the concept that we should be aware of possible surprises due to limitations in our experience.

With the problem of induction and the possibility of black swans ringing in our ears, it is easy to get embroiled in deep conceptual considerations and start questioning everything. Can we even trust that the laws of physics will continue to apply in the future?

I love such questions. But I am not a philosopher. For me it is enough to discuss them with friends over a beer or a coffee. On a practical level there are some things I am simply willing to accept. One is that the laws of physics as we understand them will indeed continue to apply on the earth at least over the next few millennia.[b] When you throw a ball in the air, I think it is more than reasonable to expect it to come down again—and to expect this today as well as in fifty years' time, wherever you are on earth: Austria, Australia, or Alaska. In this book, I am seeking confidence in climate predictions, not certainty about the behaviour of the Universe. Philosophers have much to contribute to the generation of climate predictions but we should be careful not to head down a philosophical rabbit hole that simply concludes with us not being confident about anything at all. So let's agree that there is a lot of basic science not subject to the problem of induction.

The acceptance that well-understood fundamental science will still apply in the future provides a solid foundation for the how-it-functions approach to climate change. It doesn't, however, help us very much with the look-and-see perspective. The fundamentals of science can be well understood and unchanging, while the behaviour of something as complicated as the climate system could still vary substantially over time. Indeed the behaviour of the climate system does vary naturally and substantially on timescales of decades, centuries, and millennia. We might reasonably expect it therefore to change as a consequence of increasing atmospheric greenhouse gas concentrations. For climate prediction, this suggests we have to consider carefully whether our observations contain the information we need to make predictions about what will happen when climate is pushed into a new, never-before-experienced state. Is there a reasonable chance that we'll discover new behaviour—black swans?

Of course, it is a core expectation of climate **change** that the future will not be like the past. The application of the look-and-see approach to climate prediction relies therefore on the belief that the future is like the past, only in the sense that the way climate will change in the future will be similar to the way it has changed in the past (look-and-see 2.0). We assume, for instance, that the consequences of increased carbon dioxide in a 2°C warmer world will come about through the same processes[1] as in a world which was only 0.5°C or 1.0°C warmer: we assume the changes seen with 0.5°C or 1.0°C warming provide a good basis for extrapolating to 2°C warming. Can that be relied upon? If the climate system is moving into a new state, or is being pushed in a way that it has never been pushed before, then the past may not be a good guide to the future. The processes which are most important might change. Ice sheets, ocean circulation, and land cover will all be different in the future, so the processes by which changes come about might also be different. Similar arguments hold for the impacts on society and the global economy—the past may not be a good guide. For the look-and-see approach, concerns about the problem of induction are very much alive and kicking.

10.2 Measuring climate change isn't easy

I'll return to the conceptual issues of relating the past to the future very soon but first let's consider the practical challenges of the look-and-see approach. Measuring changes in climate requires observations taken over time: that is, sequences of observations taken at regular or irregular intervals. Such observations are called 'time series'. A great deal of climate research involves the study of time series but they are also central to many other subjects as diverse as finance, astrophysics, economics, ecology, mathematics, and many more. In climate change, these time series may be of things outside anybody's individual experience, such as yearly averages of global average surface temperature, or things very familiar and local, such as daily rainfall in Lisbon. They may be of something remote to most of us, such as monthly values of the amount of sea ice in the Arctic, or something one might not consider climatic at all, like the size of the British butterfly population. Whatever the case, the starting point for the look-and-see approach to climate prediction is how we identify whether things have changed and if so, in what way and by how much.

In practice, this is often not easy because in almost all aspects of climate there is a degree of variability—even in global average temperature. This variability can have strange characteristics. It's usually not like multiple rolls of a dice but more like the number of wins that a football team has in a season. Unlike each subsequent dice roll, this year's result might be linked to last year's, even if over a long enough period, the average[2] were unchanging. When the average is changing, as it is with climate change, this natural relationship between this year and earlier years can make separating the variability from the change pretty tricky. Figure 10.1 shows some examples of how variability can make it difficult to see what is changing, even in idealized situations where the underlying change is clear and simple.

---

[1] In the same relative proportions.
[2] And the probability distribution.

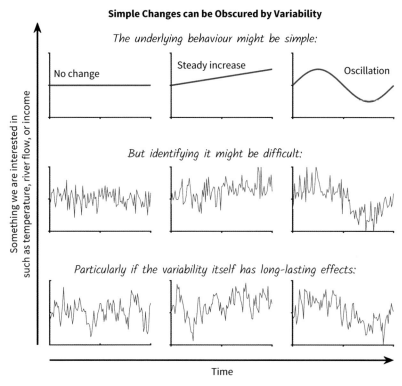

Figure 10.1 The underlying variation over time might be simple but it can nevertheless be very difficult to identify it from observations if there is also variability. The top, middle and lower plots all show the same change over time but the middle and lower ones also include different types of random variability, which obscures the signal.[c]

Climate variability tends to be larger and more complicated at small scales (say England) and smaller at large scales (say global averages)—see Figure 10.2. Thus the challenge of working out just how climate has changed locally is much harder than doing the same thing for an average over the whole planet; and it's hard enough for averages over the whole planet. The problem is knowing how much of the change we see in a climatic timeseries is due to climate as a whole changing and how much is the result of natural variability. This makes even measuring how climate has changed in the past difficult (Figure 10.3).

These difficulties lead to a whole collection of interesting challenges which span maths, statistics, and physics—but let's set aside these practical challenges for now. Instead, consider a situation where we can indeed work out how climate has changed in the past. In that case, how might we use that information to come to some conclusions as to what will happen in the future? We might try to predict temperatures in the future based on how temperatures have actually been changing in the past. Or, better still, we might try to predict future temperatures based on how temperature changes in the past have related to past changes in atmospheric greenhouse

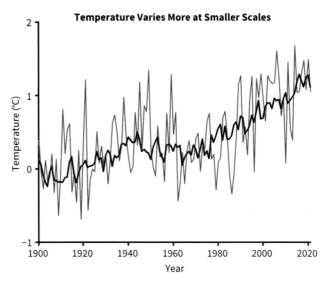

Figure 10.2 Timeseries of the change in global average temperature[d] (black) and central England temperature[e] (red). The change is with respect to the 1850–1900 average.

gas concentrations. This would provide a way to make predictions of the future based on assumptions about future concentrations of greenhouse gases, which is what we really want to do. The essential question, however, is whether we expect such an approach to be reliable; do we think the observations contain the necessary information to make such predictions?

It is tempting to assume they do because look-and-see methods are very familiar. Most of us use them all the time. We base our expectations of the future on what we see today and remember from the past. We trust our senses. It's not unreasonable therefore for our individual starting points when considering climate change to be an expectation that climate will continue to change in the way we are already seeing it change, whether that be out our windows or on the TV news. If we can see that the times are a-changin', **that's why** we accept it that soon we'll be drenched to the bone.[3] Past change **is** taken to be a good guide to future change: it's normal to think that the problem of induction is not a problem.

Imagine, though, that we had good observations of temperatures and everything else we might want to observe in the physical and social sciences, everywhere on the earth and throughout the history of the planet. Would the look-and-see perspective be able to give us a reliable picture of the twenty-first century and beyond? The answer, unfortunately, is no. The approach is fundamentally limited in what it can provide, not because of limited availability of information and statistical techniques to measure change, but because of its conceptual foundations.

---

[3] To misquote Bob Dylan.

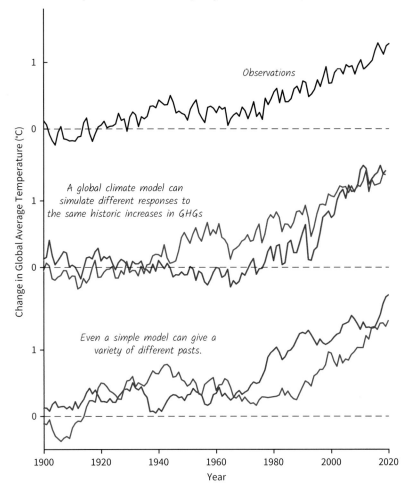

Global Temperatures are Naturally Rather Variable.
It's possible that historically they could have been quite different.

Observations

A global climate model can simulate different responses to the same historic increases in GHGs

Even a simple model can give a variety of different pasts.

Change in Global Average Temperature (°C)

Year

Figure 10.3 Upper plot shows observations. In the middle are two simulations with the same global climate model.[f] At the bottom are two simulations with the same, very simple model—one that encapsulates year-to-year and decade-to-decade variability.[g]

The thing is, we **know** that past behaviour will not be a good guide to future behaviour. We don't just accept that it might not be—as in a financial disclaimer—we **know** that it isn't. That's because key factors which influence climate, which drive its behaviour, are different today from at any point in the earth's history and we know that those factors will be even more different in the coming decades. Carbon dioxide concentrations are higher than at any point in the last million years and within the twenty-first century may well go higher than in the last 30 million years.[h] The context for such numbers is that anatomically modern humans have only been around for about 300,000 years and human civilizations for around 10,000 years.[i] That means

that we know we are entering unknown territory for our species. Observations—
however accurate and complete they may be—are of the climate in a different mode,
a different state to the one we are trying to predict. The dice has changed and we know
it will change further, but we don't have any experience of throwing this new dice.
The climate in the future will behave differently to how it has behaved in the past.
The consequence of this is that we should expect that observations do not contain
all the necessary information to make reliable predictions. Not directly at least. We
absolutely expect the problem of induction to be a real problem for us. To bring this
argument down to earth, and into the kitchen, consider something familiar—a kettle.

10.3 Kettles and climate

Think about the kettle full of water from Chapter 1. Or perhaps a different kettle of
water. Any conventional kettle will do (Figure 10.4). When you switch it on, the water
warms up. That we know. But how much will it warm up, and how fast? To answer
these questions, we can take observations of its temperature, use them to deduce
how fast it has been warming, and then predict what temperature it will reach in the
future. If we take observations over one minute and find that it warms from 30°C to
60°C then we can use them to predict that after another minute it will have warmed
by another 30°C and reached 90°C. This will be a good prediction. But if during our
observing minute it warms from 60°C to 90°C we don't expect these observations to
tell us what will happen in a further minute because the water will boil. The water in
the kettle will not reach 120°C. The observations of how it has changed in the past are
not, alone, a good basis for making the prediction and it is our knowledge of physical
science that tells us that this is the case. Here the look-and-see perspective breaks
down.

   Of course it wouldn't break down if we could repeatedly observe what actually
happens in that extra minute but climate change is about extrapolation and is a one-
shot-bet—no observations of the future, no repeats. Nevertheless, for the kettle our

Figure 10.4 A kettle.

physical understanding of water tells us not to trust a look-and-see perspective once it gets close to boiling. For climate, things are not so obvious because predictions require many different components to be combined and connected: components such as the lower atmosphere (troposphere), upper oceans, deep ocean, the upper atmosphere (stratosphere), ice sheets, ice floes, crops, forests, ecosystems, maybe even economics. Many of these components have thresholds beyond which their behaviour changes character, so the same principle applies as with the water in the kettle. A simple example is the behaviour of sea ice. Sea ice is important because it reflects more sunlight than seawater, so if the amount of ice decreases, the planet absorbs more energy from the sun and warms up more quickly. This is one of many processes that influence how much and how fast global temperatures change with increasing atmospheric greenhouse gases.

Arctic sea ice shows a lot of variability from year to year, but observations also show that the area of sea ice in the Arctic has clearly been decreasing since the mid-to-late twentieth century (Figure 10.5). The amount of sea ice also varies substantially by season, with the smallest area in September and the largest in March. One could imagine using observations to give us guidance on how sea ice will change in the future. Such predictions would be based on the observed decrease from year to year or decade to decade. But we know that when the arctic ice has gone, then it can't decrease any further. It is just like the kettle reaching boiling point. At that point assuming that changes in the future will be like those seen in the past is obviously wrong. Once September sea ice has reached zero it can't go any lower, so extrapolating from the past will fail. This is so obvious, of course, that nobody would be so foolish as to predict that the area of arctic sea ice will become negative. Imagine, however, doing the

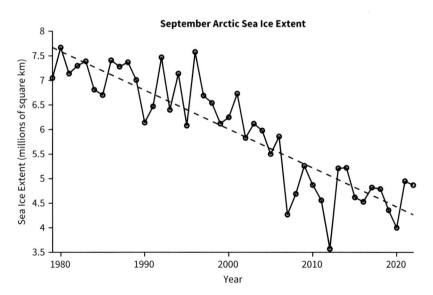

Figure 10.5 Time series of September arctic sea ice extent—the area with at least 15% sea ice cover.[j] The dashed line is the best fit straight line through the points.

same simplistic look-and-see prediction with observations of global average surface temperature over the last century or so (Figure 10.2). One might use such observations to predict how it will change in the future either by assuming the rate of change in the future will be a continuation of what has happened in the past or by deducing a relationship between temperature change and atmospheric greenhouse gas levels.[4] This may seem likely a reasonable approach, and it is one which is widely taken, but the observations of changes in temperature in the past incorporate the consequences of decreasing sea ice. We know, however, that the way arctic sea ice decreases will be different in the future because it will start to reach limits beyond which it cannot go. The prediction of global temperature will therefore be flawed beyond a certain point because the climate's response in the future will be different to that seen in the past. In this case, however, it will not be so obvious that there is anything inconsistent being done because nobody is talking about negative amounts of sea ice. Similar but more complicated arguments can be made in relation to clouds and other processes which are collectively known as climate feedbacks and which affect how much and how fast the climate warms with increasing atmospheric greenhouse gases. The point is that the change in climatic behaviour between 1.5°C and 2°C of warming will be different in character to that between 0°C and 0.5°C of warming.

Concerns of this nature represent fundamental issues for the look-and-see approach but is the how-it-functions approach any more reliable? Let's return to the problem of predicting the future temperature of the water in the kettle. A how-it-functions perspective to this prediction might look at the rating of the kettle (its wattage) and combine that information with knowledge of the properties of water[5] to deduce how fast it will warm up and at what point it will stop warming because the water is boiling. If the kettle is old fashioned, or broken, and doesn't have an automatic cut off, then it would also provide information on how long it would take for all the water to boil away, after which the structure of the kettle would warm further until at some point it would break and/or a fuse would blow. The how-it-functions approach doesn't have the problem of misleadingly predicting a temperature of 120°C because it knows that water boils at 100°C. It seems nice and accurate, but in practice I might want to supplement it with observations. Why? Because there are often specifics of the particular physical system, the kettle, which we don't know exactly. The kettle may be labelled as a 2 kW kettle but in practice it won't be 2 kW exactly. In any case, the power available from the electricity network will almost certainly be slightly different to that assumed in the manufacturer's specifications. The rate at which the water heats up will also be influenced by the temperature in the kitchen. If it's cold, like it is in the UK as I'm writing, thanks to cold winds from Siberia, then it will heat up a bit more slowly than on a hot summer's day. That's simply because more heat will be lost to the air in the kitchen while the kettle is on. Observations can therefore be of significant value in refining the physics-based prediction. Look-and-see

---

[4] Or possibly their energy trapping abilities known as radiative forcing.
[5] Its heat capacity.

and how-it-functions are not mutually exclusive approaches but they do represent very different perspectives on the problem and on where to place our confidence.

The kettle example is trivial, of course. Most of us know how water behaves. Physicists or not, we know that it boils at roughly 100°C. Furthermore, familiarity with using kettles to heat up water for a nice cup of tea or coffee means we are confident about how they behave. Consider instead then a slightly different situation where the kettle is full of an unknown liquid. In this situation, we don't have familiarity with what happens and we don't know at what temperature the liquid boils; this is much more like future climate change. Given our lack of knowledge about the behaviour of the liquid, we might be inclined to use the observations for one minute to predict the next. We might think this is the best we can do—but is it? The answer is: only to a limited extent. We don't know at what temperature the liquid boils but we do know that liquids boil. We have enough physical understanding to know it would be sensible to be cautious in our prediction because it might boil at any time. Much the same can be said about climate predictions: we don't know what will happen, but we have lots of reasons to be cautious.

The look-and-see approach to climate prediction is attractive because it appeals to a familiar way of seeing the world. We know, however, that the climate is changing a lot, so we have reasons to suspect that familiarity may not be reliable. We know that the future will not be like the past, and we have good reasons to expect that the way it will change in the future will be different to the way it has changed in the past. What this adds up to is knowledge that we need to be cautious about the conclusions from the look-and-see approach. Fortunately, the how-it-functions approach seems to hold the prospect of more reliable predictions. After all, it is founded on fundamental scientific principles and 'ye cannae change the laws of physics'.[k] However, applying this approach to climate is not easy and usually requires a computer. Ideally a big computer. As big as you can get. Or better still, a lot bigger than that.

11
Starting from scratch

11.1 The basics and a bit more

If we're going to build a picture of future climate from fundamental principles, what are the principles to start from? The most important one is the law of conservation of energy: energy cannot be created or destroyed; it can only change form, for instance when sunlight is converted from radiation into the heat of the sand on a warm summer's day. This law alone is enough to build a tremendously useful, if simple, representation of climate change. It's not, however, enough to provide reliable predictions by itself and nor can it provide many of the details we desire: local changes, rainfall information, heatwave characteristics, and so on. Nevertheless, add three more fundamental principles and we get four physical laws which can take us a long way towards these details. Together these are:

- The law of conservation of energy.
- The law of conservation of mass.
- Newton's second law of motion.
- The ideal gas law.

These laws are reliable: they have history on their side in domains way beyond climate. Conservation of energy is a profound aspect of classical physics. Conservation of mass is one of those pretty obvious aspects of physics which is nevertheless crucial to making all sorts of deductions. It simply says that mass, like energy, cannot be created or destroyed; it can move around and change form (e.g water to water vapour) but it can't just appear or disappear. These two conservation laws become combined into one in Einstein's special theory of relativity, which is important when considering objects travelling at speeds close to the speed of light—about one billion km/hour—670 million miles/hour if you're working in imperial. Such details, however, are completely and utterly irrelevant for the types of processes we're interested in when we talk about climate, so it's absolutely fine to ignore Einstein. Newton, though, cannot be ignored. Newton's second law of motion, first published in 1687, tells us how rapidly something changes its velocity when given a push. And finally, the ideal gas law, nigh on 200 years old, provides a relationship between the pressure, volume, and temperature of a gas. It's an extension of Boyle's law from Chapter 2; it makes precise certain types of behaviour which are really quite intuitive. For instance, if you try to squash a balloon full of air—nicely tied and ready for a party—you find that the balloon expands in those parts where you're not pushing; it generates nodules before eventually bursting. By squashing the balloon you are decreasing the volume of the air inside (there's the same amount of air but in a smaller space), and as a result the pressure inside increases and the air pushes out harder generating nodules and somewhat increasing the volume again. The point is that there's a relationship between pressure and volume. This relationship is one of several captured by the ideal gas law.

These laws have been thoroughly tested over hundreds, even thousands, of years. They are reliable foundations.

With these basics, and some bits and pieces about gravity and other details[1] which can also be traced back to Newton, you can build a model of the atmosphere. You start by breaking it up into millions of boxes (Figure 11.1). Each box is represented by a set of variables which describe the air within it: temperature, pressure, density,[a] speed (west to east), speed (south to north), speed (bottom to top). The four laws described above are encapsulated in six equations; Newton requires three all for himself (not surprising as he seems to have been a rather self-centred sort of chap who liked to grab the available attention).[2] The equations tell us how the variables in a box change

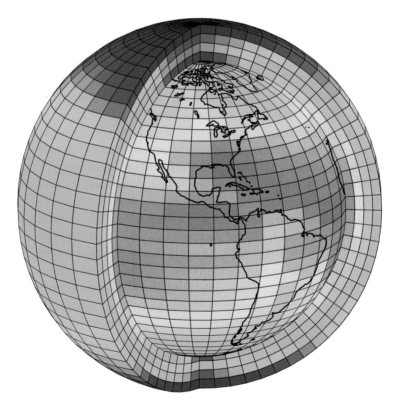

Figure 11.1 Illustration of the atmospheric part of the model grid in a global climate model. The colours represent temperature.

---

[1] Viscosity.

[2] Okay I'm being a bit mean to Newton here. He only gets three equations because his one equation handles vectors, which have direction and magnitude, rather than scalars, which just have magnitude. To solve it on a computer, however, requires breaking it down into three. But he does appear to have been a pretty self-centred sort of chap.

based on what they are now and what the values are in the surrounding boxes. That's to say, if we know what the state of the atmosphere is at the current moment, then the laws give us a good scientific basis for working out the state of the atmosphere in the future. Problem solved. Climate prediction sorted. And all based on fundamental, reliable, scientific principles. Bosh.

But of course, it's not really sorted. There are still many details to resolve. And the devil—whatever form he or she may take—resides—as is so often the case—in the details. For a start, how many equations are we actually trying to solve? The previous paragraph says six—that doesn't sound too bad, surely we can solve six equations? Unfortunately, although in each box the equations look the same, they are applied to different variables. The temperature near the ground in Bergen is not the same thing as the temperature 10 km up in the atmosphere over Delhi; both are temperatures but temperatures of different chunks of air. Thus the equations for the box near Bergen and the box above Delhi look identical but they are actually different equations because they relate to different things. As a result we need to treat each box's equations as different, unique. That means we don't have six equations to solve but rather millions of equations—six for each box—along with many millions of variables.[3] And they're all connected with one another—well each is connected with its neighbours which ends up being the same thing. That's many millions of simultaneous equations—not something you can do on the back of an envelope. It's beginning to look tricky but it's about to get much worse.

What does 'solving the equations' tell us? Solving the equations in each box tells us how the variables in that box change over a period of time. That period of time could be a decade, or a year, or a day, or 5 minutes. It's called a timestep. The process then needs to be repeated until you get to the point in the future you're interested in. Each time you solve the equations, you take one step into the future—but you might need several steps to get to the time you want to know about. With a ten-year timestep, you'd have to solve all these equations repeatedly ten times to predict what happens in 100 years. At each step the results from the last iteration are fed into the next. This process of solving equations by breaking down space and time into chunks is part of a field of study called numerical analysis, and it enables the equations to be solved (approximately) on computers. Today's global climate models (GCMs) use this approach to reformulate the aforementioned laws in a way that a computer can be programmed to solve.

Unfortunately, it turns out that you can't just choose whatever timestep you want; that would be too easy. The maximum timestep one can use is limited by the size of the boxes. In today's global climate models (GCMs), the ones used to make projections of the twenty-first century, the grid boxes are very roughly 100 km by 100 km at the surface of the Earth, with around forty to ninety layers going up through the atmosphere.[b] The boxes may be smaller or larger depending on the particular model and what it is being used for. At the typical resolution of today's models, the length of

---

[3] Mathematicians might say that the 'state space' of the problem is very high dimensional: of the order of a hundred million or higher.

the timestep is limited to a maximum of about ten minutes. That means to simulate 100 years, you have to repeatedly solve the millions of equations about five million times. That's a lot of calculations and begins to explain why climate modelling needs big computers.

This limitation on the timestep comes about because if the timestep is too long then information could jump over grid boxes and that would be unrealistic. Imagine a model simulating a hurricane which traverses the Bahamas and goes on across Miami and southern Florida and out into the northern Gulf of Mexico. If, at some point in time, the hurricane is over the Bahamas and the timestep in the model is one or two days, then it could be that at the next timestep the hurricane is already beyond Florida and in the Gulf of Mexico. The model could be saying that there would be huge damage in Nassau in the Bahamas, and no problem at all in Miami, but that the hurricane might appear again beyond Miami heading towards Tampa or New Orleans causing further damage. That would clearly be wrong. If the size of the boxes allows us to differentiate between Nassau, Florida and the north-west Gulf of Mexico then Florida is 'in' the model and a hurricane can't just jump over it. If the model resolution allows for local detail, then the timestep must be consistent with the timescales on which the local details change. If a hurricane can jump over Florida, then the model cannot be getting close to solving the equations correctly. In practice, if you try to use a timestep that is too long the computer simulation rapidly becomes unstable and starts to generate numbers that are too big for the computer to process. The model then simply stops working: it crashes, blue screen of death, 404 error—solution not found.

The study of numerical analysis provides many ways to optimize the building of GCMs but it can't get round this limitation on the timestep. The exact way it is limited is described by something called the Courant–Friedrichs–Lewy condition, which dates back to mathematical work published in 1928, long before the birth of modern computing. That's perhaps not so surprising because the concept of making weather forecasts by solving these equations using numerical analysis was first proposed by Lewis Fry Richardson in 1922. Without access to modern computing, his approach relied on humans, who he called computers, with each one solving one equation for one location at each timestep and relaying the result to those around him/her. Richardson's ideas very much reflect the way weather forecasting is approached today and provide the foundations of climate forecasting. It is an inspiring example of how progress can be made in designing good scientific approaches long before the technology exists to implement them. There are lessons here for today's climate research.

This timestep constraint is at the heart of climate forecasting because it explains why there is a balance between the geographical resolution we're interested in, the timeframe we want to predict, how soon we need our prediction, and the available computing resources. If we want high-resolution detail and we want to simulate a hundred years and we want the answer within a few months, then we might need computing resources beyond those currently available. On the other hand we can certainly simulate a hundred years in even just a few hours, but only if we're willing to accept a lower resolution, one which might not differentiate between say Trento and Verona, Rotterdam and Eindhoven, Calgary and Banff, or Santiago and Valparaiso. If we want higher resolution and still want the result in hours then that's also fine but only if we're happy to limit our outlook to a couple of years ahead; a hundred years isn't an option.

The models used in weather forecasting are built on the same foundations as climate models—indeed they are often versions of the same models—but weather forecasters only need to simulate a few days, so they are able to set up their models with very high resolution. They still have a balance to make though, because if the resolution is too high then the model runs too slowly and might take two weeks to make a five-day forecast, which isn't much use.

Every climate or weather model experiment involves compromises which are governed by their target purpose and the available equipment. Interpreting model results requires, above all else, understanding what compromises have been made.

11.2 Something's missing

After all this, though, the fundamental equations are actually just the beginning. There is much more to GCMs. What I've just described is referred to as the 'dynamical core' of the model. Its conceptual foundations are pretty much rock solid—even if interesting questions remain about the efficacy of their representation on a computer. Unfortunately though that's not enough to simulate either weather or climate. For a start, light from the sun (shortwave radiation) and radiation emitted from the earth's surface and the atmosphere (longwave radiation) interact with the gases in the atmosphere; greenhouse gases absorb longwave radiation, solar radiation is reflected by clouds and scattered by the atmosphere (which is why the sky is blue), etc. The models have to take account of these interactions if they are to be at all realistic and simulate the response to increasing greenhouse gases. Other processes are also missing from the fundamental equations: clouds and rainfall are two important examples. Without clouds and rainfall, the heating and cooling of each grid box would be unrealistic and the whole thing would not in the least resemble the real world. Indeed, without rainfall it would be missing one of the key aspects which we might want a climate model for in the first place. Yet the processes which generate clouds and lead to rainfall are not included in the core equations.

On top of the processes which are missing from the dynamical core equations, there are also processes which the equations could capture, except that they take place on scales smaller than the resolution of the models. One example is the rising air (convection) that leads to cumulonimbus clouds and thunderstorms. The versions of these models used in weather forecasting often have high enough resolution for the dynamical core equations to capture this air movement but in the climate versions the need to simulate decades rather than days means that the spatial resolution has to be much lower and hence such small scale features are missing. Another example is gravity waves. These aren't exciting waves in the fabric of space suitable for

Marvel Universe heroes to surf. No, these are like waves on a pond, except that they are created in the atmosphere near the earth's surface and travel across the planet and upwards through the atmosphere until they break very high up. This process influences the winds and circulation patterns all the way up to the mesosphere (more than 50 km above the surface) and affects the climate from one pole to the other.

Whether processes are missing because the fundamental equations don't represent them (e.g. clouds) or because the resolution at which we can run our models doesn't allow them to arise out of the equations (e.g. gravity waves and convection) doesn't matter. Whatever the origin of the problem, these processes need to be included if the behaviour of our models is to look anything like reality. So how is this issue resolved? The answer is to include the processes in the models using pieces of computer code called 'parameterizations'. These pieces of code do two things. First, they represent the impact of the missing processes (e.g. clouds, gravity waves, etc.) on the large-scale circulation patterns of the climate system—the ones captured by the fundamental equations. Second, they represent the consequences of the large-scale circulation patterns for things that we're interested in but aren't able to simulate from foundational principles (e.g. rainfall). They are critically important parts of any climate or weather model.

11.3 Beyond the atmosphere

So far, I've discussed only the atmosphere, but of course the climate system is much more than the atmosphere so in a bid to annoy my oceanography, geomorphology, surface processes, and ice sheet friends, I'll allot just a single paragraph to the rest of the climate system. If we're going to simulate climate change, then of course a model has to include the oceans, the land surface, and possibly ice sheets, glaciers, etc. The oceans can be simulated in much the same way as the atmosphere, albeit that salt plays a rather more important role and there is an even greater demand for high resolution in order to simulate many features. The principles, however, are the same: there are a set of fundamental equations that look very similar to the ones in the atmosphere, there is a trade-off between resolution and timescales,[4] and there are a range of processes which can only be included through parameterizations. Representing sea ice and the land surface—that is, rainforest, tundra, desert, agricultural land, cities, and so on—is done through further parameterizations. Ice sheet modelling is a field in its own right and involves its own fundamental equations and its own parameterizations. And then there is the carbon cycle and atmospheric chemistry—how the chemical constituents of the atmosphere interact and change over time—both of which we would expect to be important on the timescales of climate change. The bringing together of GCMs with ice sheet models, carbon cycle models, and atmospheric chemistry models is leading to a whole new range of models known as earth system models (ESMs), which include an ever wider range of processes.

---

[4] Controlled again by the Courant–Friedrichs–Lewy condition.

Although ESMs include more processes, the conceptual perspective remains the same as for GCMs: that we can predict future climate by building a computer representation of the whole earth system. This is called a reductionist approach. The idea is that we construct the whole thing from its constituent parts; build it up from the smallest building blocks. It's an attractive idea and the rapidly increasing availability of massive computer power makes it appear within our grasp. Nevertheless, we should keep our excitement in check and reflect on the reliability of this approach. The chief difficulty is the parameterizations. Well, the parameterizations and the resolution. Actually the parameterizations, the resolution and the choice of which processes to include/exclude. There are a number of chief difficulties but let's focus on the parameterizations.

The parameterizations aren't representations of well understood and robustly justified physical processes. These parameterizations are not physics. Some of them are designed to reflect physical principles but many of the processes are not well understood and even if they were, their representation in grid boxes 100 km wide doesn't come with the same sort of foundational, historical reliability that the core dynamical equations have. Consider clouds, for instance. These are critically important for how energy conservation is achieved in the atmosphere: they reflect and absorb sunlight from water and ice particles and they generate rain and snowfall which moves water and energy around. Yet cloud droplets are small. Smaller than a tenth of a millimetre. They form by condensation of water vapour on particles such as dust, aerosols, and sea salt. Individual clouds have characteristics and details that can vary on scales from centimetres to hundreds of metres. Neither individual droplets nor individual clouds can be represented in current models with grid boxes 100 km across; nor would they be if the resolution increased to 10 km or 1 km. So clouds are instead dealt with by parameterizations which describe how much of a grid box is covered.

The parameterizations are written in terms of the density of particles on which droplets can condense, the amount of water vapour in the air, how long ice particles spend in the cloud, etc., but they don't describe what's actually going on at the scale of droplets or individual clouds. This needn't be a problem so long as they are based on robust theory.[5] Unfortunately, they don't represent robust, well-tested theories for how the distribution of clouds in a grid box are related to the other variables representative of the grid box—theories that we might trust to apply in very different states of the climate. Consequently, there is no one best way to build a cloud parameterization and within any particular one there are many uncertain parameters. These parameters control, for instance, how long ice lasts in model clouds which itself has a big impact on how the clouds interact with long and shortwave radiation. Such things matter because they affect how the whole modelled climate system responds to changing concentrations of greenhouse gases, and since the values of these parameters are not constrained by theoretical understanding, they are open to

---

[5] After all, we can describe the behaviour of gases and liquids in terms of pressure, density, and temperature without representing every molecule.

selection by model developers. All of a sudden we've gone from robust physics to predictions that can vary at the whim of—or possibly the carefully considered but limited perspective of—a computer programmer.

Many parameterization schemes have the same characteristics: they don't represent fundamentally well-understood processes and they have parameters whose values are not constrained by theoretical understanding.

At this point, let's take a step back. It seems I've got embroiled in a lot of heavy—you might think tedious—detail regarding how we build models. All this detail seems largely irrelevant unless you're wanting a job building climate model parameterization schemes. Why should you care about this stuff? Why is it important? It's important for two reasons. First, it means that we can't automatically trust the models' extrapolatory predictions because although they have some elements built on robust physical understanding, they have other elements (the parameterizations) which aren't: some aspects are founded on the laws of physics, but lots are not. The application of the how-it-functions approach in terms of complicated computer models of the climate system can't claim the reliability that we'd like it to. Second, the parameterizations are tuned to enable the models to represent past observations, but that puts us in the same situation we were in with the look-and-see approach. We know climate is changing, so we know that the future will not be like the past and we have good reasons to expect that the way it will change in the future will be different to the way it has changed in the past. That's why we were suspicious of the look-and-see approach. Models tuned to represent the past suffer the same problem: they may not reflect the way things will change in the future when the state of climate will be different. They are tuned to capture how the processes have combined in the past—not how they fundamentally behave.

So the how-it-functions approach isn't the panacea that it appears to be.

Things aren't quite as bad as they seem though. Some aspects of the models do have a sound physical basis and even where they don't, they **might** capture credible behaviour. This makes them valuable research tools. The models can capture more and different information than the look-and-see approach and that makes them useful. Nevertheless, we certainly shouldn't take them as equivalent to reality, or anywhere near it. We shouldn't use them as if they generated the climate predictions we're looking for. Rather they are just another source of information. Where look-and-see makes many debatable assumptions about future drivers of change, models make more specific assumptions about, for instance, the response of landscapes or clouds being as we've described them in the model. They're different from reality but they are nevertheless potentially useful to help us understand aspects of what we might expect under climate change. They require us, however, to question hard the impact of all the assumptions we're making. We need to be very, very cautious about using these models to guide public and industrial planning, international negotiations, and government policy. Unfortunately, the structures of the research system (funding, publications, indicators of an individual's or institute's success) don't tend to reward such caution. This is a structural flaw in our research infrastructure which undermines the relationship between climate science and society, and risks undermining efforts to achieve a robust, reliable response to climate change. More on this later in the book.

11.4 More models, better predictions?

The conclusion at this point is that computer models aren't a panacea to the problems of climate prediction: the how-it-functions approach isn't the desired solution to the problems of the look-and-see approach. We seem to be in a bit of a pickle. The research challenges this raises and how we might address them are dealt with in the following chapters, but first there's a need for a bit more background on the way climate change science actually uses computer models today. Two issues are important to note: one is the diversity of different types of models out there; the other is that there are multiple examples of each type of model.

So far, I've discussed global climate models (GCMs)—computer models of the atmosphere, oceans, land surface, and cryosphere (the ice components of the climate system)—and the more comprehensive ESMs, which include details of the carbon cycle through representations of land and ocean ecology and biogeochemistry. These are reductionist models designed as a representation of the climate system built up from all the different underlying, interacting, physical processes.

ESMs and GCMs—I'll refer to them collectively as just GCMs—represent as much of the global climate system as they can[6] but in the study of climate change there are many other types of model as well. Firstly there are regional climate models (RCMs). These are the same as the global models but they only cover a particular area of the earth—say, southern Africa. They enable that area to be resolved in much greater detail but of course all parts of the earth's climate are ultimately connected with all the rest—the equations in a GCM provide a connection between everything—so we can't simulate climate change in one region independently of all the others. The RCMs get around this problem by taking the results from a GCM to provide information about what is going on elsewhere on the planet. Such an approach helps us understand regional processes better but in terms of climate prediction it creates massive conceptual problems. If the GCM's predictions are flawed then the RCM predictions will inevitably also be flawed because the GCM results are used to constrain what the RCMs can do. Despite this, RCMs can still show different behaviour to the GCM used to guide them. However, if the higher-resolution RCM results are inconsistent with the lower-resolution GCM results (for instance, they show more or less rainfall in the region) then this implies that they are simulating different physical worlds. In practice, this is often the situation with current models.[c] In this case, the GCM world has one set of rules which describe how the region behaves, and these are used to

---

[6] Neither ESMs nor GCMs are comprehensive—far from it—but ESMs allow one to study a model-future (a 'future' in the world of the model) driven by assumed greenhouse gas **emissions**, while GCMs have to be driven by assumed changes in atmospheric greenhouse gas **concentrations**. The latter involves additional assumptions about how the carbon cycle distributes the emissions around the earth system.

control the RCM world which has a different set of rules for how the same region behaves. The results are, therefore, not representative of any self-consistent world, let alone the one in which we live. It's a conceptual nightmare. What, if anything, this tells us about the future under climate change is something that needs a lot more work to understand.

Beyond RCMs and GCMs there is a range of entirely different models. Models designed to represent particular aspects of the physical and economic world. There are models of crops to study the consequences of climate change on food production, hydrological models used to study the consequences of climate change on flood risk, and what are known as 'integrated assessment models' to study the economic implications. Most of these suffer from some or all of the issues already discussed regarding the lack of reliability of parameterizations, the consequences of tuning to historic observations, and the use of questionable GCM outputs to frame and control their own simulations. The role of some of the economic ones will be touched on later in Chapter 25, which looks at how the physical changes affect us, the societies in which we live, and the economies in which we operate. In all cases, however, these models are, to some extent, driven by how we expect the global, interconnected physical climate system to change, and that information is often taken from GCMs. Consequently, the bulk of climate change research is, to some extent, conditional on the assumptions made when we build and interpret GCMs.

In addition to this plethora of different types of model, there are also multiple models of each type. Consider GCMs, for instance: the 2021 report from the Intergovernmental Panel on Climate Change (IPCC) presented results from at least forty different GCMs.[d] Some of these are different versions of the same model, while others have been developed by entirely different modelling centres around the world. Just because they were developed by different groups, however, doesn't mean that they are entirely—or even substantially—different. The modelling community shares approaches, techniques, methods, and sometimes even computer code, so different models can't be treated as independent approaches to the same problem. This lack of independence is a reason to be cautious in their interpretation: all models approach the problem in much the same way, and their shared methods and techniques mean that any consistency between models should not be interpreted as a basis for increased confidence. The often-heard claim that 'all models show' something isn't in itself a sound reason for thinking that that something is more likely to come to pass in reality.

Nevertheless, although they have links and similarities they also have differences and they generate different results (Figure 11.2). You might think that that's a problem but it most certainly is not: it is definitely a good thing because it gives us a way of thinking about—and studying—the uncertainties in model-based predictions. The greater the diversity in computer models, the better.

Comparison of models is facilitated by international collaborations known as model intercomparison projects (MIPs). A vast amount of time and effort is spent by modelling communities generating simulations which are designed to be comparable with each other in these MIPs. A vast amount of further effort is then spent

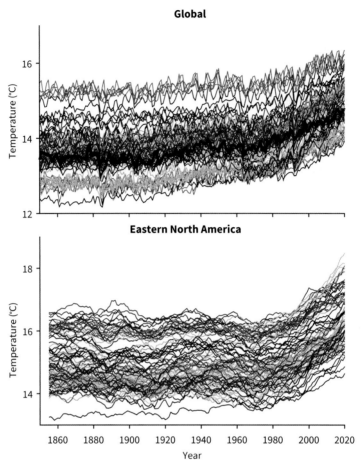

Figure 11.2 Different models are very different—even in terms of global or large regional temperature averages. This figure shows the Global (top) and Eastern North American (bottom) average temperature from multiple global climate models (different colours) and multiple simulations of the same model (same colours)—all models with only a single simulation are dark blue.[e] Global data are yearly averages. Regional data are ten-year averages. The thick black line is observations.[f] The point is that the models are all very different in the simulations they generate.

poring over the details of these simulations, trying to understand the similarities and the differences. These efforts are, in many ways, the core of climate change research. The coupled model intercomparison projects (CMIPs) are the most influential MIPs because they simulate the twentieth and twenty-first centuries under various assumptions about future emissions of greenhouse gases. There are, however, many others. There are MIPs focused on clouds, on ice sheets, on geoengineering,[g] on crop models,[h] and on impact and economic models,[i] as well as a wide variety of others. The academic study of climate change in the early twenty-first century is dominated by MIPs which absorb a vast amount of the available research funding. There are, however, no solid foundations regarding how MIPs should be interpreted since consistency of behaviour between inter-related models is simply not a good basis for confidence about the behaviour of reality.

There is huge enthusiasm for—and investment in—building bigger and more complicated climate models and running model inter-comparisons. The availability of ever-increasing computing power drives this process and hence controls how we pursue climate change knowledge. If this investment is to be in any way useful however, we need to consider much more carefully what we require of models to make confident, reliable statements about reality. Can we ever expect a model which we can just run and in whose output we can trust?

That's a big question which I'll return to later on, but before we go there it's worth first reflecting on what we expect of climate science and how that influences the reliability of what it produces. Is climate science the type of science you can ask questions of and expect robust answers suitable for guiding society, or is it still in the realm of debating approaches and methods? Is climate science pure or applied science, and are there lessons from the history of science for how we should interpret it?

12
Are scientists being asked to answer impossible questions?

12.1 First build your foundations

One might expect the foundational challenges in climate prediction to have been widely studied before anyone started building climate models. That isn't the case. This is partly because no one has ever sat down and said, 'Today I'm going to build a reductionist computer model of the climate system that I'm going to use to predict the detailed consequences of increasing atmospheric greenhouse gases'. Or even 'I'm going to write a research proposal and gather a team to spend the next ten years building a reductionist computer model of the climate system to predict the detailed consequences of increasing atmospheric greenhouse gases'. That may seem surprising given the large number of models out there and how important they are for climate policy, but the general circulation models (GCMs) we have today have mostly been adapted from models built for other purposes. The original models were built for research on particular parts of the system (the atmosphere, the ocean, etc.) or from models designed to be used in weather forecasting. They were never originally designed for climate prediction. Of course, building on what has gone before is not a bad thing to do given the complexity of the task but it does embed an implicit assumption that all we need to do is make the models more complex and higher resolution. It assumes that a weather forecasting model with added oceans and land surface processes is necessarily a good climate forecasting model. The conceptual differences between climate prediction and weather prediction aren't addressed in the rush to get an answer. The urgency of the societal issues and the availability of computational resources wash away careful consideration of whether the methods we are using are fit for the purpose of supporting society, or even of supporting scientific research on climate change. The fundamental constraints on what we are trying to do rarely get a look in. That **is** a problem.

It's not just a historical accident though that the foundations of the climate prediction problem have not been thoroughly addressed; there are also strong social drivers of science that discourage addressing the foundational issues. The demands from society—from the funders of science and the employers of scientists—are for practically useful results. Reliability is rarely a priority, perhaps because it cannot be easily assessed due to the extrapolatory and one-shot-bet characteristics. Scientific analysis has made clear why we should expect human-induced climate change; it has gathered observations which demonstrate that it is taking place; and it has illustrated the many reasons to expect it to have profound impacts on our societies. It's not surprising therefore that societies, governments, and funding bodies turn to science and ask for more detail. What will the world's climate look like in 2050? How will floods, droughts, food production, vector-borne diseases, etc. be affected? How high do we need to build our sea walls and river flood defences? How should we design our buildings and infrastructure to be resilient to future changes? What will be the impacts on economies and trade? What are we committed to, and what can we avoid?

These questions are tremendously important to ask but they are directed at climate change science as if it were an applied science: a science that you can go to and get answers to whatever questions you might have. That's to say questions about climate change are treated like engineering questions, like whether some design for a bridge will support a given amount of traffic. They assume that climate change science is about applying what is already understood to practical problems. It is not. Climate change science, and climate prediction particularly, is a young science still struggling to work out how to approach the problems and identify what works and what doesn't.

Climate change is a huge societal issue, so utilizing our best knowledge is essential. Much of climate change science is, however, pure science pretending to be applied. It's pretending to be applied in the sense that the questions it addresses are of practical importance: they are not esoteric questions relating to, for instance, what happened before the Big Bang. The questions are important, societally relevant, and even urgent, but that is not enough to make climate change science an applied science: it does not mean that climate change science can necessarily answer the questions we want it to. Many of the questions we ask of it require new scientific foundations to be built. They require us to seek out and encourage diverse approaches. They require pure research and much wider debate and discussion across disciplines and across perspectives (multidisciplinarity). It is not a matter of turning a handle.

The reliability of, and uncertainties in, predictions of the consequences of climate change should be the subject of frequent and fervent debate. Currently they are not, and this creates problems. How can a policymaker, or any non-specialist, know when predictions made by scientists are something they should trust in the way they should trust that the earth is (almost) spherical, and when they should be taken with a pinch of salt: the latest results along a winding path towards comprehension and useful guidance to society? If this question is not always at the forefront of the research agenda, then there are real risks of climate science misdirecting society. Nevertheless, remember the eleven paragraphs of Chapter 8—these concerns are not about the existence and seriousness of climate change. Rather they are about the information used to guide trillions of dollars of adaptation investments and the descriptions of our future societies which motivate personal, national, and international climate change debates, agreements, and actions.

12.2 Familiarity breeds caution

One of the consequences of climate change science being simplistically treated as an applied science is that climate models gain great prominence. They are so tempting. They're like the fanciest cake in the window of a bakery—sometimes the eating experience lives up to the looks, but more often it doesn't and you wish you'd chosen something plainer. In the same way, climate models look like they provide all the details we crave. They produce data that looks like it represents the answers we want: a detailed picture of the future. It's not surprising therefore that so much of climate change science is founded on them. Indeed it is often founded on the **same set** of global climate models because their outputs are used as inputs to other models and other types of analysis. Most climate predictions, whether of global temperature, heatwaves, health impacts or economic losses, are therefore founded in some way on the same assumptions, the assumptions inherent in climate model interpretation. That leads to a degree of comforting consensus but it's a type of consensus that we should view with caution.

There is consensus among scientists about the reality of climate change and the serious threats it poses to our societies[a] but there is always a risk that consensus is an indication that everyone has gone down the same blind alley. To ensure that this is not the case, it is important to ask about the foundations of the consensus; not just whether people agree but why they agree. In this case the consensus on the reality of the threat is reliable and is based on good understanding and strong foundations (see Chapters 8, 22 and 23). The foundations for making predictions about the details of quite what the future climate will look like are much less strong.

Some humility would be appropriate. There are many examples of overconfidence in the history of science. Scientific endeavour has often been distracted by ultimately invalid approaches and perspectives which have remained largely unchallenged for decades or centuries. That's to say, scientific progress often gets stuck in a rut. The urgency of the climate change threat requires that we should be investing significant time and effort in reflecting on what ruts we might be stuck in because we don't just want to get it right eventually, we want to get it right soon.

Climate predictions are being used to prepare our societies for the future. Get it wrong and we could be dramatically ill-prepared. If, based on a prediction, a region invests heavily in flood protection but it turns out that the real risk in that location is drought and a need for water storage and irrigation, then the prediction itself has caused damage. Furthermore, when most adaptation efforts are driven by the same family of models, multiple regions, nations, and industries are all being encouraged to prepare for the same model-informed future. If that future misrepresents reality with regard to the patterns and details of change, then vast swathes of society could all be ill-prepared **in the same way**. Large-scale adaptation to anticipated climate change is only just beginning, but the stakes are high in terms of understanding what information climate science and climate models do and do not provide.

Of course, computer models are not expected to give the 'right' answer—few, if any, who work with them think that that's the case. That, though, is not the point. The real issue is whether the uncertainty estimates we put on their projections are even remotely accurate. As with the Bank of England GDP forecast, the question is not whether the headline number is right but whether the nine out of ten probability range actually represents a nine out of ten chance of what will actually happen. A salutary lesson in this regard comes from looking at the speed of light.

In 1986 Max Henrion and Baruch Fischhoff published an analysis of how estimates of physical constants have changed over time. Their number one focus was the speed of light. The thing about the speed of light (in a vacuum) is, one, that it is very large and, two, that it is constant. If I could travel at the speed of light, I could go from London to New York and back more than twenty-five times in a second.[1] That makes the speed of light tricky to measure but the important aspect here is that it is a constant. It doesn't vary over time. The historical challenge for physics was therefore to measure it. But the measurements did vary over time. Furthermore, estimates of the measurement uncertainties also varied over time. Measurements in the late nineteenth century typically had large error bars reflecting the large uncertainties due to the difficulties of the task. By the 1930s and 1940s, the recommended value had shifted a little lower than the earlier estimates. The error bars—the measures of uncertainty— had also reduced, reflecting greater confidence in the abilities of the scientists and the methods. By 1941 it looked like there was general agreement on the correct value and one prominent physicist of the time commented that 'after a long and at times hectic history the value for [the speed of light] has at last settled down into a fairly satisfactory "steady" state'.[b] Nine years later, the recommended value increased. By a lot. The new value was higher by more than twice the previously estimated errors (Figure 12.1). The new value would have been given less than a 1% probability of being correct based on the former results. What's important for us is not that the earlier estimates were inaccurate but that they were over confident about the value they had derived. They were wrong but confident that they were right: their probabilities were unreliable.

A few years later it happened again. New measurements led to the recommended value rising again, and again it was by much more than one would have expected from the uncertainty estimates on the previous value; on this occasion it would have had less than a 2% probability of being the new value based on the previous error bars. This is just one specific example but Henrion and Fischhoff document a widespread tendency to underestimate the errors on measurements of a range of physical constants. The message is that there is a common tendency to be overconfident and to underestimate uncertainty in challenging areas of research.[c]

A related problem is that research can get stuck in particular paradigms. An existing theory is made more and more complicated so as to fit the observations rather than the theory itself being questioned. The classic example of this is the role of epicycles to account for observations of the motions of the planets. Back in the fourth century BCE Plato and Aristotle described a geocentric version of the universe in which the sun, the moon, and the planets revolved around the earth. In the third and second centuries BCE, astronomers such as Hipparchus proposed that the moon and planets moved in small circles, epicycles, which themselves moved around the earth in a large circle, called a deferent. Later, in the second century

---

[1] I'm not sure why I would want to but I could.

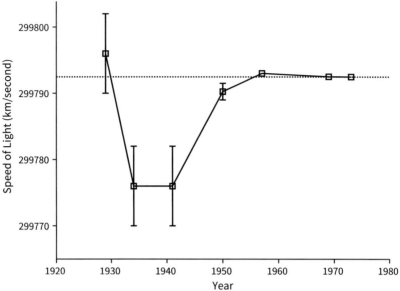

**Recommended Values for the Speed of Light
(And their uncertainty estimates)**

Figure 12.1 How the recommended values for the speed of light, and their uncertainties, changed over time.[d] The dotted line shows the value for the speed of light from 1983 onwards.

CE Ptolomy made this more complicated still by proposing that the earth wasn't quite at the centre of all these rotations but slightly offset from the centre. The introduction of epicycles enabled this geocentric model to account for oddities in the observations, in particular that the planets occasionally seem to go backwards in their orbits, something called apparent retrograde motion. The introduction of epicycles was a great success and the Ptolemaic model was taken as the standard model of the cosmos up until the Copernican revolution in the sixteenth century, when it was replaced by a heliocentric view in which the earth and the planets rotate about the sun in elliptical paths. Actually Copernicus used circular orbits but later Kepler introduced the more accurate elliptical orbits and simplified the whole description. The point for our considerations here is that Kepler's heliocentric view is much simpler than the Ptolemaic model and consistent with subsequent understanding of the laws of gravity. It is, in essence, correct. The Ptolemaic model is very good at describing most of the observations, but it doesn't describe what is actually going on, so it would be of little value when trying to extrapolate to new situations such as the behaviour of a newly observed planet. This contrasts starkly with the heliocentric model, which is able to explain historic observations AND could be applied effectively and reliably to predict the behaviour of 'new' planets or different situations.

It turns out that by adding epicycles within epicycles to the motions of a planet, one can reproduce whatever planetary orbits one wants.[2] The term 'adding epicycles' is itself used to describe the situation where a theory is made increasingly complicated in an effort to make it agree with observations. The lesson for climate modelling is to question when we are making models more realistic and more reliable for extrapolatory predictions, and when we are just making them more complicated so as to be more representative of today's climate alone. When are we using physical arguments to fit data, and when are we using physical understanding to fit reality?

There are many other examples of situations where the scientific endeavour has gone down blind alleys for decades or longer. From the late seventeenth century to the early twentieth century, it was widely accepted that light was some sort of wave in a material known as luminiferous aether but this aether was never actually detected. Eventually Einstein's special theory of relativity provided a description of light without the need for it. Sometimes, however, even non-existent objects are detected. In the first half of the nineteenth century, peculiarities in the orbit of Uranus led to the discovery of the planet Neptune. Urbain Le Verrier, who played a role in discovering Neptune, also studied oddities in the orbit of Mercury and concluded that there must be another planet even closer to the sun than Mercury itself. He called it Vulcan. In the second half of the nineteenth century a number of astronomers claimed to have observed this planet which is now known not to exist. The oddities in Mercury's orbit have instead been explained by Einstein's theory of relativity.

These examples of how the scientific community can go down the wrong track are simply examples of the process of scientific research struggling towards better and more complete explanations of the universe in which we live. They are examples of the success of science but they also suggest a need for caution in fundamentally new types of scientific endeavour, particularly if there is an urgency to that endeavour. The use of computer simulations to predict behaviour in never-seen-before situations is certainly one of these cases. History tells us that if we want answers that we can rely on, we need to invest heavily in understanding the foundations of, and in stress testing, our methods. For reliability there needs to be a large and active research community questioning the assumptions used and the claims made. Where climate science is failing society is that this hardly exists.

12.3 Being aware of our limits

'Over a twenty year period, stocks are almost certain to go up. (There is no twenty year period in history in which stocks have declined in real value, or have been out performed by bonds.)'

p. 131, *Nudge*

---

[2] At the time of writing there is a YouTube video demonstrating how one can achieve a planetary orbit which traces out the shape of Homer Simpson.

The above quote comes from the book *Nudge* by Cass Sunstein and Richard H. Thaler, the latter winner of a Nobel Memorial Prize in Economic Sciences. They go on to say: '*Many investors [. . .] invest too little of their money in stocks. We believe this qualifies as a mistake, because if the investors are shown the evidence of the risks of stocks and bonds over a long period of time, such as twenty years (the relevant horizon for many investors), they choose to invest nearly all of their money in stocks.*' This is an example of just how much confidence experts and laymen alike put in not just the look-and-see approach but in the look-and-see approach version 2.0: the future will change in the same way as we have observed it to change in the past.

Thaler and Sunstein clearly believe that stocks will continue to be a good investment on twenty-year plus timescales. They may well be right. I'm not an economist so what would I know? It might nevertheless be worth asking why they expect that look-and-see 2.0 is a reliable approach in this case. They might, for instance, believe that for this not to be the case requires a substantial change to global society and perhaps a breakdown of the market system; something that they don't believe is a credible possibility. This may or may not be a reasonable assumption, although given the scale of potential climate change it is at least open to question. Thinking about how the prediction could be wrong might be informative. It is noteworthy that they only say that stocks are '**almost certain**' to go up; they leave open the possibility of something surprising happening and stocks going down. This 'almost' makes the prediction probabilistic—they allow some small but undefined probability that the outcome could be different. An explanation of where this small probability comes from would be a significant improvement in the prediction because it would allow us to make our own judgements of how small it might be.

There are many reasons to expect future changes in climate, particularly at the scales of nations, regions, towns, and cities, to be badly represented by look-and-see 2.0 approaches. Yet the familiarity of such approaches makes them the obvious go-to method for thinking about the future. Politicians and civil servants are likely to be much more familiar with look-and-see approaches than with how-it-functions approaches,[3] and might well see statements along the lines of 'this has never been seen before' as strong evidence that it will not happen. Yet climate change implies that this is exactly the kind of statement that is not reliable, for both physical climate and also for finance systems and economics.

A better foundation for climate predictions might be a how-it-functions approach utilizing computer models, but climate models come with all their own questions of reliability; they aren't a panacea to the prediction problem. The challenges in both perspectives are deep and conceptual. They aren't about the specifics of climate or the atmosphere or the oceans, but about how the application of how-it-functions and look-and-see methods are constrained by the essential characteristics of climate predictions—the characteristics outlined in part one. The really important and interesting questions in climate change science at the moment are about the foundations of our knowledge, not about warming, heatwaves, retreating glaciers, and the like.

---

[3] Most of them are not scientists!

We should be cautious about our conclusions until we've fully examined the justifications for them. That's the message from the history of science—but the pressures on researchers today push them in the opposite direction: society and research structures want answers, or at least something that looks like an answer whether it is right or wrong. Of course, we do want answers, and we want them as soon as possible (there is urgency), but we also want reliability. If science tells us that a certain societally relevant question cannot currently be answered, then that in itself can be useful knowledge. It is certainly more useful than an answer that on reflection, and utilizing only knowledge that exists today, can be seen to be wrong or misleading. The difficulty is that the machinery of scientific research is structured to give answers, not to question what is answerable.

The lie of the land in climate change science suggests we should look out for potholes, blind alleys, and misdirections. A key aspect of climate prediction is that there is a lot of land (multidisciplinarity) and many potential blind alleys, so this is a big task.

The next few chapters get into some of the exciting issues that we need to understand to be able to approach climate predictions with confidence, and to be able to design climate science to be useful to society. The challenges concern maths, physics, economics, philosophy, history, chemistry, biology; methodological approaches to research; how we connect science with society; and structural issues about the role and drivers of researchers and experts.

If we wanted to design our first aeroplane, it might not be best to start by adapting a car or a bicycle. In the same way, we need to design and plan from scratch how we approach the climate prediction problem rather than relying on tweaks and upgrades to tools designed for other purposes.

We need to think about what we can and can't say about the future state of climate and how it interacts with society. Climate science needs to be clearer about what information is robust (and why we believe that to be the case), and also clearer about how much we don't know (and why we believe that to be the case). It needs to be at once more arrogant and more humble.

There's a lot to do, a lot of new types of problems to solve and many connections to make. The challenges of climate prediction are conceptual, fascinating, and fun. If we choose not to address them, though, climate science risks misdirecting our societies and potentially making matters worse. In an effort to help make a better world it would be easy to create many unnecessary problems. Let's be careful out there.

Challenge 2: Tying down what we mean by climate and climate change.

We think we know human-induced climate change when we see it: increasing temperatures, decreasing glaciers, increasing sea level, decreasing sea ice. What, though, do we really mean by 'climate'? What exactly are we trying to predict when we make a climate prediction? How can we tie down 'climate' in a way that is precise and yet applicable across the many different disciplines involved in the analysis of climate change? And how can we identify what uncertainties matter most for predicting climate change?

13
The essence of climate

13.1 Is this what you expected?

Welcome to Chapter 13; Chapter 13 in a book predominantly about climate change. It may seem odd to have got this far and only just be at the stage of asking: what is climate? It's actually not so strange, though because 'climate' is a slippery object. Even in the business, there is no precise, widely accepted, and widely applicable definition. Climate under climate change is slipperier still; we're talking eel-like levels of slipperiness.

This is a problem. It's difficult to have constructive debates about climate change if we can't be precise about what we mean when we use the term. It's harder still to judge what makes a reliable climate prediction if we're unsure what type of thing we're predicting. On top of that, characterizing climate change requires the bringing together of expertise from diverse disciplines (multidisciplinarity), so without a consistent expectation of what climate predictions consist of, there is little hope of getting reliable, joined-up conclusions. For all these reasons tying down climate is important, so the next few chapters are about getting rid of just enough complexity to hone in on what we are talking about when we say climate has changed. It's about what climate and climate change mean from a conceptual and mathematical perspective. These are the most mathsy chapters of the book but don't worry if that's not your thing; there are no equations, it's all about ideas.

To begin with it's important to differentiate between 'climate' and the 'climate system'. The climate system is a complicated multi-element system which includes the atmosphere, the oceans, sea ice, ice sheets, rainforests, ecosystems, and more. These elements vary over a wide range of scales: Europe vs. South America vs. Antarctica, San Francisco vs. Los Angeles vs. San Diego, the Mediterranean vs. the southern Ocean. The climate system includes so much that it's difficult to take it all in, so to get a handle on what we mean by climate we need to take a big step back and consider something that doesn't have so much complexity. A good starting point is dice—for me it usually is. In the coming chapters I'll go beyond dice to systems which have characteristics more like the real-world climate system but which nevertheless remain simple enough to play with. These will enable us to explore the consequences of the characteristics of part one for both what we mean by 'climate' and what we mean by 'climate change'. For the moment, though, let's stick to dice.

Consider a standard, fair, six-sided dice: the sort you would normally use to play Monopoly or Backgammon. What is the climate of such a dice? The best starting point for an answer comes in a quote from Robert Heinlein's 1973 science fiction novel *Time Enough for Love*: 'Climate is what we expect, weather is what we get.' So

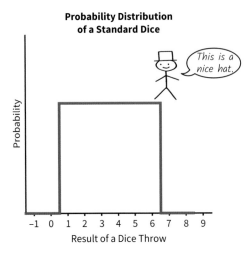

Figure 13.1 Top-hat probability distribution of a standard, fair dice.

what do I expect when I roll my dice? Well it's a standard dice so I expect the outcome to be 1, 2, 3, 4, 5, or 6, and since I said the dice was also fair, I expect each of these numbers to have equal probability of occurring. A description of 'what I expect' is a combination of what outcomes are possible and the probability of each outcome coming to pass. Figure 13.1 is a representation of this: the x-axis represents the possible outcomes and the y-axis the probability of each outcome. Figures like this are called 'probability distributions'; I'm going to be referring to them a lot. In this case it has a simple shape referred to as a 'top-hat distribution' because that's what it looks like.

When I roll my dice I get some particular outcome. Just now[a] I got a 2. I could say that the current weather of the dice is 2, while its climate is an equal probability of each number 1 to 6. Weather is what I got, climate is what I expected.

This concept that climate is the probability distribution of all possible outcomes, or all possible states of weather, is the foundation of a solid definition of climate. It's an excellent foundation but let's take it as work-in-progress. It isn't sufficient to completely tie down climate in a useful way, for reasons we'll come to shortly. Before getting to them, though, it's worth noting that even this, let's face it, simple concept is not what today's definitions focus on. There is a great desire to make it simpler still.

13.2 Averages are a poor representation of climate

Modern definitions of climate put a strong emphasis on averages. The Intergovernmental Panel on Climate Change (IPCC) defines it principally as 'average weather'. Their full definition goes:

"Climate in a narrow sense is usually defined as the average weather, or more rigorously, as the statistical description in terms of the mean[1] and variability of relevant

---

[1] The 'mean' is simply one particular type of average.

quantities over a period of time ranging from months to thousands or millions of years. The classical period for averaging these variables is 30 years [. . .]. The relevant quantities are most often surface variables such as temperature, precipitation and wind. Climate in a wider sense is the state, including a statistical description, of the climate system."

IPCC, 2021[b]

The Royal Meteorological Society says simply:

"Climate: The long term (often taken as 30 years) average weather pattern of a region."[c]

The American Meteorological Society provides a similar take:

"Climate: The slowly varying aspects of the atmosphere—hydrosphere—land surface system. It is typically characterized in terms of suitable averages of the climate system over periods of a month or more, taking into consideration the variability in time of these averaged quantities. Climatic classifications include the spatial variation of these time-averaged variables."

American Meteorological Society, 2002[d]

A serious problem with all these definitions is their focus on the average. The average[2] outcome of multiple throws of my standard, fair, six-sided dice is 3½. This is an outcome that can never actually occur. It sounds warning bells if our definition of climate could lead to climate being a type of weather which can never happen. In any case, it doesn't tell us much about what to expect: the range of behaviour that is possible. The same is true with the real-world climate system. The average temperature across the year in Chicago is roughly 11°C but typical values range from nearly 30°C on summer days down to nearly −7°C on winter nights, with much more extreme values being not uncommon.[e] The average is simply a bad summary of the variation in weather and is rarely what we should expect.

The problem is made worse by climate change. If we think about climate as an average of weather, then it is natural to consider climate change as a change in that average. Yet it is entirely possible that under climate change the average temperature or rainfall in a particular location could change little, while the distribution and the relatively uncommon extremes could change a lot. Or alternatively, the average could change a lot but the extremes very little; this would be the case, for instance, if almost all summer days become a few degrees warmer but the real scorchers stay the same or decrease. Such details matter for responding to climate change but are completely lacking when changes are presented as averages. No average can capture the diverse aspects of climate that matter across society.

---

[2] By which I mean the 'mean'.

It is interesting that this emphasis on the average is a recent move in the history of climate science. An earlier version of the American Meteorological Society's definition, from 1959, described climate as 'the synthesis of weather; the long term manifestations of weather, however they may be expressed'[f] with no mention of averages at all. Earlier, in 1938, the British climatologist W.G. Kendrew[g] had explicitly deprecated the use of averages, saying:

> "Certainly no picture of climate is at all true unless it is painted in all the colours of the constant variation of weather and the changes of season which are the really prominent features, and it is inadequate to express merely the mean conditions of any element of climate."
>
> Kendrew, 1938[h]

Defining climate as the weather we expect is more useful than an average, with or without climate change. If I visit Chicago in January, then I expect a range of rather low temperatures from cold to bitterly cold, and that helps me plan what clothes to pack. Similarly if we knew the probability distribution of Chicago temperatures for Januarys in the 2070s, then this would be useful for considering future heating needs which would be valuable for designing buildings today. In both cases, averages would be much less useful. Averages are an oversimplification.

The IPCC climate definition somewhat recognizes this when it says that a more rigorous definition involves a '*statistical description in terms of the mean and variability*'. This is better—but not much better. It allows for climate being a probability distribution, which is good, but takes it to be one which is expressed in terms of an average—in this case, the mean—and variability. This encourages researchers to think about climate as a probability distribution with a nice smooth curve—sometimes a bell curve, sometimes other simple curves that can be easily described mathematically[3] (Figure 13.2). These distributions are great if you are inclined towards statistics and want to use the many tools in a statistician's toolbox to play with climate variables. Real-world climate distributions, however, are often not represented very well by such idealizations. Even where they capture the overall shape of the distribution, they can fail to capture societally relevant details, such as the likelihood of exceeding specific temperatures or how they are changing.

We want our descriptions of climate, and of climate change, to be reliable in the sense that they give sufficiently accurate probabilities for the full range of possible outcomes. How accurate they need to be depends on how they are going to be used. This means that we can't separate climate research from the application of climate research because the latter defines what aspects of the distributions are important. Even when climate research is pure and conceptual rather than designed to be used by society, it still needs to be guided by an understanding of what would ultimately

---

[3] These are called 'parametric distributions'. They often require only two numbers to completely describe them (for instance, mean and variability) although some more complicated ones require three or four or more.

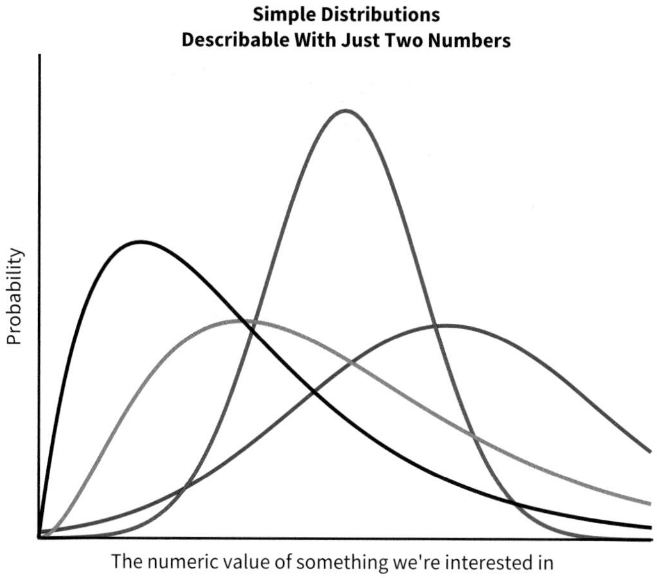

Simple Distributions
Describable With Just Two Numbers

Probability (y-axis)

The numeric value of something we're interested in

Figure 13.2 Simple distributions don't capture important details of climate probability distributions. All the distributions shown here can be described by just two numbers which tell us about their mean and their variability. The blue and red ones are examples of the classic 'bell curve' or 'normal distribution'. The black and green ones are called 'gamma distributions'—they are not symmetrical like the normal distribution but are still smooth and idealized. Distributions for many real-world climatic variables are not like this.

be useful. Relatively small deviations between reality and idealized distributions, deviations that might appear unimportant to a researcher, might be tremendously important for the practical application of predictions in terms of say health impacts or economic productivity. The focus on averages, or averages and variability, is handy for scientists but not so much for society.

This situation can be illustrated with more dice. I have a number of peculiar dice (Figure 13.3). My fair, standard, six-sided dice has a top-hat probability distribution (Figure 13.1). I have another fair, six-sided dice which is not standard. It is blue while my standard one is green, but that's not what makes it peculiar. What is odd about my blue dice is that it has no three; instead, it has an eight. The possible outcomes are 1, 2, 4, 5, 6, 8. So the average outcome is $4^{1/3}$ and the probability distribution, my work-in-progress 'climate', is shown in Figure 13.4. This distribution is badly represented by the average ($4^{1/3}$) and by any simple mathematical distribution which uses only the average and the variability about the average. The impossibility of achieving a three or a seven is an important aspect of this dice. If I'm playing backgammon then knowing that such outcomes are impossible may well change my strategy. Similarly, if I'm designing major infrastructure that

Figure 13.3 My collection of strange dice.

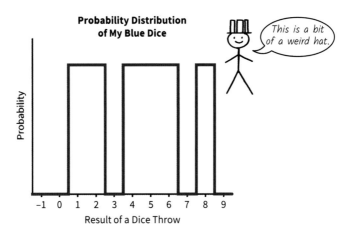

Figure 13.4 Not quite a top-hat distribution: the probability distribution of my blue dice.

I want to last for at least sixty years, I might want to know what types of potentially damaging climatic behaviour are and are not possible. That requires much more than an average and usually much more than an average and variability: it requires a characterization of the whole probability distribution, whatever shape it may take.

Assuming that climate can be usefully summarized by an average, or a simple idealized distribution, constrains our thoughts and our analysis in an unhelpful way. The very definitions of climate that we currently have in climate science work against researchers providing useful information. Our work-in-progress definition, the probability distribution representing what we expect, is already more likely to be useful.

13.3 Maybe averages aren't so bad

Having taken a dig at the major definitions, I want to row back a little. The thing about averages[4] is that there is usually less uncertainty in them than in other aspects of a probability distribution. By the end of the twentieth century a major focus in climate research had become climate **change** and in particular detecting signals of climate change: demonstrating or confirming the reality of the concern. For that particular task taking an average makes good sense.

Averages combine lots of data. The more data you have, the smaller the uncertainty in the average. Consider, for instance, the average height of American women over twenty. According to a study by the US National Center for Health Statistics the average height of American women was 161.3 cm (5′3.5″) in 2015–2018 based on a sample of 5510 individuals.[i] They give the uncertainty in this average as 0.19 cm (0.075″) and this uncertainty[5] has a precise meaning. It means[j] there is a 68% probability of the actual average being no more than the uncertainty above or below the estimated average; that's to say, there is a 68% probability that the actual average is between 161.1 cm and 161.5 cm (5′3.43″ and 5′3.58″).

What's important for the pros and cons of averages is that the uncertainty **in the average** is much smaller than the uncertainty in the height of any particular individual. The study says there is a 70% probability that a US woman would be between 154 cm (5′0.6″) and 168 cm (5′6.3″) tall—a 14 cm (5.5″) range—but a similar probability, 68%, that the average is between 161.1 cm (5′3.43″) and 161.5 cm (5′3.58″)—a 0.4 cm (0.15″) range. The point is that the uncertainty in the average is much smaller than the uncertainty about the height of any particular woman.

More important still, the uncertainty **in the average** gets smaller the larger the sample of data we have. If we only had data for fifty women rather than 5000+, then we would typically expect the uncertainty in the average to be ten times greater—maybe closer to 2 cm rather than 0.19 cm.[k] This feels right. Even intuitive. The more samples we have, the smaller our uncertainty in the average. If we have ten women and add one more who is particularly tall or short it can change the average quite a lot. But if we have a thousand women it's not going to make much difference, even if the one extra is extremely tall, say 185 cm (a little over six foot).

On the other hand even with a sample of a thousand, one extra particularly tall woman might make a big difference to the details of a probability distribution—to the probability of an individual being over 183 cm (6 foot), say. That's because there are very few of them in this category so one more can make a big difference to the probability we estimate. The uncertainty in the probability of being over 183 cm is therefore much greater than the uncertainty in the average height.

The consequence for us of these simple statistical facts is that it may be easiest to detect a difference between, or a change in, distributions by looking at their averages simply because we are likely to be most confident about what the averages actually

---

[4] Particularly the mean.
[5] It's called the 'standard error of the mean'.

are. This too can be seen in height data. If we look at data for the Netherlands we find that the average height of women over 20 in 2010–2013 was 167.5 cm[l] (5′5.9″). We can therefore be confident that the average height of Dutch women is greater than the average height of American women because 167.5 (5′5.9″) is greater than 161.3 (5′3.5″) by much more than the uncertainty in each average: 0.19 cm (0.075″). And if the averages are different, then we also know that the distributions are different.[m]

So if you simply want to identify whether two probability distributions are different, a good starting point is to look at estimates of their averages because the uncertainties in the averages will be smaller than the uncertainty in most other aspects of the distributions. Furthermore, this uncertainty decreases as we include more data. All this means we are more likely to be able to confidently detect the existence of climate change by looking at the change in thirty-year average temperatures than by looking at the change between particular years, between particular handfuls of years or between instantaneous measurements of temperature. The average daily summer temperature around Amsterdam in the thirty years 1950–1979 was about 19.7°C with an estimated uncertainty[6] of 0.06°C, while in 1991–2020 it was 21.1°C with an estimated uncertainty of 0.07°C.[n] The change is clear because the uncertainties are small. By contrast, the average Amsterdam June temperature in the period 1974 to 1976 was 19.2°C with an uncertainty of 0.39°C, while from 2004 to 2006, it was 19.8°C with an uncertainty of 0.46°C. In this particular case the data shows some warming—choose different years and one can see cooling or alternatively much greater warming—but the uncertainties are much larger which should lead us to be less confident that this change hasn't just come about by chance.

13.4 No—actually averages are bad after all

A change in an average can give us confidence that climate has changed but we do need to be cautious. If the average has not changed, it does not mean that climate has not changed. The probability distribution for the types of weather that we expect may change as a result of climate change in ways that are not captured by the average but which might still have big impacts.

This too can be illustrated by dice. It might not surprise you to learn that I have another peculiar dice. This one is black (Figure 13.3). Like the blue one it has one of the conventional numbers missing and replaced by something else. In this case it is the six replaced by a seven. Unlike the blue dice though it has another important difference. It is loaded. It is a six-sided dice but it is neither standard, nor fair. It is loaded so that the four comes up much more frequently than any of the other numbers. About seven out of ten rolls produce a four when I roll it on my carpet. The other numbers appear to come up at roughly the same frequency as each

---

[6] Standard error of the mean.

other although the three, which is opposite the four, is a little less common. So the probability distribution of outcomes, the climate of the dice, looks something like Figure 13.5.

The average outcome of my green standard, fair, six-sided green dice is 3.5. The average outcome for my peculiar, fair, six-sided, blue dice is 4⅓ and the average outcome for my peculiar, loaded, six-sided, black dice is about 3.9. Imagine that these represented the climate of Glasgow and Central Scotland: the green one as the present and the blue and black ones being different potential consequences of climate change. Consider the following parallels between the dice numbers and Glasgow weather: one cold and wet, two cold and dry, three mild and wet, four mild and dry, five warm and wet, six warm and dry, seven hot and very wet, and eight hot and very dry. The green dice doesn't have any hot or very dry or wet periods. The blue one shows the greatest change in the average: with climate change we see the removal of mild, wet days and the addition of hot, very dry days. The black one shows much less change in the average because almost all days become mild and dry, although among the remaining days, there are a few that are hot and very wet in a way never seen before. These different changes in climate would have very different consequences. The blue dice perhaps suggests a need to invest in flood protection against frequent surface flooding because rain only comes when it is warm or cold, not when it is mild, so it falls on ground which is likely to be warm and hard or cold and frozen, leading to greater run-off. The black one arguably suggests even greater impacts: the absence of many cold days, for instance, could substantially affect ecosystems and agriculture; the lack of warm days could also influence ecosystems and agriculture as well as tourism; the reduction in the overall number of days with rainfall could lead to water shortage; while the occasional hot, very wet days could lead to moderately rare but extremely heavy flooding. These very different changes are not captured in any way by the change in the average: a greater change in the average doesn't represent

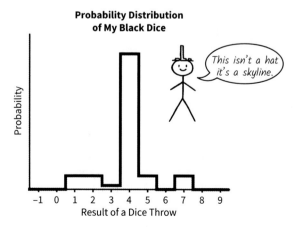

Figure 13.5 Even less of a top-hat distribution: the probability distribution of my black dice.

a necessarily more severe outcome. The black dice with the smaller change in the average could represent just as, or more, severe impacts on society as the blue dice.

The impacts of climate change are felt through changes in the details of the distribution of weather. What matters is the change in the probability of extreme temperatures, the reduction in the frequency of low temperatures, the likelihood of intense rainfall events, the longer or shorter periods with little or no rainfall, and similar details. These are the things that impact individuals and society. For the social impacts on everything from labour productivity[o] to crop yields[p] it is periods above or below particular thresholds, and the change in their probabilities, that really matter. These thresholds may be extremes or they may not—different aspects of weather matter to different parts of society—but the average weather is rarely what is most important. The focus on averages is ubiquitous in climate change assessments and is underpinned by conventional definitions of climate, but it undermines the ability of scientific research to integrate understanding across disciplines and provide useful information to society.

This takes us back to the concept that climate is what we expect: the climate of the climate system is a probability distribution of weather types, and climate change is a change in that distribution, however it may manifest itself. Thinking about climate in this way is useful and flexible: adaptable to whatever your interests and vulnerabilities are. Furthermore, we don't need to be constrained by the complexity of the real world to talk about climate and climate prediction. We can talk about the climate of a dice or the climate of anything that has some probability distribution representing what happens. Which brings us back to climate models.

At its heart a climate model is simply a set of equations that have been represented on a computer so that when we provide some starting conditions, the computer tells us what will happen next. We can study this process using much simpler systems; we don't need oceans, monsoons, rainforests, and ice sheets. Using something simpler provides a way of studying the process of climate prediction itself. It helps us identify the wood without getting distracted by the trees, and it removes the computational constraints because instead of millions of equations we can study the process using only, say, three or five. It turns out that 'climate is what we expect' doesn't entirely solve our problems—we need a bit more to a definition of climate—but to see why we first need to get a picture of what these simpler systems are. Let's take a walk.

14
A walk in three dimensions

14.1 Peculiar paths

Imagine going for a walk along a path. Maybe a hike in some mountains or a trek along ridges and around lakes. This path does not split and you can only go forward, no going back. You are also completely constrained by the path, you can't step off it on one side or the other. (This is not like rambling around a forest or a field—this is following a path.)

Or maybe imagine you're going around a museum or an exhibition. Maybe the Plantin-Moretus Museum of Printing in Antwerp or an exhibition at the National Gallery in London. Or maybe just imagine you are in a large home furnishing store—IKEA, for example—following your way around. What's important is that in the imaginary exhibition, museum, or IKEA, one room leads to the next room to the next room to the next room, and that is the only route you can follow. On top of that, you cannot go back. In reality it is never quite like that, although in IKEA it feels that way.[a]

The thing about these paths or routes is that where you go next depends entirely on where you are at the moment. If you are in room twelve of the exhibition then next you will be in room thirteen and then room fourteen and then in room fifteen. If you are in soft furnishings then next you will be in bathrooms and then bedrooms and then kitchens. There is no possibility that you could jump from room twelve to room fifteen, or from bedroom furnishings to kitchens. That just can't happen. It's not the way the world works. It's impossible.

There are usually maps for these paths—a walking map for the mountain trek, an exhibition plan, or a store outline to show where you are and, most importantly, where you go next. One feature of these maps is that they are flat—you can lay them out on a table; they are two-dimensional. The terrain of the mountains may go up and down but the map is flat. The home furnishing store may have two or three floors but the plan is two-dimensional, with the different floors represented separately and connected by symbols for stairs and lifts/elevators.

With this in mind, I want you to make a small conceptual leap. Imagine that you are in some space like a sports hall—a large, rectangular cuboid space, and also that you are not constrained by gravity so you can move wherever you want: up/down, left/right, forwards/backwards. It's not that you are floating around in space like an astronaut; you're not Sandra Bullock or George Clooney in *Gravity*, however much you might want to be. No, in this situation you have more control, you can simply 'step' to wherever you want to be next. You aren't constrained to be on the floor, you can go anywhere you want in the whole volume of the sports hall space—only you're limited by a path. You can't choose where to go next because there is a path and you must follow this path through the three-dimensional space. The path tells you what step you can take. It's a three-dimensional path like that in Figure 14.1. You start at some point and from that point onwards, the only thing you can do is move forward following the path wherever it goes. However, unlike the mountain hike or the exhibition, this path is not something you can draw on a flat, two-dimensional map because it wanders around the whole of the three-dimensional space.

To make things just a bit weirder, you can start wherever you want but from that point onwards there is only one route forward. The path appears when you choose where you're starting from. There may be a myriad of paths starting from different places but once you've chosen where you're going to begin, there is only one option for where you go next. The route forward could be the one I've made up to illustrate the point in Figure 14.1, or it could be the result of a set of rules. We might have mathematical rules that describe how the path leads on from where you currently are: they describe how your next 'step' depends on where you are at the moment. Abiding by convention, let's say we can describe your point in the sports hall space by the coordinates x, y, and z: coordinates just like longitude and latitude on a map but with a third one for height. Where you will be next (let's call that $(x,y,z)_{next}$), depends only on where you are now (let's call that $(x,y,z)_{now}$).

The rule is expressed as $(x,y,z)_{next}$ is $(x,y,z)_{now}$ plus some change in each of x, y, and z, and the change is completely governed by $(x,y,z)_{now}$. This is just saying that at every point in space there is only one step you are allowed to take and it is controlled by where you are. It feels like IKEA. If you are in bedrooms the only place you can go

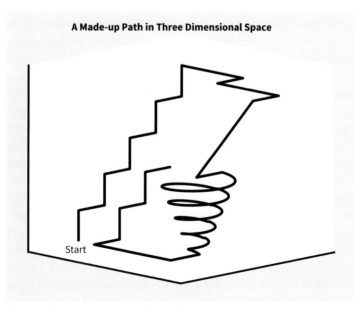

A Made-up Path in Three Dimensional Space

Start

Figure 14.1 A path in three dimensions.

next is kitchens. In the Museum of Printing, if you are in room thirteen (the Type store), the only place you can go next is room fourteen (the printing office).

14.2 Making paths with maths

There are lots of ways of making up rules for paths through a three-dimensional space but Figure 14.2 is a particularly interesting one. It is also particularly relevant to considerations of weather, climate, and climate models. The path in 10.7 wanders around two lobes and traces out something that might perhaps be seen as similar to a butterfly's wings. This is appropriate because this picture is one of the most referenced examples of the butterfly effect.

Figure 14.2 shows a path generated by the rules of what is known as the Lorenz system, or more accurately the Lorenz '63 system, because it was first described in a paper by Ed Lorenz published in 1963. With both the Lorenz '63 system and with climate models there are a set of equations which, when represented on a computer, tell us what happens next; they take us one step forward on the path, one chunk of time into the future. To go a long way into the future, and to trace a picture of the butterfly's wings' path, requires solving the equations repeatedly. This requires a tedious process of iteration but computers are good at that. The computer code to solve the Lorenz '63 system[b] and produce a path in three-dimensional space is vastly simpler than that of a climate model, but the process is essentially the same. In this

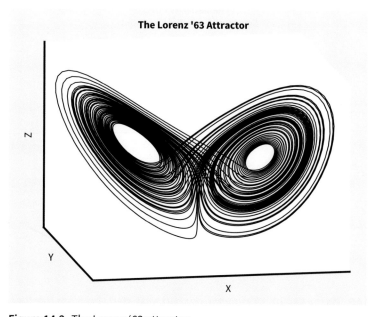

The Lorenz '63 Attractor

Figure 14.2 The Lorenz '63 attractor.

sense, the fundamental nature of the Lorenz '63 system is a good parallel for that of complicated global climate models.

There is another important reason to consider Lorenz '63 when trying to understand climate models. Both are sensitive to the way you start them off. In Lorenz '63 where the path goes in the three-dimensional space is sensitive to where you begin. Similarly what the climate model simulates is sensitive to the particular values of temperature, winds, pressure, and so on, at all the various grid points when you set it going. They are both sensitive to what are called their 'initial conditions'.

In Lorenz '63 the shape of the path which draws the butterfly's wings looks pretty much the same almost regardless of where you start from. That is, all paths end up tracing a route which looks similar to all others, even though the particular way that the path traces out the shape is very sensitive to where you start from. The shape of the path, the butterfly's wings, is called the **attractor** of the system. (This term 'attractor' is going to crop up a lot—it applies to all sorts of things including climate models.) In Lorenz '63 you might start from somewhere far from the wings, say the top left corner of the cuboid in Figure 14.2, but once you've followed the path for a little while you'll find yourself close to the wings—close to the attractor. The attractor attracts you from wherever you may start. However, if you start two paths very close to each other— even if they are close to the attractor—then where on the wings of the attractor you find yourself after some period of time will vary a lot (Figure 14.3): two paths that start close together, stay close together for a while and then—BOOM—they go off in entirely different directions. That's what is meant by sensitivity to initial conditions.

**The Butterfly Effect
Sensitivity to Initial Conditions**

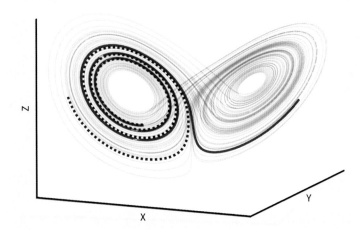

Figure 14.3 On the Lorenz '63 attractor, two paths can start very close together but after a while can do something completely different.

This characteristic of sensitivity to the starting conditions is common in nonlinear systems such as the climate system and Lorenz '63. In fact the Lorenz '63 system is related to the climate system: it represents an idealization of a type of atmospheric circulation so it is not surprising that the same sort of sensitivity is found in weather and climate models. This sensitivity is one reason why there are limitations on the reliability of weather forecasts. Even if we had a perfect model of the climate system our ability to predict the weather in, say, London in ten days' time would be limited by our ability to know the current state of the climate system, particularly the atmosphere, right now. Weather forecasting is like predicting the Lorenz '63 path when you only know approximately where you are starting from. You can do it for a while, but at some point the uncertainty blows up, becomes huge, and all you can say is that you'll be somewhere on the attractor (Figure 14.3).

But hold on a minute. The uncertainty blows up because we don't know the starting point, so why don't we just measure the starting point better? Unfortunately, **any** difference in the starting point, **however small**, will eventually blow up into a big difference in the forecast. Indeed that's how Ed Lorenz came across the peculiar behaviour of this system, which is an example of what would later be called chaos theory.[c] He and his team were studying the system by running simulations on an early computer. They decided to look at one simulation in more detail but instead of rerunning it from the beginning, they took a shortcut: they used the starting conditions, the values of x, y, and z, from a print-out from part way through an earlier simulation. To Ed's surprise, he found that it did not go on to replicate the earlier simulation but rather did something completely different. He had the insight to realize that this was because the print-out didn't include the exact values being used by the computer but rather an approximation: numbers with three decimal places rather than six decimal places. This small difference was enough to lead the simulation to do something very different. It turns out that to get the simulations to do the same thing, you have to start them from **precisely** the same starting conditions. Even the smallest possible differences will lead them to diverge drastically at some point in the future. The consequence is that accurate long-term weather forecasts are only possible if we have a perfect model of the climate system AND perfect knowledge of the conditions of the climate system today. Since neither of these are possible, weather forecasts are inherently limited. The forecast limits vary by season and location but in mid-latitude locations such as Western Europe, reliable weather forecasts are limited to a maximum of about ten to fifteen days ahead.[d] In practice weather forecasts are often good but not that good so there is room for improvement; there is however a limit to how much improvement is possible.

Nevertheless, as made clear by Robert Heinlein and the Intergovernmental Panel on Climate Change (IPCC), climate and weather are two different things. In Lorenz '63, not being able to say which path you are on, and therefore where you will be in the future in the three-dimensional sports hall, is not the same as saying you have no information at all about where you will be in the future. Quite the contrary. Whatever path you happen to be on, you will be somewhere close to the butterfly wings of Figure 14.2. That's why it's called an attractor—wherever you start from, you're going

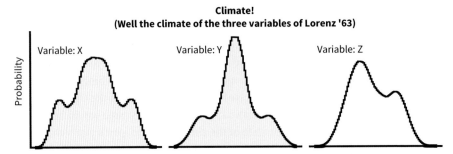

Figure 14.4 The climate—the probability distributions—for the Lorenz '63 variables.

to end up close to it. If you know your starting point approximately you can make a good prediction of where you'll be for a few steps ahead but beyond some point the uncertainty becomes so great that you could be anywhere, but only anywhere near the attractor. In Heinlein's terms, you don't know what you'll get but you still know something about what to expect. This suggests that for Lorenz '63 the attractor itself—the butterfly's wings pattern—is somewhat related to what we mean by 'climate'. That's why the concept of the path generated by a model is so important to understanding what we need to do to tie down the concept of climate. If we run a computer simulation of a very long path over a very long time, we can build up the probability of each coordinate (x, y, and z)—the probability distributions for the variables of Lorenz '63 (Figure 14.4). These distributions describe the values we should expect[1]: they are essentially the climate of each coordinate of the system.

14.3 The climate of a model (part 1)

The coordinates of each point on the Lorenz '63 path in Figure 14.2 represent the variables of the mathematical system. For Lorenz '63 they relate to physical quantities: y and z to temperature variations and x to the movement of air, in an idealized representation of convection in the atmosphere. Plotting them as coordinates in Figure 14.2, shows us how the rules of the system constrain the variables as they change over time: it illustrates which combinations of values are possible and which are impossible.

   Chaotic systems like this have fascinating and peculiar properties and there are many excellent books which describe their behaviour.[e] This, however, is not one of them. This book is about climate predictions and the challenges of describing the future. The intriguing properties of chaotic systems are only relevant here to the extent they have consequences for climate predictions, and the physical meaning of the Lorenz '63 variables is entirely unimportant. What's important for us is that

---

[1] Like in weather forecasting we can predict what happens a few steps ahead but beyond that point all we can say is that we will be on the attractor. The distributions describe the values we should expect beyond the point that predicting a particular path becomes impossible.

Lorenz '63 is a system of equations which when implemented on a computer take a collection of related variables and step them forward in time, and that what the variables do—the path that they follow—is highly sensitive to their initial values. Both these characteristics also apply to global climate models, so we can use Lorenz '63 as a tool for thinking about climate and climate models.

We can think about the x, y, z coordinates of Lorenz '63 as like any three climate variables; let's say temperature in Chennai, humidity in Chennai, and temperature in Bangalore. The probability distributions of Figure 14.4 show the climate of each variable individually. This is one way to view climate. For some purposes though, you might be interested in how two variables relate to each other. For instance, the possibility of high temperatures in Chennai and Bangalore at the same time might be important for electricity demand and production in the region. In that case, two separate probability distributions don't tell us what we want to know about the impact of climate on the regional electricity system. With our model this is not a problem. A long simulation of the path traces out the shape of the attractor which shows how the variables relate to each other and enables us to build up the probability distribution for two variables together—something called a joint distribution. Figure 14.5 shows this for x and y: a different, more complex picture of the climate of Lorenz '63. One could do the same for all three variables together, although then it becomes difficult to draw the picture.

We could go further still and consider the **consequences** of the variables rather than the variables themselves. For instance, temperature and humidity both influence comfort and health risks in humans and animals. We could create a measure of

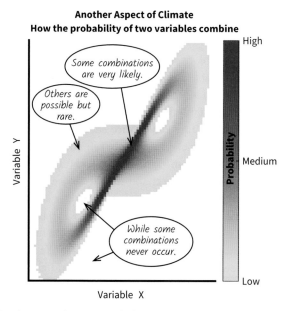

Figure 14.5 The climate—the joint probability distribution—for two of the Lorenz '63 variables together.

comfort by combining these two variables in various ways; indeed, a range of thermal comfort indices exist based on these and other climate variables. At each point on the path we could calculate this measure and in that way create a probability distribution for comfort. This too would be a version of the climate of the Lorenz '63 system—in this case, the climate of a derived or dependent quantity. Having long paths which explore and map out the attractor enables us to create the climate—the probability distributions—for whatever we want, so long as it is dependent on the variables of the system. They are all versions of its climate and we can do the same with the output of climate models.

This relationship between climate and the path of coordinates exploring an attractor is extremely valuable. It allows us to think about the climate of a mathematical system and hence of a climate model. It isn't yet, however, sufficient to tie down what we mean by 'climate' in a way that reflects what we are actually interested in. It isn't yet a useful basis for measuring climate, even within a climate model. The problem is that in reality, and in global climate models, we don't want 'climate' to encompass all possible states that the climate system can be in; we don't want climate to be the distribution across the whole attractor. To see why, consider Edinburgh in Scotland. At the height of the last ice age, between twenty-six and twenty thousand years ago, Edinburgh—or rather where Edinburgh now is—was under an ice sheet. Such a situation is a state of the climate system which is clearly possible, but most of us would consider it misleading to include it within a description of Edinburgh's climate. A tourist website that said 'in summer Edinburgh gets on average over five hours of sunshine per day but in some years it can be under many metres of ice' would not only be considered bad advertising but also actively misleading. Yet the probability-distribution-from-the-attractor-of-a-mathematical-system perspective would include such a possibility. After all, it has happened in the past.

We want our models to be able to simulate such icy conditions so that we can study paleo-climate and relate data about the world in the distant past to our understanding of climate today. But when we talk about the climate of today, or the climate of fifty years in the future, we don't want to include such possibilities, **even if the attractor of reality[2] includes them**. There are no ice sheets in Scotland today, so even without human-induced climate change, Edinburgh could not possibly be under ice in fifty years' time. Ice sheets can't form that quickly. That means the under-ice possibility is not part of what we expect and should not be part of our probability distribution for climate in Edinburgh in the next century.

To see how we can get a probability distribution from a global climate model that reflects the behaviour we would expect in fifty years' time (in a model) we need something simple but a little more complicated than Lorenz '63. We need a mathematical system that captures a few more features of the real-world climate system and of global climate models.

---

[2] If such a thing exists, which is something open to mathematical, physical, and philosophical debate.

15

A walk in three dimensions over a two-dimensional sea

15.1 Key features of the climate system

The climate system consists of the atmosphere, oceans, land surface, and cryosphere (the ice components), but what are the features of the system that are fundamental for tying down its future climate? It's not about the actual processes but about how they behave. One fundamental feature is simply that the system can't do just anything: there are physical laws, physical relationships that restrict what it can do. Another is that these relationships are often nonlinear leading to a sensitivity of future behaviour to uncertainty in the initial state—to where we start it off from. So far, Lorenz '63 is a good parallel: it encapsulates nonlinearity, initial-condition sensitivity, and a set of rules which provides a constraint on the behaviour of the variables.

But there are three—well, two and a half—further features that we need to capture. One is that different aspects of the climate system vary on different timescales—some are faster and some are slower. Another is that the climate system is naturally driven by constantly varying inputs of energy from the sun, which drives the daily cycle of day and night, the annual cycle of the changing seasons, and, to some extent the paleo-cycles of ice ages and inter-glacial periods. The final one—or half because it is closely related to the second—is that the climate system is changing due to mankind's emissions of greenhouse gases which are changing the flows of energy into and within it.

The first additional feature—differing timescales—is related to the differences between the atmosphere, the oceans, the cryosphere and the land surface (forests, vegetation, ecosystems, etc.). Think about, for instance, the maximum daily temperature in Boston, Massachusetts. This varies substantially from one day to the next but the temperature of the seawater off the coast of New England varies much more slowly: one day is much like the previous one despite there being gradual changes with the seasons and from year to year. This is because the oceans tend to vary on slower timescales than the atmosphere.

This doesn't mean, however, that we can treat the different components of the climate system as if they were not connected. The oceans affect the atmosphere and the atmosphere affects the oceans. The slow processes set the context for the fast ones but the fast ones also affect the slow ones. It is all one system and all the different components interact with and affect each other. The sensitivity of the climate system to initial conditions means that the future state of the atmosphere is sensitive to the details of the present state of the atmosphere **and the oceans**, while the future state of the ocean is sensitive to the details of the current state of the oceans **and the atmosphere**. It's all one system and that's how we need to think about it—particularly in the context of climate change.

The second and third additional features encapsulate the fact that the climate system is not isolated: it is influenced by variations which are external to it. It receives energy from the sun which varies from one season to the next and from day to night, while human emissions of greenhouse gases are trapping more energy in the lower atmosphere and driving long-term changes in behaviour.

To study what we mean by climate in a way that is useful for understanding the real world, we need something that is like Lorenz '63 but that includes these extra features: variables that vary on different timescales, a seasonal cycle, and the possibility of driving wholesale change in behaviour which is akin to climate change. One such system is the Lorenz–Stommel system.

15.2 Thinking in five dimensions

The Lorenz–Stommel system is similar to Lorenz '63, except that it has three variables which vary relatively rapidly and two others which vary relatively slowly. The fast variables can be thought of as a parallel for the atmosphere in a global climate model, and the slow ones as a parallel for the oceans.[a] All the variables and processes in the Lorenz–Stommel system have a basis in physical reality but as with Lorenz '63 their physical meaning is not important for us.

In addition to its atmosphere-like and ocean-like variables, the Lorenz–Stommel system has a parameter which partially controls the difference in atmospheric temperature between the equator and the pole. This is useful because in reality this temperature difference varies with the seasons and is also expected to change as a result of global warming. By varying this parameter, we can include a seasonal cycle as well as long-term change on timescales of decades or centuries—in other words, we can include something which is a parallel for the consequences of varying solar energy inputs over the year and also the driven changes resulting from increasing atmospheric greenhouse gas concentrations. Finally, the whole system is nonlinear—it is sensitive to initial conditions—and has an attractor which constrains what values each variable can take—just like Lorenz '63. The Lorenz–Stommel model is not a model of the climate system, but it has all the features necessary to study what 'climate' is and what makes predicting it uncertain. We can use it to explore what we would need to do in order to predict climate in a fully fledged climate model.

Like the Lorenz '63 system, and also global climate models, the equations of the Lorenz–Stommel system can be solved on a computer in steps—each step taking us one timestep into the future. In this way it creates a path just like that of the Lorenz '63 system (Figure 14.2), except that it doesn't have three variables, it has five, and that makes it hard to visualize—we can't draw a picture of it in three dimensions, we can't visualize it in our minds.

This problem of visualization is actually quite important. A key aspect of all these models is that the rules of the equations create a link between **all** the variables. For Lorenz '63 there are three variables so to see what happens next we need a path in three dimensions. We couldn't show it properly in two dimensions—by drawing the behaviour of just two variables—because for any particular values of each of those two variables there are a myriad of paths going forward, dependent on the third. The path is only unique when all three variables are known. Remember, where these paths go next is completely controlled by where they are at the moment—just like in IKEA or the exhibition—but that doesn't help us if, in Lorenz '63, we only know the values for two variables: the rules of the system only tell us what will happen next when we know all three. Fortunately, we are familiar with visualizing things in three-dimensional space. We can even draw the three-dimensional path on a two-dimensional page because we have lots of experience of interpreting a two-dimensional image as a three-dimensional object.

For Lorenz–Stommel, things aren't so easy. There is a problem of visualization because it has five variables so the path of the Lorenz–Stommel system is a path in five dimensions, not three. It's important to realize though that this is just semantics. There is nothing clever going on here about the state of the universe. There are not two hidden dimensions that you've been missing all your life. You don't need to squint and tug your mind around to some never-before-considered perspective. This is just a way of talking about these systems: the Lorenz–Stommel system has five interconnected variables, so we need to imagine the shape of the path in five dimensions. Any representation in our conventional three dimensions will be missing some important aspects of the connections between the variables. The trick, if there is one, is not to try to visualize what's going on at all. 'Visualize' suggests you can see it in a conventional way, but you can't visualize something in five dimensions. All you can do is understand what is going on and be aware of the limitations of what you can visualize and what we can draw.

Mathematicians have the advantage here since many of them do this sort of thing every day. They come across equations describing three variables all the time and they get familiar with translating those equations into a behaviour in space. In this way, they might 'see' what the equations describe. When they have equations that are similar to the three-dimensional ones but with more variables, it is a handy conceptual hop to talk about positions and behaviour in more dimensions. It can be a really useful hop in discussing the behaviour of complicated systems and may help them go skipping and jumping to valuable new insights. But that hop can be quite a leap for those not familiar with it.

The thing to remember is that thinking about five dimensions is simply a way of describing the behaviour of, and the relationship between, five variables. The variables might relate to the real world; for instance, temperatures in different cities in California or economic output and human welfare in different regions of Poland. Or they may be abstract mathematical quantities. Whatever the case, they are described by our equations. Don't get distracted by trying to 'see' extra dimensions but do

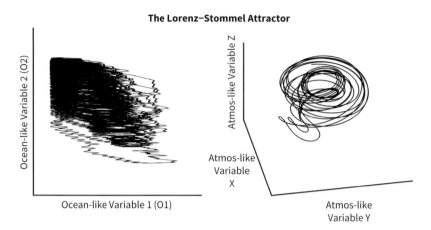

Figure 15.1 The attractor of the Lorenz–Stommel system. On the left is a 5000-year-long path of the ocean-like variables. On the right is a one-year-long path of the atmosphere-like variables.

remember that the path is actually moving about in a 'space' that is 'bigger' than anything we can draw.

Because the Lorenz–Stommel path is in five dimensions, I can only draw some aspects of it. Figure 15.1 shows a three-dimensional path for the three atmospheric variables and a two-dimensional path for the two ocean variables; together they represent the attractor of the system. Where the path goes is sensitive to the starting conditions, so if I set off two paths from identical values for the ocean-like variables but slightly different ones for the atmosphere-like variables then after some time the path—even the aspect of the path represented by the ocean-like variables—will look very different between the two (Figure 15.2). The same is true for climate models, so this is a powerful tool for getting our minds around climate, weather, and climate prediction; even more so than Lorenz '63 because of the extra features it has.

15.3   The climate of a model (part 2)

What, then is the climate of the Lorenz–Stommel system? If climate is what we expect, as represented by a probability distribution, then one answer comes from doing exactly the same thing as I did with Lorenz '63: take the probability distribution from a very long path. Figure 15.3 shows such distributions.

To deduce these distributions for the Lorenz–Stommel system we have to use a path which is many tens of thousands of years long (in model time) because the ocean-like variables only change rather slowly so it takes a long time for the path to explore the possible states of the system. Nevertheless, such a path paints an almost complete picture of the attractor, and it provides a good estimate of the probabilities

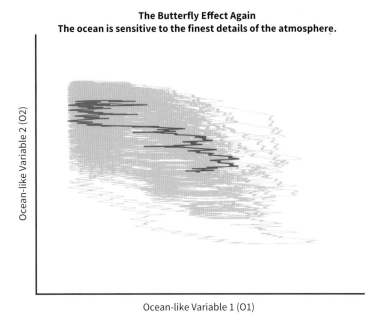

The Butterfly Effect Again
The ocean is sensitive to the finest details of the atmosphere.

Figure 15.2 In Lorenz–Stommel, the future path of the ocean-like variables can be highly sensitive to even the slightest differences in the initial conditions of the atmosphere-like variables, even though they vary much more slowly. This shows the attractor of the ocean-like variables (grey) along with two paths (red and blue) starting from almost the same initial conditions but with very small differences in the atmosphere-like variables.

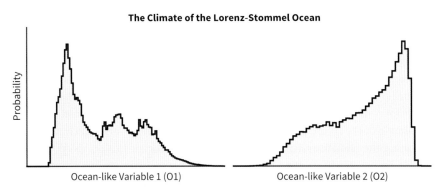

The Climate of the Lorenz-Stommel Ocean

Figure 15.3 The probability distribution for O2 and O1 in Lorenz–Stommel from a 100,000-year simulation.

for what can happen in Lorenz–Stommel. The equivalent in reality, however, would include ice ages which isn't what we're looking for when we're making climate predictions of the twenty-first century. These distributions are all very well, but they don't solve the Edinburgh-under-ice problem (see section 14.3).

To get a more useful and relevant interpretation of climate, we need to think about weather. If climate is what we expect and weather is what we get, then we need to interpret 'weather' as the state of the climate system at some point in time. This includes everything: winds over the north sea, temperatures in Arizona, ocean currents in the Indian ocean, glaciers in the Andes, everything. It's not just about whether it's cold and windy, or hot and humid outside.

Nowadays we have observations of many climatic variables so we do know a lot about today's 'weather': to some extent we know what we've got. In the Lorenz–Stommel system the parallel would be that at some point in time we know the values for the five variables; in other words, we know where we are in our five-dimensional space (Figure 15.1). In this case, there is no uncertainty in the state of the system at present; it just is what it is. In the real world, however, observations are limited so we only know the weather approximately; nevertheless, to some extent we still know what the weather is today. The question is, what did we expect it to be?

If I were sitting in Oxford 600,000 years ago—which I wasn't—and I asked myself what I expected the weather to be in Oxford on 6th January 2021, then I might reasonably have included the possibility that the world would be in an ice age today.[b] The earth has gone in and out of ice ages many times over the last million years, so that would be a reasonable possibility: conditions consistent with an ice age should be part of the probability distribution for what I expect. But if I were sitting in Oxford yesterday—which I was—I would not include such a possibility in my prediction for today. The probability distribution of weather types that I expect—the climate—fundamentally depends on **when** I am looking from.

Thinking instead about the future, there are lots of states of weather in the future that are consistent with observations of the weather today. That's to say, uncertainties in our knowledge of today's weather—uncertainties in our initial conditions—lead to many possible futures.[1] If climate is the weather we expect, then a climate prediction is the probability distribution of weather in the future that is consistent with what we know it to be today. It would be ridiculous to 'expect' something in the future that was not possible given what we can see out of our windows or can monitor with our satellites today, so an ice age cannot be part of the climate distribution that we predict for tomorrow or even for 100 years' time—with or without global warming. The climate of the future is a probability distribution but not one which extends to everything the climate system can do, only those things it can do given the best-possible knowledge of its state today. When we make a climate prediction, this is what we are aiming for: it is the best we can conceive of achieving.

The Lorenz–Stommel system provides a tool to get our minds round what's going on here. In discussing it, it is useful to remember that the Lorenz–Stommel system describes how variables change over time; it even includes a seasonal cycle. As a result we can talk about 'time in the model', and even months and years 'in the model'; I've

---

[1] This is true regardless of other uncertainties. It would be true even if we understood perfectly how the climate system worked and what greenhouse gases mankind will go on to emit.

done this a couple of times already. To help make the distinction between model time and real-world time clear, from here on I'll *italicize* words which refer to model time.

In the Lorenz–Stommel system if we know the precise values of the variables *now* then there is no uncertainty where the path will go in the *future*, and therefore what values the variables will have in, say, *30 years' time*. As in Lorenz '63 the path only goes one way and it never splits. It just wanders around in its five-dimensional space. The system is said to be 'deterministic' because if you know exactly where you are in its five-dimensional space *at some point in time* then where you will be *in the future* is completely determined. Put another way, if you know the weather now, you know the weather in the future. But in the real world, we can never know the precise values of all climatic variables (temperature, humidity, pressure, etc.) now. Observations are never exact. You never know exactly what the weather is at the moment.

The consequences of this can be seen in Lorenz–Stommel by simulating lots of paths all starting from initial conditions (values for the five variables) that are **almost but not quite** the same as each other. In this way, we can capture the consequences of only knowing the weather approximately: only knowing approximately what values the variables have *at some point in time*. Such a collection of paths is called an initial-condition ensemble. If we know approximately but not precisely what the values of the variables are *now* then we can generate lots of possible paths going forward (see Figure 15.2) and we can build a probability distribution from those paths of all the possible values in *30 years' time*. Such a distribution tells us what we should expect those variables to be in *30 years' time* given that we know approximately but not precisely what they are *today*. It is a probability distribution—a climate prediction—but it is not the same as the distribution built up over *thousands of years* of simulation (Figure 15.4). It's a probability distribution founded on the state of the system *today*: a much better representation of what we actually expect and therefore a better representation of *future* climate.

With these paths we can create probability distributions for each *year* going *forward*. We can draw these probability distributions and watch the distribution widen

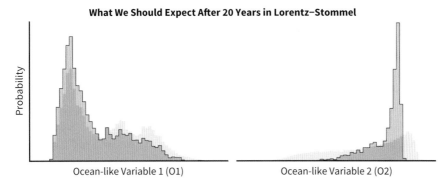

Figure 15.4 The probability distribution for O2 and O1 in Lorenz–Stommel from a 1000-member ensemble after twenty years of simulations (blue) and from a single 100,000-year simulation (pale red).

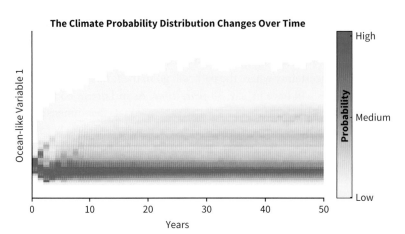

Figure 15.5 The probability distribution for O1 in Lorenz–Stommel expanding and changing over fifty years.

and change shape as we get further into the *future* (see Figure 15.5). It expands away from the almost certainty of the conditions *now*. One might even say that climate doesn't exist *now*, only weather. In Lorenz–Stommel there is some very small uncertainty in the variables *now* which parallels the unavoidable uncertainty in our observations of the present day atmosphere and ocean in reality. Future climate, however, captures the uncertainty in the path we are on as we move forward—the uncertainty which arises from even the smallest uncertainty in the current state. The future climate expands and broadens as we move away from the fixed point of *now*.

15.4 Studying the future is different from studying the past

These considerations of Lorenz–Stommel show us that a useful definition of the climate of the future is the probability distribution of future climatic behaviours in the light of our knowledge of the state of the climate system today: a more specific version of 'climate is what we expect'.

Interpreting climate this way however is problematic for climate research because it is at odds with the way climate has previously been understood and defined—particularly in studies of the past. When we study the climate of the past we focus on identifying what actually happened. This involves fascinating research using corals, tree rings, long tubes of ice extracted from ice sheets in Greenland and Antarctica, and many other wonderful and challenging techniques, but its goal is always tying down what actually happened. Thought about in terms of the Lorenz systems, it's about tying down where the path in *the past* came from and how it took us to where we are *today*. We may consider multiple possible paths going forward into the future but we focus on only a single path in the past.

The real-world climate in the past may still be thought of as a distribution but one built up from that single path. We might talk, for instance, about the climate of Vienna in the mid-twentieth century as being the probability distribution constructed from observations between 1940 and 1970. Or we might talk about the climate at the height of the last ice age as being the distribution of temperatures in some locations throughout that period.

Whether we consider climate to be a distribution or just an average, the study of the past leads to the idea that it is calculated over a period of time—something which is built into the conventional definitions of climate from the Intergovernmental Panel on Climate Change (IPCC) and others. This encourages us to think about climate in a deterministic way: to think that what happened was the only thing that could have happened. Probability distributions, when they are used at all, describe segments of the path rather than different pathways: they capture the idea that there is natural variability within any single path but not that there were many possible, potentially very different, paths.

This history-based view of climate is problematic. It holds back our ability to make societally useful climate predictions. Within its paradigm the aim of climate change research is essentially one of finding the right, or approximately right, path to describe the future. Unfortunately that's not the universe we live in. In the real world, as in the Lorenz systems, there are many potential paths going forward and the aim of climate change research should be to describe this plethora of paths and to evaluate the probability distributions resulting from them.[2]

The friction between a single path in the past and multiple paths in the future also leaves us in limbo regarding the climate of the present day. What is the climate of today? At the moment the climate system is simply in whatever state it is in, the weather is what it is and that is that. Arguably climate doesn't even exist in the present. But that isn't a satisfactory answer because we all have an idea of the climate of at least the place we live: we know the types of weather that we expect. So what is the probability distribution for what we expect in the present day? We could use a history-based approach to describe it, in which case we might take the distribution from the last thirty years. But climate change means we know that this is an inaccurate description of what we should expect today because we know today's climate is different from that thirty years ago—the history-based approach blurs the changes over time and thus does not represent today. Alternatively we could use the multiple paths definition and say it is the distribution of possibilities today arising from the state of the climate system at some point in the past. But in that case, when? Thirty years ago? A hundred years ago? Or 100,000 years ago?

---

[2] In practice most researchers accept that there are multiple potential paths going forward but the history-based paradigm leads to a common assumption that if you look over *thirty years* they will all look much the same in terms of their average and probability distribution. If that were the case then one, or a few, paths would be enough to tie down the climate of the future. Unfortunately, there is every reason to expect that not to be the case. Different paths could have very different thirty-*year* distributions. Distributions over thirty *years* are not the same as distributions over many paths.

There is also a different way to view the climate of the past. Given our knowledge of the state of the climate system today, there are many possibilities for how it could have been. There are many different paths that lead to today's conditions—imagine Figure 15.2, but with the two paths leading to the same point rather than away from it. We can think of the climate of the past, therefore, as being a distribution that broadens and expands as we go backwards in time from the present day, just like the climate of the future expands as we go forward. This sounds a bit wacky, but it is an important concept for climate modelling. The climate models used for climate predictions are often evaluated against past observations, but we need to be very careful with such tests: the past could have been different and still lead us to today. We need to evaluate our models against not just what was but what could have been because even a perfect model will be unlikely to reproduce the actual path that was reality. This creates all sorts of tremendously interesting challenges at the intersection of statistics, maths, philosophy, and physics.

Getting back to the main issue, though—what about human-induced climate change? The discussion so far has been mostly in the context of an unchanging climate but the multiple paths interpretation of climate works just as well when we include human-induced climate change. And the Lorenz–Stommel system provides a tool for studying that situation.

15.5 Giving climate a push

If you want to know how climate change will affect the place where you live, or if you want to build infrastructure that will last a hundred years and not be inundated by floods or collapse due to instabilities caused by droughts and heat waves, then you want to know what to expect. The most useful climate forecast will tell you what to expect so that you can plan accordingly: it will give you a probability distribution for future weather or whatever else it is that you're interested in.

In the illustrative mathematical systems of the previous pages, the distribution is built up from paths in three-dimensional space (in Lorenz '63) or five-dimensional space (in Lorenz–Stommel). In a fully fledged climate model, it is . . . well, we'll come to that in the next chapter, but it is also generated from paths: a sequence of different states of temperature, humidity, and other climate variables. The particular distribution we look at could therefore be the distribution for a variable such as one of the Lorenz–Stommel variables or for daily maximum temperature in Austin, Texas in a climate model. It could also be something more complex. It could be a distribution of some combination of temperature and wind strength to account for wind chill to give a perceived temperature in Edmonton, Canada. Or it could be a combination of temperature and humidity to give an indication of comfort in Jakarta, Indonesia. Or it could be a combination of multiple days or weeks of temperature and humidity representing the health risks associated with heat waves. Or it could even be some combined measure of the consequences of temperature, humidity, precipitation, and soil moisture over several years, which combine to influence the agricultural production in the Western Cape in South Africa. Since the probability distributions are generated from multiple possible paths, they can represent quantities that are influenced by sustained periods of heat or drought or whatever aspect of climate is important for you. The use of multiple paths allows us to construct the climate distributions for whatever we are interested in.

Whatever it is that we're looking at, the probability distributions broaden out from the current state of the system as we move into the future. The distributions will be broader in a hundred years than in thirty years, simply because the different paths will have time to explore more of the natural behaviours of the climate system—more of the attractor

When thinking about the next century though we also need to account for the big push we're giving the climate system by emitting large quantities of greenhouse gases. This is changing the range of potential climatic behaviour: it is changing the climate attractor. How do we describe how this interacts with the paths spreading out across the wide range of natural behaviours?

The answer is pretty obvious: we simply include the drivers of change in our model simulations. In the Lorenz–Stommel system, the change comes about by varying a parameter which we take as a parallel for atmospheric concentrations of greenhouse gases. In a climate model it comes about by changing the atmospheric concentration of greenhouse gases in the model. In each case, the processes of the system—the physical processes represented by the equations—remain the same; it is only this parameter that changes. It's like the laws governing how a ball moves when you throw it to a friend on Mars. The laws are the same as those applicable on earth, it's just that the strength of the gravitational pull is different; it's a simple change in a parameter. This means we can evaluate climate under climate change in exactly the same way as within a non-changing climate. We create lots of paths going forward from where we are now and generate probability distributions from those paths. The only difference is that when we use our equations, our model, to create those paths we allow some parameter—our greenhouse gas concentrations parameter or something equivalent—to change gradually over time.

For our mathematical systems—Lorenz '63 and Lorenz–Stommel—the shape of the attractor (Figures 14.2 and 15.1) will change as the parameter changes but the rules of the system still create a path in three- or five-dimensional space. Like in the unchanging system, the path never splits and where you go next is completely controlled by where you are in the system's space. Also as in the unchanging system, paths which start off very close together can potentially, after a while, diverge and head off in very different directions. That's because these systems are nonlinear. These characteristics are also characteristics of climate models, which is why studying simple systems can help us understand how to design the most useful experiments with big complicated climate models.

Figure 15.6 shows two attractors for the Lorenz–Stommel system. You can think of them as representing different states of the climate system—different states but each stable and unchanging. One can be thought of as a parallel for the climate of the earth in the seventeenth century, before the industrial revolution, and the other

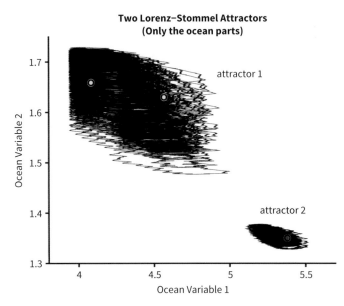

Figure 15.6 Two different attractors for the Lorenz–Stommel system.

for the climate of the earth in, for instance, the thirty-third century when perhaps it has become 3°C warmer but is again stable and unchanging. One represents climate before anthropogenic climate change and the other after it has occurred. Which is which? It doesn't matter. This is just a conceptual game of getting our minds round how to think about the problem of what climate is under climate change so either can be either. I'm only using Lorenz–Stommel to get a handle on the issues.

Climate change in this system involves starting at some particular point on one attractor and then gradually changing the climate-change parameter until it gets to a new value representative of the other attractor; it reflects a situation of, say, increasing atmospheric carbon dioxide concentrations from those seen in pre-industrial times (280 parts per million) to twice those seen in pre-industrial times (560 parts per million), after which they are kept constant. The system starts off on one attractor and ultimately ends up on a new attractor but while it is being driven to change, it might not be any attractor at all. How to study and describe this situation mathematically is one of those exciting areas in need of further research. For the practical purposes of climate predictions though, we simply start a collection of paths from approximately the same point on one attractor and watch as they move apart from each other but generally towards the other attractor.

At any point in *time* we can construct the climate—the probability distribution— from the collection—the ensemble—of paths. Figure 15.7 shows exactly such a situation. It shows the two ocean-like variables in the Lorenz–Stommel system (variables O1 and O2) but if you want to keep your mind on the real-world climate prediction problem, you could think of them as, for instance, average summer temperature (O2) and rainfall (O1) in, let's say, Oslo. They aren't actually anything to

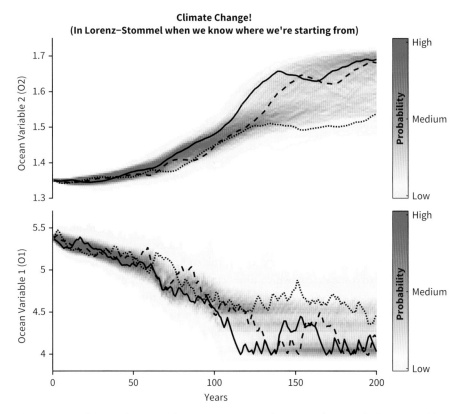

Figure 15.7 Climate change in the Lorenz–Stommel system, along with three possible paths through the changing probability distributions.

do with Oslo or rainfall, but seasonal averages change slowly like the oceans, and the parallel is useful for getting our heads around how we should be thinking about climate under climate change. The paths all start within the small red dot in Figure 15.6 but the system as a whole is moving from attractor 2 to attractor 1; the climate change parameter takes a hundred years to make this change.

In this case, Oslo temperature (O2) increases steadily until after a while the distribution spreads out. At the same time Oslo rainfall (O1) gradually decreases and again spreads out. These changing probability distributions show how the climate in this system changes over time due partly to the nonlinearities in the system and partly to the fact that the whole system is being driven to change. Founded on where we are starting from—the red dot in Figure 15.6—and the changing value of a parameter to reflect climate change, these are the changing climate distributions which show what we should expect at any *future* point in time.

So this is climate prediction within the Lorenz–Stommel system with climate change. It tells us what we expect. In reality, though, we will only experience the twenty-first century once: the one-shot-bet nature of the problem comes in to play. Figure 15.7 also shows three possible pathways through the distributions. In two (solid and dashed) the Oslo temperature (O2) rises gradually and then rapidly around years 110 to 130, while in the other (dotted) we don't see the final rise within the 200-year simulation. Similarly, Oslo rainfall (O1) falls rapidly to a very low value in the first two but hovers around some medium decrease in the other. The point is that individual pathways can look quite different from the impression given by the distribution of possibilities. This rings a bell of caution: even if we have good climate predictions—good probability distributions—how should we use them to understand the risks we face?

The physical processes of the climate system, combined with the way we are giving the system a big push through greenhouse gas emissions, control the probability distributions for different types of weather in the future. Different levels of greenhouse gas emissions lead to different changing probability distributions over the next century and beyond, and this can be illustrated and studied with simple nonlinear systems. Greenhouse gas emissions are of course something that people, businesses, and governments have some ability to control, so human societies have the potential to choose, to some extent, which time-varying probability distributions we want the future to arise from. The particular pathway we take through those probabilities is, however, not something we can control (Figure 15.7). That depends on happenstance. We can choose what sort of dice to roll and maybe the numbers on it, but we can't choose the outcome. This is the essence of the human-induced climate change problem. It raises deep and difficult questions about how we, as societies, respond to risk and uncertainty. More on this in Chapter 25. For the moment though, there is one more fundamental aspect of climate prediction under climate change to consider.

15.6 We don't know where we are—not even roughly

The concept behind the multiple-paths-into-the-future approach is that the paths are all based on the same physical system, are all equally consistent with the state of the world *today,* and all share the same assumptions about how humanity will behave in the future—most importantly regarding greenhouse gas emissions. To apply this future-climate-is-a-probability-distribution-conditioned-on-the-state-now approach, we need to know approximately what the state is now. Unfortunately in reality, despite what I said earlier, we don't know the state now at all well. We don't know what the state of the system is today, not just in terms of not knowing the fine details but in terms of whacking great details. Our satellite observing systems are fantastic demonstrations of today's technology but there are many gaps in our knowledge, particularly with regard to the state of the oceans, and especially the deep oceans. We know we are not in an ice age but there are still major features that we don't know well. The problem of prediction is therefore not just about the consequences of the smallest uncertainties.

The good old Lorenz–Stommel system again provides a handy tool for illustrating the consequences and understanding how to explore possibilities. Climate prediction is about how the probability distribution changes going *forward* under some steady push representing climate change—a push implemented by changing the value of some parameter over *time* (Figure 15.7). We can repeat that process from a different starting point to see the effect of not knowing where we are starting from—not knowing in a substantial way rather than not knowing the fine detail. Earlier I chose to move from attractor 2 to attractor 1 but now let's consider the alternative case of moving from attractor 1 to attractor 2. Remember, this is all about exploring what's involved in climate prediction, not about actually making predictions, so I am free to choose whatever best illustrates some plausible concerns. In the same way as before I can apply climate change by changing a parameter, in this case pushing the system from attractor 1 to attractor 2, but I'm going to start off two groups of paths, one from each of the yellow dots in Figure 15.6. The result is two different changing probability distributions over time (Figure 15.8)—two different climate forecasts. In one the variable O1 distribution is pretty much always increasing, while in the other there is a period in the *first few decades* when a significant number of paths show it decreasing. To think in terms of parallels with the real world we might, again, think

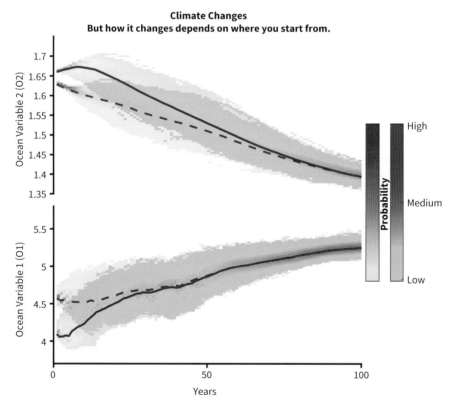

Figure 15.8 Climate change in the Lorenz–Stommel system based on two different starting points on the initial attractor. The lines show the average (median) of each distribution. It can take more than fifty years for the average to stop depending on the starting macro-initial conditions.

about the variable as if it were the average summer rainfall in Oslo. In that case, we see an almost steady increase when starting from one point but a significant chance of an initial decrease for a number of years to decades in the other case, even though longer-term rainfall is still increasing. If such a situation occurred in reality, we can imagine getting severely misled by observations and the look-and-see approach to climate change.

So it turns out that there are two different sorts of sensitivity to starting conditions. The first is due to the fundamentally nonlinear nature of the system and the butterfly effect. This is sensitivity to the finest details of the starting conditions. We don't know **exactly** what the initial conditions should be—something called 'micro-initial-condition uncertainty'—so we don't know exactly what will happen in the future. This gives rise to climate predictions being intrinsically and unavoidably probability distributions.

The second is messier, less theoretical, and about the practical application of models in predicting climate. We don't know many large-scale features of the initial conditions—something called 'macro-initial-condition uncertainty'[3]—so we don't know what the most relevant probability distribution is. Typically macro-initial-condition uncertainty is mostly associated with slowly varying components of the climate system, such as oceans, ice, the land surface, and parts of the upper atmosphere.

Uncertainty expressed by a probability distribution and uncertainty in what that distribution should be are both inherent to the climate prediction problem. The former is simply what constitutes a climate prediction while the latter is uncertainty in that prediction. Both uncertainties would exist even if we had a perfect climate model.

An important difference between micro- and macro-initial-condition uncertainty is that the latter is potentially reducible while the former is not: we could, in principle, get better observations and perhaps know better where to start our models from. In that way we could reduce macro-initial-condition uncertainty and thus reduce uncertainty in what the future probability distributions will be. This is in stark contrast to micro-initial-condition uncertainty and the probability distributions which arise from it. The uncertainty in future behaviour represented by those distributions cannot be reduced or removed by better observations or by any other means: they are intrinsic to the climate system.

There are deep and fascinating questions regarding **how** we might use observations to reduce macro-initial-condition uncertainty in model-based predictions. This is not a trivial task—not practically, not theoretically, not even conceptually. Models are different from reality and their attractors are therefore different from reality's. As a consequence, we shouldn't expect the best forecasts to come from starting off

---

[3] In practice macro-initial-condition uncertainty is not simply a matter of how big the uncertainty is but also about how rapidly the uncertainty in a variable can affect its future behaviour and the behaviour of other variables.

our models from observations of the real world; rather, we want a starting point that in some way reflects the state of reality, but in the world of the model—a starting point that parallels reality rather than is the same as reality.

The fundamental challenges here haven't even begun to be addressed and tend to be brushed under the carpet in the rush to buy the next big computer and to give an answer—any answer, reliable or not—to questions of changing flood risks, heatwaves, agricultural production, and so on. There is an urgent need to understand how to explore these sensitivities so we can make better and more useful predictions. And so we can understand what meaning there is in the information we currently have.

15.7 Challenges from an idealized world

When we talk about climate and climate change, most of us leap immediately to thoughts of temperatures, rainfall, floods, droughts, ecosystems, ice sheets, etc. Predicting the response of the climate system to increasing atmospheric greenhouse gases, however, requires some more conceptual considerations. Being clearer about quite what it is we are trying to predict involves fascinating challenges associated with understanding the essence of climate—and the processes of climate prediction—better. We need to know what we are trying to do, and we need to understand how to identify the sensitivities and interactions that can most affect the outcome. This requires a new focus in climate research: one centred on understanding the problem rather than jumping straight to answers. It's not, at least at first, about the temperature distribution in Chicago in the 2070s but about the factors that could influence that distribution: both real-world factors and factors related to how we use computer models.

Simple mathematical systems can help us look past the complexity of the real world and of complicated climate models, to focus on the core issues without being distracted by our favourite landscape, ecosystem, or historical period. Understanding the diverse behaviour of such systems can guide our efforts to improve climate model experiments and the methods we use for processing observational and model datasets into descriptions of climate change. We need to be clear though about our aims. We need to ensure that the expertise that resides in the mathematical sciences is applied to problems that have practical application in the climate sciences. It is essential to tie down the essence of climate and climate prediction if different disciplines are to share an understanding of the fundamental goal.

There are many mathematical challenges related to how we formalize descriptions of the behaviour of a system under climate change, particularly for a system with variability on timescales from seconds to centuries. How do we identify which variables have the greatest impact on future climate distributions? Given the difference between models and reality, how do we use observations to inform us about the best initial conditions from which to start our models?

There's still lots to do in the mathematical sciences and also in identifying what is relevant and useful as information and data are passed from one discipline to another. For the moment though, it's time to get real—okay, not real but a *bit* more real—by considering what these issues mean for the interpretation of fully fledged climate models.

Challenge 3: When is a study with a climate model a study of climate change?

If experiments with climate models are to be studies of climate change, then we need to be able to measure climate change within the models. How do we design experiments that enable us to study climate change in the world of a climate model? When are the assumptions we make necessary and useful to help us understand the problem and when are they unhelpful disclaimers to cover the fact that the experiment doesn't address the question we're interested in? Is there any point in running a climate model experiment within which it is not possible to measure climate change?

16
Climate change in climate models

16.1 What you see depends on how you look

What would you think if I told you thieves had broken into my property and stolen my discovery? Years ago, I met someone at a business meeting who, during small talk over coffee, made such a statement. Well, not quite. The statement was about a colleague of his. He told me that certain neighbourhoods of the city in which he lived, and which I was visiting, were dangerous. He gave an example of a colleague who had been the victim of thieves who had 'stolen his discovery'. This is a friend-of-a-friend type tale; its accuracy is completely unverifiable. Fortunately, the truth about the theft doesn't matter. What's of interest here is my response. My mind whirred. I'd never before been presented with such a peculiar statement. I could only imagine a situation like 'Back to the Future' or 'Ant Man': a wacky scientist working in a basement. For me the big question was: what was the discovery that had been stolen? It must have been important for it to have been worth stealing and for thieves to have targeted it. But how do you steal a discovery? Was it simply his notes, or was it something physical like Doc Brown's time travelling DeLorean or Hank Pym's shrinking suit? The situation felt particularly peculiar because the person I was talking to—like everyone in the organization—was a diplomat or manager. It seemed strange—implausible indeed—that anyone in this social setting would be making a discovery worth stealing. The world felt confusing, unreal, and a lot more exciting.

Lacking confidence, I held back from doing the sensible thing and simply asking what the discovery was. Instead, I obfuscated and made sounds of general interest, and the individual went on to describe the incident in more detail. After several minutes—a surprisingly long time—it became apparent that the discovery was not some sort of scientific breakthrough, as I had assumed, but rather some sort of car: a Land Rover Discovery.

The Land Rover Discovery had been launched about eight years beforehand but I've never been very interested in cars, and in any case the Discovery was way out of my financial league so its existence hadn't impinged on my consciousness. I breathed a sigh of relief that I hadn't revealed my stupidity. It was clearly ridiculous to have thought that 'discovery' referred to a scientific breakthrough. Who makes scientific discoveries at home? Particularly of the sort that can be stolen? The mere idea is ridiculous—solidly in the realm of fiction. Yet I had no other context for understanding the term 'stolen my discovery'. I knew my interpretation was crazy but I had no tools to understand it any other way. The individual telling the tale lived in a different world to me: a world where discussion of cars was common. Presumably he couldn't imagine anyone not understanding that 'Discovery' referred to a high-end vehicle.

Our perspectives were completely different and posed a barrier to communication. Fortunately, it was an issue of no import.

What a climate model is, is subject to perspective in a similar way. To many climate modellers, climate physicists, and environmental scientists, a global climate model is a representation of reality. It's as simple as that. They view the models as being built, to a large extent, on physical laws, and because of this they see them inherently as representations of the real world. Even those bits that are not built on physical laws—that's to say all those parameterizations discussed in Chapter 11—are seen as being motivated by an understanding of the physical and natural world and hence inherently representations of that physical and natural world. They know that their models are not perfect—there are processes missing and errors of representation—but nevertheless, the models are seen as, in essence, representations of reality. Furthermore, for them, the more we include in the models—the more complicated we make them—the closer to reality they will become. The process of model development is one of getting steadily closer and closer to an almost perfect representation of the real-world climate system. Moreover, this is perceived as being synonymous with the models providing ever more reliable predictions. This perspective takes it to be undoubtedly the case that if a model is a good approximation of reality, then its simulations of the twenty-first century (under some assumptions about future greenhouse gas emissions) are good approximations of what will actually happen. The modellers know that there is unavoidable uncertainty due to natural variability but this is not seen as very important.[1] Their perspective is that current models are good and hence their simulations of the future are reliable[2]; future models will be better, and hence their simulations of the future will be even more reliable. That's the modellers', climate physicists', environmental scientists' view.

Many mathematicians, philosophers, specialists in chaos and dynamical systems, and even many non-climate physicists, have quite a different view of what is going on. For them, a global climate model is simply a set of simultaneous equations: a complex set of relationships. There are a large number of variables, and the computer code represents the equations which take the current values for these variables and maps them onto a new set of values representing some point in the future. It steps the variables forwards in chunks of time—our timesteps. The models are in essence just like Lorenz '63 or Lorenz–Stommel except that they have more variables—many, many more variables. Fundamentally though, for these specialists a climate model is simply a set of simultaneous, nonlinear equations.

You see a set of simultaneous equations; I see an almost complete representation of reality—let's call the whole thing off?

No, let's not call the whole thing off. Both these perspectives are important to help us understand climate change and make climate predictions. They both have something

---

[1] This is related to the view discussed in the last chapter that a single path gives a reasonably good approximation to climate distributions in the past and therefore also going forward—that's to say a a good representation of what we should expect.

[2] By which I mean close to representing what will actually happen.

to add but they also both have flaws and limitations. The mathematical perspective can help us better use and study computer models of climate and also of many other systems: economies, floods, agriculture, space weather, brains, galaxies, etc. Yet it may also take us ultimately to a dead-end by telling us that there is no climate model or climate model experiment that provides 'the answer': the prediction of all aspects of future climate that we think society seeks. Meanwhile, the climate physicist's perspective can help us use the models to better understand climatic processes, and from that understanding we could build descriptions of more or less likely futures. It could therefore side-step the problem of there being no modelling system that directly provides the answer to climate prediction questions. Yet often, that is not how the physicist's—modeller's—perspective is applied. Rather the models are widely interpreted as directly providing what people are looking for: they are taken as providing predictions.

This idea that models generate climate predictions, or even that the next generation of models might provide something closer to climate predictions, has the rug well and truly pulled out from under it by the mathematician's perspective. How? Well, what the mathematician's perspective brings centre stage is that the model is an object in its own right. It is different to reality—a fundamentally different object. Like reality, everything in a climate model can influence everything else within it and these relationships are often nonlinear, but a model doesn't include the same processes as reality, so their combined effect will be different.

If we add new processes or increase the model resolution then the model changes. It becomes a different object: what it contains is different; how the processes interact and the consequences of nonlinearity are different; its attractor will be different. It's a leap of faith that improving some aspects of the model will necessarily lead to an overall better simulation of the whole real-world climate system. It may not. It's like following a path around a lake to the town on the opposite side—for much of the time you aren't getting closer to the town, indeed you may be getting further away. Similarly, making parts of the model better could make the whole thing worse. That's the case until we reach perfection (reach the town), or perhaps, optimistically, some sufficient level of adequacy to make predictions. But how would we know when we had reached that level, how would we know when we had reached our destination? When is a model good enough? It's not an easy question to answer because of the extrapolation characteristic. We have no way of testing it and assessing the aspects that matter for climate predictions under human-induced climate change.

The climate modeller's perspective of wanting bigger and bigger models, based on the assumption that they will generate better and better predictions, is built on foundations of sand. We don't yet have any idea of what is necessary and sufficient to make reliable predictions, so a 'better' model (i.e. one with more processes or more realistic representations of some processes) is not synonymous with a 'better' or 'more reliable' tool for prediction. A 'better model' only implies that some processes are represented in a more realistic way than before, not that those processes are actually realistic or that the combination of processes is sufficient or that the model as a whole is likely to produce simulations that could be taken as reliable predictions.

But I'm getting ahead of myself. These issues are at the heart of the climate prediction problem but I'll deal with them—head-on—in Chapter 18. The essence of this chapter is not the relationship between climate models and reality but what it means to talk about climate change, even just within a model.

Climate modelling experiments today tend to be designed by climate physicists and hence almost exclusively from a climate physicist's perspective. Embracing a more multidisciplinary view of what's necessary in a modelling experiment would help us design them far better. Even if they couldn't provide reliable probability distributions for the real-world future they could nevertheless much better support research and in that way better inform society.

As a first step, if model simulations of climate change are going to be at all informative we need to be able to measure climate change within the model. The mathematical perspective of the last few chapters provides a foundation for understanding what is necessary in order to do that, and in a few pages' time I'll describe some climate modelling experiments that have taken similar approaches. First, though, let's take a moment to consider how the concepts raised in the context of simple nonlinear models apply to complicated global climate models.

16.2 More dimensions please

A climate model typically has more than ten million variables but in essence it is the same as the Lorenz–Stommel system. The equations step each variable forward in chunks of time—timesteps—and as in Lorenz–Stommel and in IKEA there is only one path. The only significant difference is that the path is not in an easy-to-visualize three-dimensional space, or a more-tricky-to-visualize five-dimensional space, but rather an unimaginable ten-million-dimensional space: there is one dimension for each separate variable. Don't worry about the bizarreness of this concept—it is just like a path moving around in a three-dimensional sports hall: where it goes depends on where you start from along with the rules encapsulated within the model which restrict the path to a limited domain in the ten-million-dimensional space. That's to say there may be an attractor—or something similar to an attractor[a]—that limits what values the variables can take. With climate change, the domain of possible values will change, as was the case in the experiments with Lorenz–Stommel (Figure 15.7). The only question is how.

Each variable has one axis in this unimaginable ten-million-dimensional space but you don't need to visualize it—just think about it in three dimensions. Be aware, however, that there are more than another 9,999,997 dimensions that you're not looking at. If your position in this space at the beginning of the path is different, then what happens to the path in the future will be different. That's to say, if the starting conditions of any one of the variables is changed at the beginning of a climate simulation, then the subsequent values of potentially all the variables will be different after some period of time.

Measuring climate change in a climate model is about quantifying the consequences of this sensitivity. When we study climate change in the world of a model, we want to know what *future*[3] climate (in the model) would look like under some scenario of *future* greenhouse gas emissions (in the model), while assuming that we know very well (but not perfectly) the starting conditions. Those starting conditions are the temperature, humidity, pressure, winds, ocean salinity, etc. in all the model grid boxes. We can make the climate predictions we want by generating many paths going forward from some particular starting conditions. If we know very well—but not perfectly—where to start all the paths from, then we are exploring micro-initial-condition uncertainty. Collections of climate model simulations that generate such paths are called micro-initial-condition ensembles: they are like the collection of paths originating from within one of the red or yellow dots for Lorenz–Stommel in Figure 15.6. The probability distributions from the many paths of a micro-initial-condition ensemble represent climate in the *future*. If most or all paths show decreasing rainfall over southern Africa then we know that the probability of that happening in our simulated *future* is high; count the proportion of simulated paths that show wetter *futures* and the proportion that show drier *futures* and we have the probability of each. The probabilities tell us what to expect. That's climate predicted. In our model. For a particular choice of greenhouse gas emissions. And for the particular, but still approximate, starting conditions.

With enough simulations—a large enough ensemble of this nature—we could provide everything from the probability of global average temperature exceeding the Paris 2°C warming target in *2100*, to the probability of December temperatures exceeding 30°C in Johannesburg in *2062*, to the probability of rainfall exceeding 300 mm in September in Bangalore in *2047*. These predictions would be conditional predictions because they would be based on the assumed scenario for future greenhouse gas emissions, and they would only apply within our particular model, and for our particular—if approximate—starting conditions. Nevertheless, they would be a good start.

In practice we don't know very well where to start the models from. This is partly due to limited observations, particularly in the sub-surface oceans, but mainly because a model is a different object to reality, so even if we knew the state of reality perfectly we would still have deep questions to answer regarding the best—the most informative—starting point for our model. This is what a mathematician's perspective tells us. And these issues don't just apply to climate models: they are just as important for economic modelling of climate change impacts and for modelling the consequences for ecosystems, health, infrastructure, and so on. They are also highly relevant outside the field of climate—in astrophysics, medicine, and beyond. They are fundamental issues about how we use computer simulations to tell us about reality and they influence the reliability of a wide range of science—and increasingly social science—in the early twenty-first century.

---

[3] Remember: italicised descriptions of time are time in a model not time in reality.

Since we don't know the starting point well, we also have macro-initial-condition uncertainty—which we saw in Lorenz–Stommel in the choice of which yellow dot to start an ensemble from (see Figure 15.6 and Figure 15.8). So we have two very different sources of uncertainty that have to be evaluated and interpreted in very different ways. Unfortunately studies of these uncertainties in the climate modelling community are few and far between.[4] I can, however, tell you about three experiments that have taken tentative steps towards studying these issues. Doing so seems worthwhile at this point not only because this is a book about climate change—so I really think it's time to include some results from climate models—but also because introducing these experiments will shine a light on the difficult compromises that have to be made in any climate modelling experiment.

16.3 We know where we are but we're not sure where we're going.

One of the earliest, large micro-initial-condition ensembles with a global climate model was run by a project called climate*prediction*.net. This was a computing experiment, launched out of Oxford University in 2003, which asked volunteers to run climate model simulations on their home PCs. Thanks to the generosity of the volunteers, it was able to run much larger ensembles (collections of simulations) than any other project. It was originally designed, by Prof Myles Allen and myself, as an exploration of the consequences of errors in climate models,[5] but it was also used to explore their sensitivity to the finest details of their starting conditions. In 2007 my colleagues and I published results from a micro-initial-condition ensemble with forty-nine simulations[b] which was very large for the time and is still large now; we later expanded the work to several such ensembles, each with around 500 simulations, one of which is used in Figure 16.1.

This was an early attempt to address the issue of measuring climate change in a climate model—but it was, to say the least, idealized. The compromises we had to make were significant. If you're interested in making climate predictions, then you might want a collection of simulations that seem somewhat realistic and have vaguely plausible scenarios for future greenhouse gas emissions or their atmospheric concentrations. Well, climate*prediction*.net didn't do that—at least, not in its early stages. For a start the model we used was a global climate model with a full representation of the atmosphere but only a very simplified representation of the ocean. On top of that the scenario we explored was not meant to represent a plausible scenario of the future. It considered a situation in which atmospheric carbon dioxide concentrations were instantaneously doubled. Did you read that right? Yes. It took a model world which was somewhat reminiscent of pre-industrial climate and simply doubled the atmo-

---

[4] Although assessment of micro-initial-condition uncertainty is beginning to gain in popularity.
[5] More about that aspect of the project in Chapters 19 and 20.

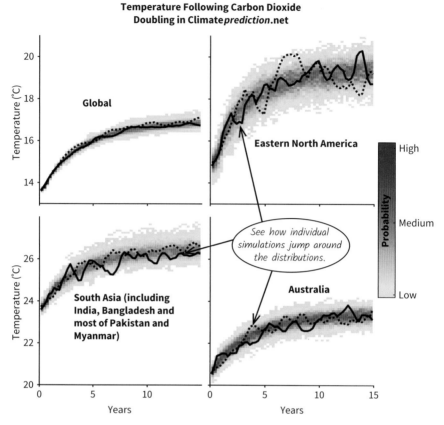

Figure 16.1 Changing temperature probability distributions from a large initial condition ensemble produced by climate*prediction*.net. Colours show the distribution; lines show two individual simulations. They represent the response to a sudden doubling of carbon dioxide using a model that allows the world to get to its final temperature much more quickly than could happen in reality.[c]

spheric concentrations of carbon dioxide. All of a sudden. Overnight. Why did we do it this way? We had our reasons. Honestly. To mention just one, home PCs at the time were much faster than my ZX Spectrum but would still only be able to simulate a handful of decades in a month, and we couldn't expect most volunteers to stay with us for more than a month or so. We were interested in how sensitive the climate system might be to increasing greenhouse gas concentrations, and with such short simulations this would be most easily deducible from a scenario which doubled atmospheric concentrations.[d] So that's what we did.

What's important here, though, is that we were able to run lots of these simulations—many, many simulations, all starting from almost the same starting state. To be precise all the temperatures, pressures, winds, humidities, etc. in all the grid boxes—that is, all the million or so variables[6]—were **exactly** the same in each simulation **except** in one grid box in the atmosphere over the middle of the pacific ocean where the temperature was altered. It was altered, but only by a bit. No more than a couple of degrees. That's to say, all the simulations started from exactly the same point in the million or so dimensional space of the model variables, except in regard to one dimension where it was very slightly tweaked. Tweaked to a degree much less than the variation between day and night.

So what happened? Well, Figure 16.1 shows how the climate changed going *forward*. The simulations warmed up rapidly *over the coming years* and then levelled off at a new temperature. The warming was much more rapid than would be the case in reality because the model didn't have a full ocean but a simplified one which was only 50 m deep—so it took much less heat to warm it up. Some regions warmed up more (for instance, Eastern North America) and some less (for instance, South Asia). Each simulation generated a different path *going forward*, with each variable responding differently, but all simulations showed global warming.[7]

The collection of simulations taken together gave the probability distribution for anything from global average temperature to Australian average temperature to the temperature in the grid box around Sydney or Seattle. These distributions are the temperature component of *future* climate. Unfortunately, the limitations of the experiment also meant we couldn't get all the data we would have liked back from the participants, so we can't see how the climate in specific places such as Seattle or Sydney changed over time but Figure 16.1 shows the kind of results we found for some large regions.

There are a couple of interesting messages from these results. One is that there isn't much uncertainty in what happens to the average global temperature. That's to say, within this model we know pretty well what to expect: the changing probability distribution going *forward* is narrow, so the prediction is quite precise—unlike the Bank of England's twelve-month GDP forecast. Another is that there is nevertheless substantial uncertainty regarding what happens in any particular region and no doubt larger uncertainty still for specific locations such as Seattle or Sydney. The consequences of initial-condition uncertainty for global average temperature are small but for local changes, they are large. This turns out to be a general conclusion for climate predictions.

For me this climate*prediction*.net experiment has two very significant advantages over all other similar ensembles. First, it is—even today[e]—the largest initial-condition ensemble with a global climate model, and size matters because it affects how well we can tie down the probability distributions. Second, I spent more than five years of my life working on it so it's dear to my heart. Nevertheless, we had to

---

[6] The model used was lower resolution than is typical today so the number of variables was more like a million than ten million.

[7] That's to say the average of the temperatures across all the grid boxes at the planet's surface increased in all simulations over the fifteen years.

make severe compromises in terms of the type of model and the greenhouse gas scenario.

Another experiment carried out more recently used a better model and a scenario for *future* greenhouse gas concentrations that is more likely to reflect a plausible future reality. Compromises must always be made though; in this case the biggest compromise was that computational capacity limited the number of simulations to a much smaller forty. The experiment was carried out by Dr Clara Deser and her group at the National Center for Atmospheric Sciences in Colorado. They used their model to simulate the period 2000–2060 and they published their results in journal papers in 2012.

The key messages from this experiment align with the results from the climate*prediction*.net ensemble. When averaged across the whole globe the probability distribution was pretty narrow: the different simulations generated warming of between roughly 2 and 2.5°C over the 2005 to 2060 period.[8] When looking at smaller regions the distribution was much wider: 0.8 to 3.4°C warming between 2005 and 2060 for the average temperature across the continental United States.[9]

Take a moment to think about these numbers. What they are saying is that under ideal conditions for prediction—that is, working only in the world of their model[f] with no worries about model error and assuming we know the starting state very well indeed—the rate of warming over the United States over the 2005 to 2060 period could be anywhere from under 1°C to over 3°C. These different possibilities would have significantly different impacts on agriculture, health, energy demand, frequency of heat waves, and many other things, so the uncertainty is itself important. Yet this uncertainty is also unavoidable: it is intrinsic to the type of nonlinear system we are dealing with. There is no getting away from it. That's a valuable piece of information and it comes from a climate model.

In case you're inclined to climate change scepticism though, remember that all the simulations warmed. A lot.

The work of climate*prediction*.net and of groups such as Clara Deser's illustrates that we can produce probability distributions for what 'we expect' in our models: we can make climate predictions in model worlds. It also tells us that even in the best possible situation, there will always be significant, unavoidable uncertainty as to what will actually happen. This is important beyond climate science because it means that societal choices regarding how we respond to the threats from climate change have to be framed in ways that acknowledge significant uncertainty in what the consequences will be.

We need to be careful when using these uncertainties though. Remember that the uncertainty at the global scale is small. This means that if you 'get lucky' locally, somewhere else has to get unlucky: if the warming over the continental United States turns

---

[8] Only land regions were included in this global average while both land and sea are included in the climate*prediction*.net results in Figure 16.1.

[9] These numbers are for winter but the summer figures are very similar: 1.3 to 3.4°C.

out to be at the low end of the distribution then the warming over some other region has to be at the high end of its distribution.

Of course, we are not actually in this best possible situation for prediction. We also need to consider that our knowledge of the starting conditions is not almost perfect.

16.4 We don't know where we are but we still want to know where we're going

I've already described what macro-initial-condition uncertainty is in terms of a mathematical system, but let's have an analogy to make it more familiar. Imagine you're taking a summer road trip through Europe. You're leaving in a few minutes to drive to the Netherlands but you don't know whether you're setting off from Toulouse in southwest France or Venice in northeast Italy. You know roughly where you're going to be this time tomorrow—somewhere in the Netherlands—but where you're going to be in three or six hours' time is very, very uncertain. The ultimate destination might be fairly clear but even so the journey is extremely unclear.

Okay, I admit this is a pretty implausible situation but for the purposes of exposition let me strain the analogy further. Consider that you're not the driver but a young child passenger. You have little knowledge of geography and location, and of French and Italian architecture—you're just on holiday. This perhaps somewhat explains your lack of knowledge about where you are. You are, however, quite aware of the temperature of the wind blowing in through the open car window. In that case your observations of driving through the outskirts of Venice or Toulouse would be largely indistinguishable. Average summer daytime temperatures are much the same in both locations: taking only temperature observations wouldn't help you very much in distinguishing between the two. And, of course, you end up in roughly the same place: somewhere in the Netherlands. Yet the sensations you feel during the rather long drive would be quite different. From Toulouse, the temperature you experience would be dominated by the daily cycle and a gradual, moderate cooling related to going north. From Venice, sometime in the first half of the journey you would cross the Alps and experience an hour or so of significantly cooler temperatures. The view would also be quite different. The journey is different.

It might be a strained analogy to imagine that you don't know what country you are driving in, but for climate models the parallel is quite plausible. In climate simulations the starting details are open to question. They are not just a little uncertain—they are very uncertain.

The difference between rapidly varying variables and slowly varying variables comes into play here. In the climate system there are some aspects which vary much more rapidly than others. By and large, aspects of the lower atmosphere can vary very quickly: one night, you could be experiencing temperatures of minus 5°C and the next it could be ten degrees warmer; one day could be calm and sunny and the next blowing a gale with torrential rainfall. The variations are greater in some locations than others but generally speaking, the conditions in the lower atmosphere can change rapidly. By contrast, other aspects change only slowly. The ocean temperatures and circulation patterns vary much more slowly, particularly the deep oceans. But it's not just the oceans. The land surface (rivers, forests, etc.) and even aspects of the atmosphere (the stratosphere in particular) can vary much more slowly than the lower atmosphere. This characteristic of the climate system is captured in the Lorenz–Stommel system by it having atmosphere-like rapidly-varying variables and ocean-like slowly varying variables.

Wholesale differences in the starting conditions of rapidly-varying variables have much the same impact on future climate as tiny differences to a single variable, but wholesale differences in slowly varying variables can change the complete picture of what we should expect in the *future*. Bear with me while I unpack this statement. Remember the butterfly effect? Well thanks to the butterfly effect small uncertainties in **any** variable rapidly lead to a different path—a different simulation of the *future*. So far so good but what about whacking great uncertainties? Well what happens then depends on which variables we're talking about: do they describe air or water or ice or trees? If you start a model simulation in June[10] but from conditions in the whole lower atmosphere that correspond to December, then the model might well rapidly revert to northern hemisphere summer conditions. Getting the conditions for the whole of the lower atmosphere substantially wrong wouldn't really matter[11] because those conditions would change quickly to come into line with the energy inputs from the sun and the state of the ocean, the cryosphere, and the land surface. After a few *days* or *weeks* of simulation, it would be similar to having started the model from conditions that correspond to June.[12] As a consequence, a June start with June initial conditions and June start with December initial conditions would be much the same as two simulations with a June start and June initial conditions where one has a slight change in the temperature in just one grid box as in climate*prediction*.net. The consequence of wholesale changes to the state of the lower atmosphere would be much the same as making only the smallest of changes: it would be just like invoking the butterfly effect. As it turns out Dr Deser and her colleagues used this approach in their experiments. They used different starting conditions for the whole of the model atmosphere as means of generating different paths—different simulations—although they didn't go as far as using winter conditions for starting a simulation in summer, only different possible days during the same season from a pre-existing simulation.

What they didn't do, however, was change the state of the ocean. That they kept the same. Changing the state of the ocean can lead to much longer lasting consequences because the ocean variables tend to vary more slowly. If the overall state of some region of the ocean is different then because it can't change very quickly the whole distribution of *future* climate could be different. Consider, for instance, a circulation pattern in the north Atlantic known as the Atlantic Meridional Overturning

---

[10] That's to say with solar input consistent with northern hemisphere summer.
[11] Up to a point.
[12] I must admit that I haven't done this experiment but this is what I would expect—so long as the model doesn't crash.

Circulation which carries heat north from the equator to the Arctic and plays a role in keeping northern Europe warmer than it would otherwise be. If the characteristics of this circulation were different at the start of a climate simulation—perhaps its overall strength or the particular pattern of sub-surface flows—then it could potentially affect how the whole climate system responds to changing atmospheric greenhouse gases not just for *days* or *weeks* but for *decades*. It could change substantially what we should 'expect'. It could lead to wholesale changes to the probability distributions that represent *future* climate change (in our model).

Large and coherent differences in the starting conditions of a simulation can lead to large and long-lasting differences in the subsequent behaviour of the modelled climate system. This is only the case, however, if those changes are in slowly varying aspects such as the patterns or strength of ocean currents or the amount of moisture held in the soils of a region. Uncertainties in the overall starting state of such slowly varying components are what I mean by macro-initial-condition uncertainties. Change the macro initial conditions and what you expect in the *future* changes: you get a whole different climate going *forward*. It is like starting your trip to the Netherlands from Toulouse instead of Venice. It's like taking a different yellow dot in Lorenz–Stommel as your starting point (Figures 15.6 and 15.8)

Prof Ed Hawkins at the University of Reading, in collaboration with colleagues Robin Smith, Jonathan Gregory, and myself, has taken a few tentative steps towards illustrating this effect in climate models. Our experiment used a more complex model than the early climate*prediction*.net simulations: it had a full representation of the ocean. It was however a model with less detail than that used by Dr Deser and colleagues. Its scenario for future greenhouse gases was the rather idealized situation of carbon dioxide concentrations increasing by 1% each year. That's more idealized than Dr Deser's but less idealized than climate*prediction*.net. These were some of the compromises we had to make.

Our experiment came in two parts. First was to repeat something similar to what the Colorado team and climate*prediction*.net had done: roughly a hundred simulations, all beginning from exactly the same starting conditions in the ocean and almost but not quite the same starting conditions in the atmosphere.[13] As in the earlier experiments, the multiple simulations provided a probability distribution which changed over time and represented what we should expect within the model world given the starting conditions; that is, it represented climate in the *future* (see Figure 16.2). At the time of doubled carbon dioxide concentrations, it gave more than a 95% probability that the global average surface warming was between 2.3°C and 2.6°C. That's to say, like the other experiments it showed some intrinsic uncertainty in the global average surface warming, but it was small. At regional and seasonal scales, however, the uncertainty was much bigger—Figure 16.2 shows the June/July/August season for the Amazon Basin.

---

[13] In this case a change of about one thousandth of a degree in the surface temperature of the ocean at a randomly chosen grid point somewhere on the model earth.

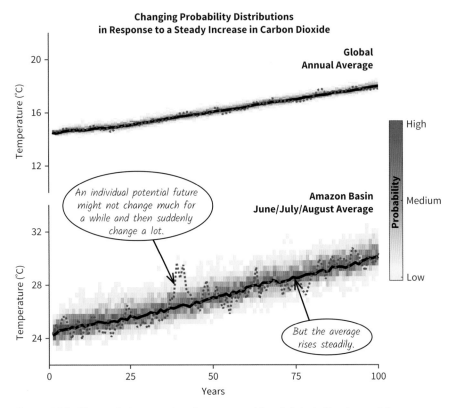

Figure 16.2 Change in temperature in an ensemble of global climate model simulations.[h] Colours show the changing distributions; the solid black lines show the steadily increasing average (median) over the multiple possibilities; the dotted blue lines are an example of one particular simulation.

The next part of the experiment was to repeat the first part but with different ocean conditions at the beginning.[g] Starting from a different ocean state gave different probability distributions for the *future* (in the model) (see Figure 16.3). The differences weren't big for the global average but for particular regions, they were significant. In the first collection of simulations, the winter warming rate in the first twenty years over Europe had been, with 95% probability, between −1°C and +0.7°C (Figure 16.3—top) but in the second collection the numbers were between +0.5°C and +2.3°C (Figure 16.3—bottom). These ranges hardly even overlap! In the second set, all the simulations showed short-term warming over Europe and for most of them it was at a level higher than **any** seen in the first set. The two climate predictions were extremely different for a number of decades. That leaves us with a problem of interpretation. Should we expect, in this model with this scenario, a high chance of initial cooling over Europe before the warming kicks in, or should we expect warming immediately?

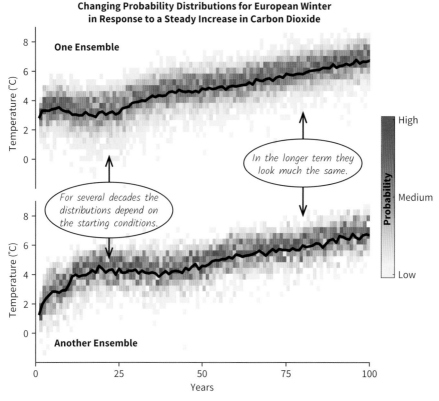

Figure 16.3 Change in the probability distribution for average winter temperature across Europe from two ensembles of model simulations.[i] Colours show the changing distributions. The black line shows the average (median) over the multiple possibilities in each ensemble.

A third collection gave numbers between −0.8°C and +0.6°C. These are moderately similar to the first set but that only raises new questions. Does its similarity mean that the first results are more likely to be right? What would a fourth or fifth or hundredth collection of simulations reveal? How should we interpret these multiple collections of simulations? How does the predicted distribution—the predicted climate—itself depend on the macro-initial conditions? We are left with some deep questions— deep uncertainties—regarding how to interpret the model results, even when we are only trying to understand what climate change means within the confines of the model.

Just as in the Lorenz–Stommel system, there is intrinsic uncertainty about what the future state of the modelled climate system will be—uncertainty in what to expect— which arises simply because we can never perfectly know the starting conditions. This uncertainty can be captured by a probability distribution representing what to expect: representing *future* climate given good approximate knowledge of where we are starting from. On top of that uncertainty though, there is also further significant uncertainty concerning what that probability distribution actually looks like. The distribution itself can depend on the large-scale features of the starting state in ways that we don't yet understand. That's what the Hawkins results demonstrate. In terms of my implausible analogy, the probability distribution for the temperature of air blowing in through the car window on our journey to the Netherlands, two hours after setting off, would be completely different if we were coming from Toulouse rather than Venice. This would be the case despite the fact that the probability distribution for the air as we set off would be much the same, and the probability distribution for where we end up would be much the same.

16.4 Designing climate model experiments

Building and running climate models is complicated, expensive, and time intensive, so it is not surprising that once you've built one all you want to do is run it and see what happens. For the study of climate change, though, that is simply not enough. There is a huge, largely unaddressed challenge of working out how to use these models effectively: what is a useful experiment to run?

There are two overarching challenges here. One is concerned with the relationship between a model and reality: what collection of simulations with a non-perfect model would be most useful for telling us about the future behaviour of reality? That's the subject of Chapters 18, 19 and 20. The other is a consideration of what experiments we would need to run to be able to quantify climate change, even if we were only interested in climate change within our model.

Similar questions arise in many other areas of research. They apply to models of hydrology, of health systems and disease propagation, of ecosystems and agricultural production, and potentially to economic models of society and parts of society. These are not just questions for climate change but rather for how we utilize computer simulation models. They are questions for almost every area of research but I'm going to stick with the issues for climate models because those are the ones I know best.

The modellers' and climate physicists' perspective—that in essence the model simply represents reality—is fundamentally unhelpful. It tempts us to think that making predictions is simply a matter of running the model and seeing what happens. One of the difficulties with this is that running a single or even a handful of simulations of a climate model doesn't provide the information we want regarding what we should expect under some scenario of climate change. Single simulations, or small numbers of simulations, don't enable us to measure climate change even within the world of the model. They can't really be considered studies of climate change at all. Yet the vast majority of climate change modelling experiments are of this nature. This is a problem.

Large numbers of simulations are needed to be able to measure climate change, so all modelling experiments which claim to be studying climate change should involve such collections as a matter of course. That they don't raises serious questions regarding their suitability for informing scientific understanding let alone societal planning.

Of course, limited computational capacity restricts the number of simulations that are possible but that shouldn't be allowed as an excuse; there are always compromises to be made so the real question is whether we are making the right ones. If we want to study climate change as opposed to simply climate processes then it may well be more useful to have large ensembles with less complex models than just a few simulations with the latest top-of-the-range model. This though is a hard sell to modelling institutions and their funders. Everybody wants to use the latest, the best model, but what's the point of using the latest model if you can't measure climate change within it? In car terms, it's all very well having the latest Discovery but if as a result you can only afford fuel or electricity to drive ten kilometres, then maybe you'd be better with a Fiat 500.

It's a matter of urgency to design future model experiments that can address specific societal questions rather than simply running the latest, best model and seeing what happens. It's also a matter of urgency to better understand and communicate the limitations which affect the usefulness of today's modelling experiments as a source of information to support societal decisions.

Climate model outputs are commonly provided for use by wider society. They are often accompanied by information to indicate that they depend on the particular models used, the experimental design, and sometimes the way the model outputs have been processed. Nevertheless, they look like climate predictions, and they are presented as if they should be used as climate predictions, so what are those who need to plan for the future going to do? This is not dissimilar to Shakespeare's witches telling Macbeth that 'none of woman born shall harm' him. Macbeth would inevitably misinterpret this as an indication that he is safe to go into battle. The prediction was reliable but also misleading because of the condition attached. There is a responsibility on the modelling community to ensure that the assumptions and dependencies of modelling results are not presented simply as a disclaimer but rather that their implications are communicated clearly so as to avoid misinterpretation and misuse.

To understand the uncertainties and dependencies in modelling results—and hence to communicate them—we need suitably designed ensembles. Collections of simulations in which the differences are only in the starting conditions are known as Initial-Condition Ensembles (ICE), or in the most recent jargon Single Model Initial-condition Large Ensembles (SMILE) (climate scientists do love their pronounceable acronyms). At the moment, there aren't many of these but their numbers are growing. There are, however, some serious challenges regarding how such experiments should be designed and how big they need to be to give robust answers.

First and foremost there is simply a need to recognize the fundamental differences between the two types of initial-condition-ensemble experiments. An initial-condition ensemble which is founded on starting states that are almost but not quite identical explores intrinsic uncertainty in the climate system: the consequences of nonlinearities which lead to unavoidable uncertainty in what to expect. Such ensembles provide a measure of *future* climate itself (within the model). These are the micro-initial-condition ensembles. In these the differences in the starting state are either small or are only in variables that rapidly adjust to the conditions of the rest of the system, which means, essentially, differences in the lower atmosphere. (The uncertainty in the starting states for these ensembles is called micro-initial-condition uncertainty.) If we look at a collection of such simulations after each has simulated, say, *20 years*, we can construct a probability distribution of behaviour in the model. It might be, for instance, that 95% of simulations show a temperature change of between 0.5 and 2.3°C over Europe. Or 70% show five years of drought conditions in East Africa between years fifteen and twenty-five. These distributions represent the probability of different types of future behaviour within the model. They are conditional on the assumptions made about future greenhouse gas concentrations **and also** on the overall pattern of the starting conditions, but they are still probabilities. The reliability of the probabilities—as a description of model behaviour—depends on how many simulations are run: the bigger the ensemble, the more reliable the probabilities. How big an ensemble you need depends on the questions you want to ask. If you're interested in changes in average temperatures over southern South America, then maybe a few hundred are enough. If you're interested in changes in snowfall in a particular town in the Canadian Rockies, then you might need thousands. This is one of those questions in climate prediction which has not yet been studied. We don't know how big an ensemble we need nor how that relates to the questions we're asking. That's a serious problem for using climate modelling to support society. It's one of the many outstanding challenges of climate prediction.

We can be confident, however, that the bigger the micro-initial-condition ensemble the more reliable the probabilities we get from it. That's to say, we expect that as the number of simulations in a micro-initial-condition ensemble increases, the probabilities built from the simulations get closer and closer to reflecting the actual probability of the behaviour of one more simulation. Put another way, we expect that the ensemble-based probabilities are close to actual probabilities of *future* behaviour within the model; they tell us about what any randomly drawn simulation would do. Put another way still, we expect that the probabilities from one large micro-initial-condition ensemble will be the same as those from another large micro-initial-condition ensemble so long as both use the same model, have the same scenario of greenhouse gas emissions, and take small variations (but not the same variations) from the same starting state. The point is that the results of a micro-initial-condition ensemble can, so long as it is big enough, be interpreted as probabilities.

That's nice not only because probabilities are useful, but also because we have many tools for handling them. They are familiar. Researchers can plough them into all sorts of more and less complex analysis procedures. Non-researchers also usually have a good intuition for how to interpret them. Whether you're a statistician, an economist, or a person who catches buses, plays Monopoly, or cycles to the station, we all kinda know what to do with probabilities.

The second type of ensemble is founded on different states of the ocean or other large-scale differences in relatively slowly varying parts of the system, such as soil moisture levels. These are the macro-initial-condition ensembles and the difference in the starting state is macro-initial-condition uncertainty. The character and interpretation of macro-initial-condition ensembles are completely different to that of micro-initial-condition ensembles, in two very important ways. First, unlike micro-initial-condition ensembles, macro-initial-condition ensembles don't produce probabilities. Second, the uncertainties inherent in macro-initial-condition ensembles are potentially reducible, while those in micro-initial-condition ensembles are not; these uncertainties relate to the starting state in each kind of ensemble.

Think about the issue of probabilities first. Why don't macro-initial-condition ensembles generate probabilities? To answer that think about why micro-initial-condition ensembles do. In micro-initial-condition ensembles each simulation is like a new throw of a dice or a new measurement of the height of a randomly selected adult male volunteer in the city centre of Utrecht. With enough dice throws we can deduce the probability for the outcome of the next throw of that dice. With enough volunteers, the resulting distribution is likely to be a good representation of the probability distribution of the height of men in Utrecht—or at the very least of the probability of the next randomly chosen volunteer in Utrecht city centre. In the same way, micro-initial-condition ensembles provide a good representation of the probability of different states of *future* climate based on the particular large-scale features of the starting conditions. How good depends simply on the number of simulations, just as it would depend on the number of throws of the dice or the number of volunteers measured. Each simulation is like a random draw of the possible outcomes.

So far, so good, but what could we deduce from observations of adult male heights if the observations consisted of 1000 randomly selected volunteers from each of the capitals of the European Union—27,000 volunteers in total? The probability distribution constructed from this data set would not represent the probability distribution of adult male heights across the European Union because different European countries have vastly different populations—3.7% of the observations would be from Malta but less than 0.2% of the population is Maltese. Consequently, the distribution wouldn't represent the probability of the height of a randomly chosen adult male European. With suitable adjustments for population sizes we might be able to extract a good estimate of such a probability distribution but it doesn't just jump out of the data set; we would need more information.

If we only had one volunteer from each capital then the problem would be worse still because we would have such a small sample. That would be like a macro-initial-condition ensemble where each simulation starts from a different pattern of circulation in the world's ocean. It wouldn't represent the probability of anything.

We could instead construct a micro-initial-condition ensemble for each ocean state and hence get a probability distribution for future climate conditional on each ocean state. That would take us to a situation similar to having 1000 volunteers from each country but we still couldn't generate a probability distribution which reflected the behaviour of a randomly chosen ocean state. The problem is that we don't, at the moment, have a means of saying when one ocean state is significantly different from another. Or by how much. Or how likely different ocean states are in the climate system. A macro-initial-condition ensemble consisting of micro-initial-condition ensembles for each ocean state would be like a situation where we had 1000 volunteers from each of twenty countries but in a situation where we didn't know which countries were selected, or indeed how many countries were in the European Union or what their populations were. In this case processing the data to give a probability distribution for the European Union as a whole would be impossible. For climate we don't know how many different ocean states we need to examine (a parallel with how many countries there are in the European Union) nor how likely they are to occur (a parallel with the population of each country). This problem is a challenge: it may be solvable, but with the state of scientific understanding today this lack of knowledge undermines our ability to produce reliable probabilities for future climate even within the world of a model.

All is not lost though. Regardless of our ability to generate probability distributions, macro-initial-condition ensembles can nevertheless paint a picture of what is plausible under scenarios of future greenhouse gases within our model. They can map out domains of possible future behaviour. They can help us get a handle on the scale of uncertainties in future climatic behaviour in the model, even if they don't give us probabilities. Furthermore, with well-directed research there is the possibility of learning better how to relate observations of the real world to particular states in our model. That might enable us to limit what states to include in our macro-initial-condition ensembles: it would be like saying that while we don't know how many European Union countries there are or what their populations are, we do know that we're only interested in the Scandinavian ones, or the Mediterranean ones, or the central European ones. That would provide a route to reducing uncertainty in climate predictions by limiting what initial states are of interest to us.

I've reached the point of speculating on what we might be able to achieve in climate modelling if we got some clever people with the right skills working on the most useful problems to solve. At the moment though, we are a long way from that. Few climate modelling experiments have even micro-initial-condition ensembles of sufficient size to measure climate change within their scenarios and almost none have macro-initial-condition ensembles.

There are a whole host of challenges arising out of these issues: challenges for the design, implementation, and interpretation of climate model experiments. We certainly need more experiments that involve micro-initial-condition ensembles; only with these can we measure climate change in a model. We also need lots of micro-initial-condition ensembles for different large-scale starting conditions of the slowly varying components of the climate system; that's to say we need macro-initial-condition ensembles in which each macro state has a micro-initial-condition ensemble. Only with these can we start to evaluate the impact of different ocean, ice, and land-surface states on the subsequent climatic probability distributions in our models. Such simulations could help us understand the significance of different types of changes to the starting conditions. What are the **aspects** of the ocean that lead to different future climate distributions? When does a different ocean state make little difference, like selecting a distribution of adult males from Amsterdam rather than Utrecht, and when does it make a lot, like selecting a distribution of adult males from Lisbon rather than Utrecht?

Related to this is that macro-initial-condition uncertainty is potentially reducible, whereas the uncertainty explored by micro-initial-condition ensembles is intrinsic and unavoidable: no plausible observing system could tie down the surface temperatures to less than one-thousandth of a degree and even if it could, errors at the one ten-thousandth of a degree level would have the same consequences. Perfect observations are impossible, hence the unavoidable consequences of micro-initial-condition uncertainty. However, in terms of macro-initial-condition uncertainty it is quite plausible that we could get better observations of the current large-scale ocean state and if we understood what aspects of the ocean state were most important for climate predictions then we could focus on getting better observations of those aspects. That would limit the range of ocean states we need to consider and could reduce our overall uncertainty regarding what to expect. The same can be said regarding other aspects of the climate system: ice and soil and vegetation. It's a route to better climate predictions.

There are definitely reasons for optimism but there are also significant and fascinating challenges. Remember, models are different to reality. If we could visualize the attractor of a climate model in its 10 million dimensional space it would look different to whatever the equivalent might be for the real world. The model starting conditions which are most informative about real-world, multi-decadal climate change are likely to be quite different to real-world observations. Even really detailed and good observations of reality will only be of limited use in guiding our model experiments because we shouldn't expect that starting our model from the observed real-world state provides the best prediction. What we need is to understand the most important characteristics that influence future change and to create a starting state in our model which is both consistent with the way the model behaves and reflects the state of these characteristics in reality. Achieving this is one of the many mathematical, physical, conceptual, and modelling challenges awaiting the next generation of scientists.

The challenges in this chapter are intrinsically linked to those of the previous chapter and the study of nonlinear systems. They are also linked to questions of the relationship between computer models and reality, something philosophers will also want to get involved in. That relationship is the subject of Chapter 18 but first I propose we take some time off models and think a bit more about the inherent limitations of reality. If models and the how-it-functions approach are so problematic, maybe it's worth returning to the look-and-see approach and thinking about what we can deduce from observations. Climate change is fundamentally about climate prediction but we nevertheless want to know what the climate is today and how it has changed. That would at least provide a good starting point—so what can our climate observation systems tell us about the shape of climate distributions in the present?

Challenge 4: How can we measure what climate is now and how it has changed?

We don't experience climate: we experience, and we measure, weather—the state of the climate system at some point in time. To what extent is today's climate knowable from observations of weather? How do the timescales of human-induced climate change limit our ability to measure climate and climate change? How can we look at observations through the lens of the things we care about and get information that is relevant for making practical decisions? How can we know when changes in a climatic distribution reflect the consequences of human emissions rather than natural variability? How can we relate local climate to large-scale climatic processes whose responses to increasing greenhouse gases are more open to speculation based on physical understanding? How can we use physical understanding to speculate about changing local climate?

17
Measuring climate change

17.1 If only there were more realities

Journalists covering climate change have often had a hard time. The scientific community has been telling them that climate change is real and serious for decades. Yet when faced with a huge story which seems to demonstrate the point, scientists have often become woolly and circumspect. There have been heatwaves in Russia and Western Europe, wildfires in Australia and California, floods in Pakistan and Central Europe, to name just a few examples. All would seem to provide examples of climate change, but when a journalist tries to get the all-important headline quote that climate change **caused** this flood or that wildfire, they have often found themselves faced with a seemingly peculiar reticence to commit. The event is said to be consistent with what we would expect, it's indicative of climate change, it has been made more likely as a consequence of climate change, but never it has been caused by climate change. This may seem peculiar but it arises from an understandable desire to be accurate, to not make claims beyond those that can be scientifically justified. It is understandable and indeed appropriate but it nevertheless hands over to the journalists the difficult job of communicating a robust and simple message fished out of a sea of complexity and uncertainty.

Climate change communication is inherently difficult because climate change is all about probabilities. Changing probabilities. Uncertainty and ambiguity in changing probabilities. It's not exactly the mainstay of popular journalism or of easy communication.

It's difficult to make absolute claims about floods, heatwaves, and wildfires in the same way that it's difficult to make absolute claims about all sorts of things. Did you get the job you wanted because you're brilliant or because you were lucky? Did I lose that game of Backgammon because I've lost my knack at Backgammon, or because I was unlucky? Did that house get flooded because of the collective acts of the human population in emitting greenhouse gases over the last 30, or maybe 270, years, or was it just unlucky? The answer might be one or the other or, in many cases, both.

If climate is what we expect, then climate is a probability distribution and climate change is a change in that distribution. A flood or a heatwave might have been very unlikely without the historic emissions of greenhouse gases but with them it may have become likely. The question is—how do we know how much more or less likely an event has become?

The essential barrier to answering this question is that we've only got one earth, one reality. Using the methods of the last few chapters, evaluating the likelihood of such events today, given twentieth-century greenhouse gas emissions, would require many realities which were all the same up to, say, 1750 or 1850, at which point they each experienced the smallest of adjustments to climatic conditions before they continued to today. The nonlinearities in the climate system would then lead to many different but equally likely twentieth and twenty-first centuries and hence probability distributions for today. It would be a large micro-initial-condition ensemble of reality, not of a model. The fraction of those realities that have the particular flood or heatwave this year would be the likelihood of that flood or heatwave this year. The **change** in that fraction between 1900 and this year would be a measure of climate change over the last 122 years, because I'm writing in 2022.

But even this wouldn't be enough to assign the change to human influence because there are natural variations over centuries and millennia, so the change in likelihood over the last 122 years is not the same as the change in likelihood as a consequence of human greenhouse gas emissions. For that, we need another ensemble of realities of the twentieth century but without those emissions. We need two, very large, micro-initial-condition ensembles of reality, one with and one without human greenhouse gas emissions (like Figure 15.8 vs. Figure 15.5 for Lorenz–Stommel). This would enable us to say, with confidence, that the chances of this flood, or that heatwave this year had changed by such-and-such an amount: say by nine percentage points from a 1% probability to a 10% probability. We still wouldn't be able to simply say that the particular event was 'due to climate change' but we would be able to say that the probability of the event had increased or decreased by some specific amount as a result of mankind's activities. Communication would still be difficult because the concept of changing probabilities is awkward and unfamiliar. Nevertheless, that would be the best we could possibly hope to be able to say.

Unfortunately we don't have lots of different versions of reality; no jumping between parallel universes for us, I'm afraid. We do, however, have climate models and we can do something similar with them. Indeed there are several modelling groups around the world who are routinely running modelling experiments of this nature.[1] They don't simulate hundreds of years but they do run ensembles of short simulations that are roughly similar to the present day. They run them twice. They run them with and without the changes in atmospheric greenhouse gas concentrations which have resulted from human activities. The differences between these ensembles are often seen as providing the change in probability of specific events as a result of climate change. They have consequently become central to climate change communication efforts and are widely picked up by the media. But do they really reflect the changing probabilities in reality? The substantial differences between models and the real world provide reasons to doubt such an interpretation.

Setting aside the issues of model interpretation though, the whole idea that we can assign the change in the probability of a specific event to human-induced climate change is founded on the basis that we know what the climate probability distributions are in the first place. But what can we know about climate and climate change

---

[1] They're known as Detection and Attribution (D&A) experiments.

in reality, on local scales?[2] How can we measure climate and climate change for particular aspects of the climate system and for particular locations? How can we measure it in ways that are useful for supporting practical decisions: infrastructure management, investment decisions, disaster preparedness, wildlife protection, and the like? To what extent can we measure specific aspects of climate rather than simply demonstrating that climate overall has changed?

These are important questions. Measuring climate and climate change in a specific, local way is important for at least four reasons. First, it is a crucial component in societal debates over our response to the wider threats of climate change, and therefore to public engagement with emission reduction policies. We all have cares and concerns that are local, even if we also feel like we're national or global citizens. Climate change debates would therefore benefit from being frame-able in a local context: from being able to show what climate is locally and how it has changed. It's about making the direct effects of climate change real and personal.

Second, we want to understand how climate has changed so far in order to support the building of climate-resilient systems. Climate change is all about prediction but given the huge challenges in making climate predictions, a fall-back position would be to design our infrastructure to at least be robust to what we should expect today. We ought to ask whether our current systems, and all the new infrastructure we're currently building, is up to the challenges of the present, let alone the future. The practical challenges of climate change are not just about preparing for the future—they're also about preparing for the now. It would be tremendously helpful therefore to know what the climate is, now.

Third, measurements of how local climate distributions have changed are important inputs into research. To understand and study how non-climatic systems such as groundwater, ecosystems, and economies respond to climate (the distribution of weather), we want to know what the climate is today and how it has changed in the recent past. We also want such information to be able to assess the model-based studies mentioned earlier—those that attribute the change in probability of events to human-induced climate change. For this, we want to know the extent to which our models can capture climate distributions, so of course, we want to know what those distributions are.

Fourth, these measurements turn out to be a valuable component in building the best possible information about future climate. They are one route to non-model-based approaches to societally useful climate predictions. The how-it-functions approach can potentially give us useful speculations about how large-scale components of the climate system might change in the future, so if observations can relate local climate to how those components have behaved in the past, then we have a way of providing useful speculations (conditional-predictions) at local scales: scales that are applicable for planning infrastructure and adaptation initiatives. This is about

---

[2] It is certainly possible to say that climate has changed on large scales and that over a large region the frequency of certain types of events has changed—the Intergovernmental Panel on Climate Change (IPCC) documents many such instances. That, however, is quite different to talking about specific events or specific locations.

combining look-and-see with how-it-functions approaches to get a non-model route to climate predictions that are nevertheless of value to society. Understanding what observations tell us is consequently a crucial part of prediction.

The problem, though, with measuring climate is that unlike weather you can't see climate out your window. Statements about climate require analysis and processing of data. Climate may seem like a familiar concept but it is actually rather unfamiliar. Without data analysis it is subject to the vagaries of personal memory and experience; we may have a sense that it used to snow in April and now it never snows beyond March—or some equivalent for wherever you live—but this sort of assessment doesn't provide a solid foundation for the practical management of our societies or even for the social and political debates we need to have to respond to climate change. It would be tremendously helpful if we could tie down the reality of climate.

What we need is a way of processing observations of weather into assessments of climate (and climate change) in a way that can be looked at through the lens of what affects us and what influences practical societal decisions. We want flexibility for different users of climate information to extract whatever it is they are interested in or vulnerable to. And it needs to come with some assessment of robustness: if the probability of an event is said to have changed by twenty percentage points, to what extent do we believe that number—how uncertain is it, how reliable is it?

It turns out that the timescales of human-induced climate change fundamentally limit what can and can't be said with confidence about observed climate change. To begin to see why, let's focus on something relatively easy. Temperature. Not global average temperature, local temperature. How would we measure changes in the climatic distribution of temperature in some place such as Perpignan, France? Or better still, how could we measure an aspect of that distribution that matters for some practical decision (because it's often not the whole distribution that matters—only parts of it)? It may be temperatures above some threshold that influences tourism, labour productivity, air conditioning energy demand, health risks, etc. How would we measure how the probability of exceeding some particular temperature has changed for a summer's day in Perpignan?

17.2 How to measure a distribution when the distribution is changing

The present-day temperature aspect of local climate is the probability of experiencing different daytime temperatures in some particular location today: the probability distribution that in some sense represents what we should 'expect' at the moment for daytime temperatures in that location. This isn't a precise definition but it captures the general sense that we feel we know what type of weather to expect in a particular location. How then might we deduce this from historic observations and how could we tell if and how it has been changing?

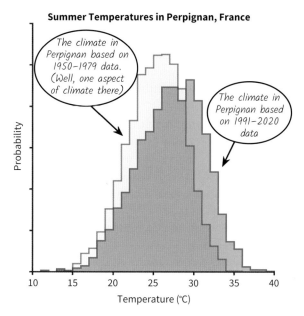

Figure 17.1 Summer climate distributions for Perpignan, France based on thirty years of data.[3]

A starting point might be to construct the probability distribution from the time series of weather observations—an approach discussed in section 15.4. We could, for instance, take the last 30 years of temperature observations near Perpignan in France. Take one value for each day: the maximum temperature reached that day. Select only data for summer—June, July, and August. Count the number of days with temperatures in each of a selection of categories (20–21°C, 21–22°C, and so on) and hey presto we have a probability distribution for summer temperatures around Perpignan (see Figure 17.1). We can do the same for 30 years starting in 1950 and from this we can tell whether the temperature aspect of climate in Perpignan has changed (see Figure 17.1). It has. It's got warmer.

This is a start and a useful indication of climate and of climate change. There are, nevertheless, some concerns.

Concern 1

The most obvious problem is that a distribution from data over the last 30 years is unlikely to represent today's climate. In 1990 atmospheric carbon dioxide levels were less than 355 ppm while in 2020 they were over 410 ppm: a 15% increase.[3] That means the amount of energy trapped in the lower atmosphere each day has

---

[3] If we allow for changes in other greenhouse gases and aerosols rather than just carbon dioxide then it is equivalent to a more than 30% increase in carbon dioxide.

increased through this period and therefore the attractor of the climate system—and in many cases the local climate probability distributions—have inevitably changed. These 30 years of data cannot therefore be relied on to represent the probability distribution for this year—or indeed for any year—because the climate has been changing rapidly through this period. This is an intrinsic barrier to observing and measuring climate under climate change.

We know that climate is changing at a rate that is fast in the context of thirty years not only because of the rate of increase in atmospheric greenhouse gases, but also because of associated observed changes such as sea level rise, ice loss from glaciers and ice sheets, warmer days and nights over most land areas,[b] and, of course, changing global average temperature. Changes at both global and continental scales have been large over the last thirty years (and longer)—much larger than we would expect from natural variability—so we should expect that the distributions of many local variables in many locations will also have seen large changes. A climatic distribution built from data over thirty years mixes the variability inherent in the meaning of climate—a distribution that reflects what we should expect—with the forced change aspect of climate change. It blurs the measurement of climate. It represents a mixture of lots of different climatic distributions—each representative of a different time.

Concern 2

This blurring of distributions over time is one problem. Another is not knowing whether thirty years would be long enough to represent a particular climatic distribution in any case.

Consider a situation where the climate of a location changes, either due to human-induced climate change or natural variations on timescales of centuries or millennia. This situation is like changing from throwing one type of dice to throwing a different one—for instance, from throwing my standard, fair green dice to my peculiar, loaded black one. If we want to know the distribution representative of a particular period—say the last 11,000 years (the Holocene), or perhaps the seventeenth century[4]—then we just need throws of the particular dice that's representative of that period. Nevertheless some period of time would still be too short: you can't assess the probability distribution of a standard dice with only ten throws. The question, therefore, is whether you need five or ten or thirty or a hundred years to assess the climate distribution for temperature—or rainfall or wind speeds or whatever—in some particular location. Thirty years is the period used by the World Meteorological Organization but the answer is likely to depend on the variable you're interested in and the particular location.

Regardless of climate change therefore, we still need to ask whether 30-year distributions like those of Figure 17.1 describe the variety of behaviour that **could** have

---

[4] A period before human-induced climate change.

been experienced through that thirty-year period or only what **was** actually experienced? Are they like the underlying probabilities for a dice throw (one in six for each of the numbers 1 to 6 for a standard, fair dice) or are they like the distribution from ten or twenty actual throws? In terms of the climate system if reality had been different due to the flap of a butterfly's wings thirty years ago, would the distribution nevertheless look much the same or might the distribution itself look quite different? Either case is possible. The particular sequence of temperatures would certainly have been different thanks to the butterfly but the distribution could nevertheless be much the same—or alternatively, the sequence of temperatures could have been different and the distribution different as well. In the first case, the distribution can be thought of as fairly robust: it represents what we should have expected given the state of the climate system thirty years ago and the changing greenhouse gases. In the second, it is not and is likely to be misleading as a representation of this period.

The problem for climate is that we don't know how many sides the dice has, so we don't know how many observations we might need to produce a robust description of a local probability distribution. And we expect the number to vary by variable and by location.

Concern 3

Finally we should remember that the underlying climate distributions representative of the last thirty or a hundred years could themselves have been different because of natural variability within the climate system. That's to say the last thirty or a hundred years could have looked different as a result of a butterfly flapping its wings five hundred years ago, thanks to the consequences that could have had on the slow components of the climate system—the ocean circulation patterns, etc. Greenhouse gas emissions since the industrial revolution would almost certainly still have led to global warming but local climate distributions over recent periods could have been quite different. This is the issue discussed in section 15.4 and illustrated in Figure 10.1.

Handling the concerns:

In terms of interpreting observations there's not a great deal we can do about the third concern, apart from being cautious to avoid over-interpreting observational data— it's not the only way the past could have panned out. The first and second concerns, however, are fundamental to how we go about interpreting observations of climate. And they pull in opposite directions. Addressing the first implies that we should build our climate distributions from as short a period as possible in order to best represent the climate of particular points in time. This in turn would enable us to see how climate (as a distribution) is changing over time and thereby to 'measure' climate change. Addressing the second concern however implies we should build our climate distributions from as long a period as possible in order to best resolve what each distribution is and avoid being misled into thinking that what actually happened is the same as what could have happened during a particular, relatively-short period[5]: that just ten rolls of a dice gives us the probability for the next roll of the dice. We want the distribution to be robust.

If we look at the data knowing that we want an assessment of robustness to accompany the distribution, we can choose to approach the challenge in a slightly different way to Figure 17.1. Indeed we can use the observations themselves to give us an indication of whether the distributions we generate are robust: whether we have enough rolls of the dice.

Before going into this though, take a moment to think about why it's important. Like the model results, information of this type can influence how we manage our societies and invest in infrastructure worth billions of . . . well, whatever currency you fancy. We should really put a high value on understanding the reliability of our information sources: not just getting an answer but having a measure of how reliable that answer is. This is just as important when we look at observations as when we analyse models.

17.3 When to trust measurements of climate

For observation-based climate distributions, we can get an idea of their robustness simply by shortening the number of years we use to build a distribution and then getting multiple versions of it within a few decades. Let me elaborate on how this works. If instead of taking our distributions over thirty years, we choose a much shorter period—say eleven years—then we can construct not just a single distribution but a collection of distributions—say ten—within a twenty-year period. We can build a distribution based on years 1–11, another on years 2–12, and so on up to years 10–20. That means we're looking at a more precise period overall (20, not 30 years), which is therefore less susceptible to the blurring from climate change (concern 1). By looking at ten distributions rather than one, we are also getting an idea of the variability in the distribution: that's to say of its robustness (concern 2). It sounds like a win-win situation but the price we pay is that these distributions are all less likely to actually be robust because they contain less data. Nevertheless we are more likely to know whether they are robust or not and this knowledge is itself worth a lot.

Figure 17.2 shows just such distributions constructed using data from 2001-2020 for a collection of places: Perpignan in France, Lisbon in Portugal, Arezzo in Italy, and Oxford in the United Kingdom. Instead of the usual probability curves though, these are shown as exceedance distributions: they give the probability, on the y-axis, of the temperature on a random day being greater than the temperature shown on the x-axis. It turns out that presenting the distributions this way will enable us to see more easily how much the probability of days being above—or below—a particular

_____

[5] A few decades or so.

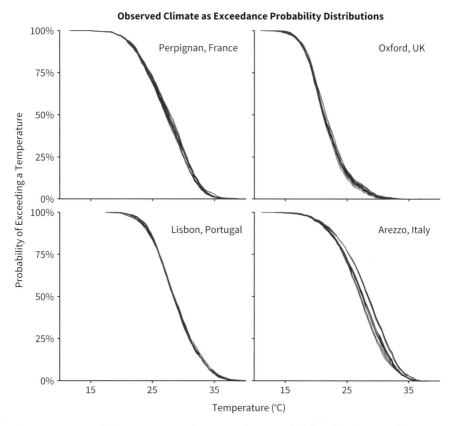

Figure 17.2 Local climate presented as exceedance probability distributions for summer temperatures. Each line is constructed from observations for 11 consecutive summers during the 2001–2020 period.[c]

temperature changes when the distribution changes. That's useful—more on it in a moment—but first note that in Perpignan and Lisbon, the different distributions are all pretty much the same; in Perpignan the probability of the temperature exceeding 28°C is about 45%. By contrast, in Arezzo they are all over the place. The message is that for Perpignan and Lisbon the distributions are indeed quite robust while for Arezzo the distributions themselves experience a lot of variability. One of the benefits of using a shorter time period is that we can see this.

We now have an indication of which distributions are reliable and which are not.

We can do the same for some point in the past so long as we have the data. Figure 17.3. shows collections of distributions for 1950–1969 alongside the ones for 2001–2020. For Perpignan, each set of distributions has little variability within it but is quite separated from the other set. From this we can say with high confidence that the climate of Perpignan is different in the early twenty-first century from that in the mid-twentieth century. In fact we can go further and pick out a threshold that might matter to us. We can say, for instance, that the likelihood of experiencing days above

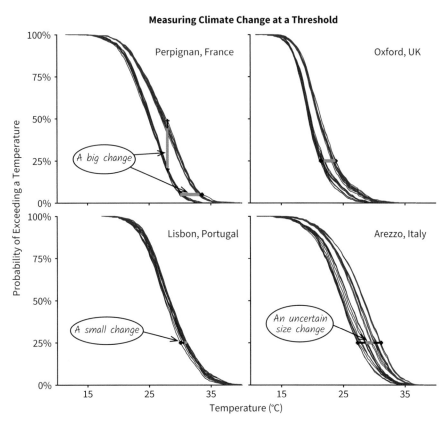

Figure 17.3 Measuring changes in summer European climate. We can measure climate change at particular locations and thresholds by comparing the exceedance probability distributions from an early period (1950–1969: blue) with more recent ones (2001–2020: red). The vertical line shows the change at the 28°C threshold; the horizontal lines show the change in the temperature which 25% of days are above (5% for Perpignan); the green lines show the smallest measured change over 51 years, the black lines the largest measured change.[d]

28°C has risen from about 21% to about 45%. We're at the point of saying we have measured local climate change: measured local climate change through the lens of an aspect that might arguably be considered relevant for decisions and planning. Woo!

The figure also shows that the climate of Lisbon—well the aspect associated with daily temperatures—has not changed much: the collections of distributions in each period are closely clustered and they are much the same as they were roughly sixty years ago. There appears to be a robust message of little change. In Arezzo in both periods the distributions fluctuate a lot: all the distributions in the later period are warmer than in the earlier period suggesting that Arezzo's climate has indeed warmed, but it is difficult to say by how much. The intrinsic nature of the variability of the climate system makes it difficult to say by how much because this aspect of climate in this location is itself not very steady. What's useful in terms of connecting climate science and societal planning is that this is something we can identify from the observations: the observations themselves tell us whether our measurement of climate change is potentially robust. The look-and-see perspective is starting to provide some results which might actually be quantitatively useful rather than just generally informative.

17.4 Mapping the changing shape of local climate

When faced with making practical decisions, we are often vulnerable to particular parts of climatic distributions. Human productivity appears to decline when temperatures go above about 28°C[e] and the number of deaths increases above and below certain temperatures; the relevant thresholds vary by country and location.[f] Climate distributions enable us to look at specific thresholds and to measure how they change between pairs of distributions that are decades apart. They provide a route to describing quantitatively the changes in risk and impacts associated with changes in local climate. To do so though, we need to consider how we're going to interpret these multiple distributions. As in so much of climate change science, there are many routes forward but few have been studied and there are conceptual issues at play.

One approach is to pair up the distributions so that each pair portrays a change over fifty-one years: the 1950–1960 distribution with the 2001–2011 distribution, the 1951–1961 one with the 2002–2012 one, and so on. Each pair provides a measure of the change in the probability of exceeding, for instance, 28°C or any number of other details that may be important for climate-vulnerable decisions. So far, so good, but what should we do with those ten measures of climate change? The problem is that they're not independent of each other in the way ten rolls of a dice would be: each distribution contains observations that are also used in other distributions. It is also plausible—even likely in some locations—that the distributions will themselves go through cycles of variation—shifting higher for some years or decades before shifting lower again. These natural, longer timescale variations may lead to eight or nine measurements that reflect one type of climatic behaviour and only one or two for the alternative but it may be the case that both types of behaviour are actually equally likely. The upshot is that although your inner statistician may desperately want to take the average of our different measurements of climate change, that would be the wrong thing to do. We have to resist the temptation to use statistical tools just because we have them.[6] In this case averaging them would lead to potentially misleading results because we have good reason to expect they aren't random samples of possibilities. In any case, averaging would mean throwing away the information they contain about climatic variability and uncertainty—exactly the information that should be at the

---

[6] Resist until we can justify their applicability—something we'll come across again in the coming chapters when looking at models.

forefront of good decision-making and the reason for using multiple distributions in the first place.

Instead of taking the average then, we can pick out the largest and smallest change at our threshold of interest, over 51 years. These values provide the most meaningful information from the set of ten assessments. For the 28°C threshold in Perpignan the two are similar and large—a change of about twenty-five percentage points but with increases ranging from twenty-two to twenty-eight percentage points—which means we can be confident that there has been significant change in this aspect of climate (because even the smallest change here is large) and also confident about how much it has changed (because the smallest and largest change are similar). In Lisbon they are much smaller (ranging from practically zero to about seven percentage points) so we can be confident that there has been relatively little change in this aspect of climate. In Arezzo the smallest change is quite small (an increase of about seven percentage points) but the largest is quite large (an increase of over 33 percentage points) which tells us that we can't say whether it has changed by a lot or a little.

The message is that sometimes we can measure climate and how it has changed in the recent past and sometimes we can't. It depends where geographically you are looking and what part of the distribution is of interest to you. That's just the way of things in a complex system. However, in this approach the observations themselves tell us about what we can and can't say regarding historic climate change and the aspects we're interested in. It also provides flexibility to focus on the part of the climate distribution that affects whatever decisions we're facing, while at the same time intrinsically presenting information about robustness.

We can go further with this perspective and map how the shape of climate has changed in different ways in different regions. Consider, for instance, how the hottest days in summer have changed. We can pick this out from our distributions by looking at the temperature that is exceeded on only 5% of summer days—the hottest summer days—and seeing how that has changed in the observational record: the horizontal lines in Figure 17.3. If we do this at every location across Western Europe, some will have changed a lot and some by not very much. Some will have a lot of variability and some not very much. To identify a consistent—and arguably the most robust— message available from the data, we focus on the smallest observed change. If this is big, we know there has been substantial change. Of course it may be an underestimate, as a consequence of variability, but if it is large then we know that this aspect of climate change is large. On the other hand if this number is small, then we know that either there has been little change or there is so much variability that it masks a clear signal. This is therefore a conservative approach because places identified with small changes could actually be seeing big changes—we can't be sure—but in places showing big changes, we know with relative confidence that there have indeed been big changes. It's all about picking out what we can say with some confidence. Taking this 'smallest observed change at a threshold' approach, we can make a map—Figure 17.4—which shows how the hottest days in summer have been changing. In northern France and across the Low countries and northern Germany we see that the hottest summer days have warmed by at least 2°C in many places.

Climate Change Affects Different Types of Day Differently
The warmer than average days in a typical summer:

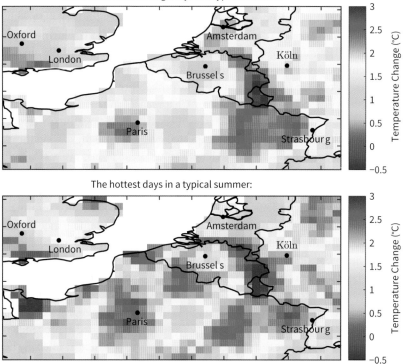

The hottest days in a typical summer:

Figure 17.4 Maps of the 'at least' warming level for different types of days in Northern France, Southern UK, Belgium, the Netherlands, and western Germany. For each location, these are the equivalent of the green horizontal lines in Figure 17.3—the smallest change over multiple samples. They show the change between the 1950s/1960s and the 2000s/2010s. The upper plot shows the change in the warmer than average day in a typical summer—the temperature for which a quarter of days (25%) are hotter. The lower plot shows the change in the hottest days in a typical summer—the temperature for which one in twenty (5%) are hotter.[g]

By contrast if we look at temperatures for the warmer-than-average-but-not-extremely-hot days in summer—the temperature exceeded on a quarter of days—we find that they too have risen but by less than the hottest days, except in the eastern part of this region—Figure 17.4. The pattern is different from the very hottest days, which shows that the shape of the climate distributions is changing in different ways across Europe.

This is the kind of information we're looking for. It uses weather observations to begin to paint a picture of changing climate distributions in a way that can be tailored and personalized to your particular interests, locations, and vulnerabilities. It is, of course, only a start. Impacts and vulnerabilities are connected with many diverse aspects of climate. Even if we stick with temperature, it is not only individual days but sustained periods that can be important: heatwaves in summer or perhaps long periods of warm temperatures in winter.[h] Changes in the frequency of heatwaves and similar events are difficult to assess outside models because they are relatively rare occurrences so we don't have many examples of them from which to build a picture of how they might be changing.[7] It turns out, however, that there are mathematical techniques we can use to extract information from distributions of daily temperatures, about the length, intensity, and probability of heatwaves in specific locations. Observation-based work by Prof Chapman of Warwick University, and colleagues, showed that in England whereas a six-day period with daily temperatures over 28°C degrees might have been expected every six to eight years in the early twentieth century, now an event of that length would be expected every two to four years. Such an event is now two to three times more common. This research was based on data for central England. It could be applied to higher temperature thresholds but the point is that it demonstrates the potential to provide such information from the type of observational weather data that we have. It demonstrates another approach to measuring climate change, again designed to focus on particular aspects that might be interesting and relevant to society—this time heatwaves—and again identifying how weather observations can reveal changing climate while letting the uncertainties shine through.

The context for measuring climate change is that different people need or want different information. Health planners might be interested in the probability of exceeding some temperature threshold or the likelihood of a particular intensity heatwave, whereas ecosystems might be vulnerable to the reduced likelihood of freezing during winter or the change in the likelihood of a certain level of rainfall in a season. Indeed, for many things we are less interested in temperature and much more interested in rainfall. It turns out, fortunately, that the approaches of this chapter can also be applied to the analysis of rainfall observations. In fact I originally wrote a section on rainfall for inclusion here but rainfall brings with it many complexities which are a distraction from the main focus of the challenges of climate change. So instead of getting distracted by them, let's assume we have some measures of local climate and how it has changed—whether they be in terms of temperature or rainfall or some other climate system variable—and consider how they can be used for predictions.

17.5 Observations as a route to predictions

When we measure changing local climate distributions, the results will inevitably be affected by how much the planet as a whole has warmed, by variations in the regional patterns of warming and by natural cycles in the climate system. Measuring

---

[7] It's worth noting that this is the case when looking at the situation in a particular location but if we simply want to know whether there is a tendency for climate change to increase the likelihood of heatwaves more generally then we can look across a wide region, or a continent, in which case we might have sufficient examples to be able to identify that things are changing even though we might still not know how they are changing in a particular place.

changing local distributions alongside how they relate to the wider state of the climate system provides a route to using observations to provide more useful climate predictions. That's because large-scale features of the climate system are more amenable to prediction—or at least informed speculation—based on a physical understanding of how they behave. If observational data can measure not just how a particular aspect of local climate is changing but also how those changes are related to some larger-scale climate system behaviour, then there is a route to saying much more about what to expect in the future for that aspect of local climate.

Consider, for instance, the North Atlantic Oscillation (NAO). This is something that substantially affects European weather, particularly in winter. It also affects weather systems more widely. It comes in two phases: a positive phase, in which the difference in atmospheric pressure between Iceland and the Azores is relatively high, and a negative phase where it is relatively low. The types of weather in the North Atlantic and Europe tend to be quite different in each phase and the climate system often stays in one phase or the other for several years at a time. Furthermore, it is plausible that climate change might lead to a change in the relative frequency of the two phases: one might become much more common. If a local temperature or rainfall distribution is changing during the positive phase but not during the negative one, then we arguably have information on which to base informed speculations about future local temperature or rainfall distributions using our understanding of the NAO.

Research insights into the mechanisms of the integrated climate system, including aspects such as the NAO, may provide theoretically based predictions of change in large-scale, relatively slowly varying, components of the system. If we can predict such changes by understanding these aspects of the system, then the connections found in observations might sometimes provide a route to translating them into local-scale predictions. In this way, effectively targeted analysis of observations could provide a means to translate scientific understanding into usable and relevant products—conditional local predictions—that could help us build more resilient societies. It's a route to connecting the look-and-see approach with the how-it-functions approach to provide better and more useful climate predictions. In this way, observations themselves become tools for predictions.

17.6 Getting the most out of our one reality

Climate-sensitive decisions are made throughout our societies every day and these decisions are often sensitive to particular thresholds in climate distributions. They may also be sensitive to how climate variables vary over time and how they combine. One can easily think of examples. The flood risk associated with increased-intensity rainfall might depend on whether it is more likely to follow a period of drought and heat which has left the soil dry and therefore increased the rate of runoff. An insurance company or government might be sensitive to changing combined risks such as the consequences of a heatwave across multiple regions simultaneously, something which might lead to an overstretched medical system. Or they might be sensitive to windstorm damage in one region followed by flood damage in another within a short timeframe, something which might lead to exceptionally high damages, insurance claims, and economic impacts. Industries with international supply chains and/or portfolios may even be vulnerable to multiple events in very different parts of the world.

The huge relevance of climate change for society and the urgency of the issue means that we would all benefit if more research—both pure and applied—were approached through the lens of what might be useful. A key challenge for the social sciences is directing physical science research in a societally useful way. This is particularly difficult for the pure and theoretical work that is required to underpin the applied analysis but even these would benefit from being carried out in the context of the social aspects of the problem. A key challenge for the physical sciences is achieving a more nuanced understanding of how their understanding can support social science and society—of really engaging in certain interconnected, multidisciplinary issues.

There would be substantial value in being much clearer about which aspects of climate influence our infrastructure, planning and design decisions, and how those aspects have changed over the last century or so. That's why measuring climate and climate change is one of the key challenges in climate change. Of course, making decisions without perfect information is something we all do all the time, but to respond most effectively to climate change, it would be helpful to understand what can be known and what cannot: what is understood now, what might be quantified in the future and what is likely to remain unknown or at least very uncertain for a long time. Good decisions can be made even if the current or future state of climate is ambiguous, but there is significant value in being clear not just about what can be said but also about what cannot be said. To say, 'we've looked and understood what's going on with the variables you're vulnerable to, but the conclusion is that it's not possible to tell you how they are changing' is quite different to saying 'we don't know because we haven't looked'. Information about what we cannot know is still information, and potentially useful. This may sound obvious but it is a point often overlooked by those producing climate information from both observations and models. There is a danger of miscommunicating the state of knowledge in the dash to be helpful and provide answers.

The challenges in processing weather observations into measurements of climate change reflect some of the wider challenges in climate prediction. There are hard questions still to be answered regarding what we can and cannot conclude from a set of observations given the characteristics of climate change: extrapolation, one-shot-bet, nonlinearity. There are also practical issues; observational analysis needs to be carried out from the perspective of what we want to know because climate change is *not just of academic interest*. Academic researchers are unfortunately rarely engaged with the actual vulnerabilities and conditions under which industrial and societal decisions are made, so it is difficult for them to focus analysis on what might be most useful; this is an aspect of the multidisciplinarity characteristic.

The timescales of climate change limit what aspects of climate change can be measured directly, and raise challenges regarding how the data should be interpreted. Looking at the distributions for Arezzo in Figure 17.3, the amateur statistician in us all wants to combine the distributions from the 1950s, and also those from the 2000s, giving us just one representative distribution for each period; we want to take some sort of average and make it simple. That way, we could say clearly that this is how climate has changed. But while it's certainly true that the distributions have changed between these periods, it is unclear how and by how much. It's unclear quite what is going on. The underlying distributions representative of each year could be varying following some natural cycle or some relationship with other parts of the climate system. The distributions themselves are not independent—they are not like separate rolls of a dice. And on top of this there are unavoidable limitations in measuring climate change which arise from the rate it is changing. The theoretical challenges in observational climate analysis are about understanding when an evaluation is robust and could be informative and when it could instead be misleading about the future. They are about reflecting on what sort of statistical or econometric analysis is meaningful for the questions we are interested in. Asking how an assessment could be wrong or uninformative is how we build confidence that in fact it is right and useful.

The complexities of climate and the relative youthfulness of climate science mean we need to be careful to avoid over interpretation of observational data: to avoid inferring or implying they provide more information than they do. We need to be asking what scales and types of change might we possibly be able to measure and when might changes be real and impactful but simply not identifiable in the historical record due to variability?

There is a vast amount of work on the statistical analysis of physical observations but not so much on their translation into time-varying climate distributions which reflect climate change. What is also missing is the study of how physical climate processes relate to local climate distributions rather than just climatic events. There is much study of the statistics of physical climate but not so much of the physics of climate statistics. Such analysis would be extremely valuable when we are trying to generate physically-justified speculations about potential future climatic behaviour, an approach known as 'narratives' or 'tales' (more on these in Chapter 21). These approaches are all about relating aspects of climate change that might be predictable with climate science, to the distributions representative of local behaviour which impact society directly. Making these connections is a crucial element in assessing climate risks: physical risks, financial risks, economic risks, whatever.

The analysis of observations needs to shift towards prioritizing the information that might be useful to society, on the one hand, and the methods for assessing robustness on the other. And speaking of robustness, it's time to return to models and the elephant among these pages: the relationship between climate models and reality.

Challenge 5: How can we relate what happens in a model to what will happen in reality?

The explosion of computational capacity in the early twenty-first century has led to an explosion of computer models of natural and social systems. To what extent can the outputs of a climate model be considered a prediction of the reality it is meant to represent? How close to the real world does a model have to be to provide reliable predictions? How should we design experiments to explore the consequences of model error? What can such experiments tell us about reality? In what way do multiple models provide more information about the real world than a single model? When is a model too bad to be included in our considerations? When do models inform, when do they mislead, and how can we tell the difference?

18

Can climate models be realistic?

18.1 The extrapolation barrier to trusting models

Computer-based representations of reality are ubiquitous in academic research and throughout society. They have substantial influence across many domains but especially when it comes to assessing climate change. It is therefore important to understand what they can and what they cannot tell us.

This and the next few chapters are a journey through the inherent challenges we face when trying to use computer models to make predictions about the real world. They raise conceptual—sometimes philosophical—issues about what represents reliable evidence in science and social science. The challenges are immense. As you read on, you might feel increasingly pessimistic about our current abilities and even perhaps about the prospect of making progress. But don't despair—this challenge and the next are a pair. They work together as bad cop, good cop. This one is about the deep and in my view fascinating challenges that we need to address if we are to use computer models to make climate predictions—predictions we might reasonably describe as reliable and which can be taken at face value. The next challenge focuses on how we can nevertheless get valuable information from today's models and today's experiments, despite all the unresolved conceptual issues.

So to begin with, imagine you have a climate model and you want to use it to make climate predictions of reality under some assumptions about future greenhouse gas emissions. How do you decide if the results from your model are reliable?

Let me be more specific. Imagine you have a global climate model that you think is sufficiently realistic that it is worth starting it off from conditions that reflect the state of the real climate system today (we don't) and you have really good detailed observations of the state of the climate system today in all its multifarious aspects (we don't). Furthermore, you have enough computational capacity to run a huge ensemble—a micro-initial-condition ensemble where in each simulation you tweak the starting conditions just a tiny amount to get a different simulation of the twenty-first century (we don't). In each simulation you make the same assumptions about atmospheric greenhouse gas concentrations through the twenty-first century.[1] You would then have model-based climate predictions which are expressable as probability distributions for whatever you are interested in: the average June temperature in Nairobi in 2064, the average August rainfall in Buenos Aires in the 2060s, the maximum temperature in Shanghai on 25th January 2087, the temperature in Melbourne which

---

[1] If your model is an earth system model, then you might make assumptions regarding emissions rather than concentration because your model includes a carbon cycle and can translate one into the other.

will be exceeded on only 5% of February days in the 2070s. You would have whatever you want for any time in the twenty-first century and for any location. You would have very precise climate predictions of everything you might be interested in, anywhere in the world, under the assumed future greenhouse gas concentrations. The predictions would look great and they would be accurate predictions of what happens **in your model**. They would be robust in the sense that you couldn't do better with this model. How, though, would you know whether they were reliable predictions of reality?

We can assess the reliability of probability forecasts, such as weather forecasts, by looking at outcomes. If we have a collection of forecasts and subsequent outcomes (forecast-outcome pairs) then we can measure whether the probabilities, and therefore the forecasts, are reliable. If they are and they are made by a model, then we don't care how the model has been built. We don't care if it accurately reflects the processes at play. We can treat it as a black box that simply spits out a forecast and we trust that forecast because we can see it repeatedly doing a good job. This is the look-and-see approach, not to forecasting but to forecast evaluation.

Model-based climate predictions can't be assessed this way due to the characteristics of climate change, specifically extrapolation made worse by the one-shot-bet. The extrapolatory characteristic means we inherently don't have any forecast-outcome pairs for the type of prediction we're trying to make so we can't evaluate it as we could a weather forecast or a GDP forecast. The one-shot-bet characteristic means that even after the fact there will only ever have been one outcome, one twenty-first century, and you can't evaluate a probability forecast with only one outcome.

The upshot is that we can't use a cycle of forecasts and outcomes to assess a model's predictive capability. Instead our confidence has to come from an assessment that the model realistically captures all the relevant physical processes at play: the reason we might trust a climate model forecast is because of the model's 'how-it-functions' origins. An assessment of a climate model for climate forecasting should therefore involve evaluating whether it accurately represents everything that is going on—not everything in the universe but everything that we think could possibly matter for our prediction.[2]

How, though can we do that? The answer is—we don't know. So instead, we tend to look at whether a model can simulate historic observations on the assumption that if it can do that it must be getting the processes right. This chapter explains why a good comparison with historic observations wouldn't give us the confidence we desire, as well as showing that in fact today's models do not compare at all well with observations. It ends with some intriguing conceptual issues regarding just how close to reality a model would need to be to be able to make reliable probability forecasts. Here goes.

---

[2] Which leads us to thorny questions about what could possibly matter for multi-decadal predictions, but that is a question for another time or another book.

18.2 Our goal should be the future but it's actually the past

To begin with we know that an ability to realistically simulate the past is not an indication that a model can realistically simulate the future simply because we know that the future will be substantially different to the past and we confidently expect that the behaviour of climatic processes—along with their nonlinear interactions—will be different in the future. Think back to the kettle of section 10.3—observations of the kettle between 60°C and 90°C don't tell us what will happen subsequently and in the same way showing that a model can simulate the 60°C to 90°C warming does not indicate that it can simulate the process of boiling that will become important as it continues to heat up.

Nevertheless if we built a model from basic physical principles, then we might feel that its ability to represent past observations would be a good indication of its ability to represent the future because it would seem to suggest that we'd done a good job at representing reality. Unfortunately, we are not in such a rosy situation. There are good reasons to expect our models to represent the past much better than the future.

These reasons arise from the fact that even though physical principles are used in building climate models, so are historic observations. They are used in a number of more and less obvious ways. For a start, global climate models are actively tuned[a] to represent observations of the past. This acknowledged tuning is only for a few handfuls of variables, which are often global or regional averages, but this is just the tip of the iceberg. Beyond these tuning processes there will inevitably be unacknowledged or even subconscious use of observations. Climate model developers work in the field of climate so they are inevitably aware of a wide range of observations of the climate system; there is thus the possibility of contamination: a risk that they will make modelling choices to make their component look more 'realistic', even outside any agreed tuning process. Furthermore, there are sociological pressures on modelling centres. A modelling centre that found their model was significantly worse than other models at simulating past observations would be under pressure to go back and look at it again in a way that wouldn't happen in relation to processes that are only dominant in the future. These factors combine to make it much more likely that a model will realistically represent behaviours that have been seen in the past than those that will become dominant in the future.

These issues reflect an inherent conflict in model construction. On the one hand, the aim of climate model development is to try to get a good representation of the climate system so it would be ridiculous not to compare the model results with observations and use that information to help improve the model; it's a valuable way to understand model deficiencies and target improvements. But on the other hand, doing this inevitably prioritises representing processes and interactions which have been dominant in the past, over the ones which may dominate in the future. If our ultimate aim is to generate information to guide society regarding the future, then that's a serious problem: it prioritizes what we can measure over what we're interested in.

This is an example of the 'in-sample' problem: data used in creating a prediction system is of limited value in assessing the reliability of that prediction system. Let me unpack that.

If we make a forecast of tomorrow's weather, next year's GDP, or the average January temperature in Nairobi in two years' time, there's no way the predictions could have been influenced by knowing the outcome. The prediction is made for a situation that is said to be 'out-of-sample' which means that whoever made the prediction had not seen the result. The subsequent outcome is a good test of the forecasting system's reliability **because** there is no way it could have been adjusted, consciously or subconsciously, to get the right answer specifically on this occasion. If it does well[3] then it is likely to indicate that the prediction system is generally applicable for such forecasts. But when we assess a climate model against past observations, we make these assessments using data which was already available to those building the models: the data is potentially 'in-sample'. The quality assessments are therefore much less useful as an indication that the model can simulate the future because the model could have been adjusted to get those particular observations right without capturing the inherent characteristics of the system. Unfortunately, there is no alternative. The extrapolation and one-shot-bet characteristics of climate change mean that we can't get access to relevant out-of-sample data. All we can do is be aware that this is a problem and remember that any success in replicating past observations provides only a limited indication of reliability for simulating the climate of the future, or even for simulating alternative paths of climate in the past.

This concept of 'alternative paths of climate in the past' came up earlier in the context of multiple potential paths in the Lorenz–Stommel system. It's important in the assessment of climate models because historical climate could have been different. I'm not talking about there not being global warming, nor about day-to-day variations in weather, but about the regional patterns of change over decades, and even the particular time series of global average temperatures. If historical climate change could have been different, then it is not clear how well we want our models to simulate the observed past anyway. Our observations only reflect what actually happened: different behaviour would certainly have been possible[4]—natural variability and its interaction with climate change could have played out differently. Assessments of model 'hindcasts', that is simulations of the past, suffer from having only one example of reality with which to compare. Until we have a good understanding of the characteristics of climate variability over decades, centuries and longer, it is difficult to know how different the past could have been and therefore how closely we want our models to align with observations. There is research being pursued on this subject but its relevance and importance for climate predictions and their evaluation, as well as for how we respond to climate change, is largely overlooked.

The consequence of this is that using historic observations to tune our models runs the risk of making them over-constrained—insufficiently variable—which makes

---

[3] Or rather well on multiple occasions so that we can assess its probabilities.
[4] This relates mainly to concerns 2 and 3 in section 17.2.

them of limited value in representing both potential pasts[5] and potential futures. Worse still, it encourages us to think that the climate system is less variable than it might well be because the only way of studying alternative twentieth centuries is by using these same models. We have an unhelpful feedback loop—climate models have less variability than reality because of the way we build them and this leads us to believe that reality may in fact be less variable than it probably is.[6]

18.3 Are climate models close to or far from reality

Despite all these concerns, a starting point for assessing the quality of climate models has to be to look at simulations of the past. What else can we do? If a model is good at simulating the past, that doesn't mean we should necessarily trust its predictions of the future but if it is not good at simulating the past—'not good' after allowing for an assessment of how the past could have been plausibly different—then we have reason to think that the model and reality are very different beasts.

It's important at this point to say that global climate models are fantastic achievements of modern computer-based science. Building up from a wide variety of different physical elements, they are able to simulate reasonably realistic-looking versions of extremely complex climatic behaviour such as the El Nino Southern Oscillation, the North Atlantic Oscillation, and even just the spatial and seasonal patterns of surface temperatures across the planet. They are amazing achievements and if used wisely, these models are tremendous tools for understanding climate. But what about prediction? They are widely treated as if they provide climate predictions; experiments are designed to simulate prediction scenarios. Do their simulations of the past give us confidence that they can be interpreted that way?

Let's have a look at global average surface temperature: the average temperature across the surface of the planet, averaged again over a whole year. It is common to look at how this has varied over recent decades and centuries and to compare observations[c] of it with results from climate models: climate models driven by concentrations of greenhouse gases that reflect the twentieth century. In order to focus on climate change these comparisons are usually presented as changes with respect to some particular period, say the average temperature over the 1850–1900 period, as in Figure 18.1. All today's models produce year-to-year and decade-to-decade variability, but they all also simulate substantial global warming since the middle of the twentieth century, warming not dissimilar to that seen in the observations.

---

[5] This is one element that undermines the reliability of the detection and attribution studies referred to in section 17.1

[6] Analysis of long observational time series going back thousands and millions of years can counter this tendency a little, but not much because many researchers are very familiar with climate models while far fewer are familiar with the literature on long-term climate variability[b]—another illustration of the consequences of limited perspectives and the importance of multidisciplinarity.

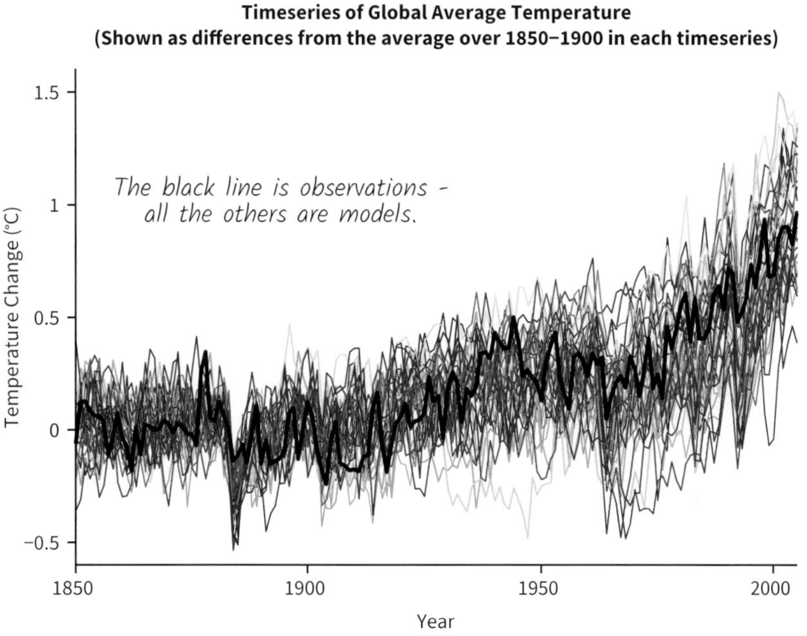

Timeseries of Global Average Temperature
(Shown as differences from the average over 1850–1900 in each timeseries)

Figure 18.1 Time series of change in global, annual average temperature from observations (black line) and from the CMIP6 ensemble (colours). Temperatures are shown as the difference from the average over the 1850–1900 period.[d]

So far, so good. It's certainly reassuring that all the models show global warming, as this is a basic physical consequence of increasing greenhouse gas concentrations. However, if instead of looking at the change from some particular period—the 1850–1900 average in Figure 18.1—we looked at the actual global average temperatures, as in Figure 18.2, the plot tells a different story. Now it is clear that while all the models show warming, they also each have quite different values for the global average temperature. In fact, they have quite a wide range: some models are a degree or so cooler than observations and others a degree or so warmer.

It's not just a matter of global average temperature, though. If we look at regional data instead, we see the same thing. In many regions, the range of behaviour is even greater than in the global average (Figure 18.3).

These differences demonstrate that our models are actually very far from being consistent with reality. Global annual average surface temperatures are often a degree or so different from observations and regional annual average surface temperatures can vary from each other by 6°C or more. Such differences are large in three ways. First, they are large by comparison with the year-to-year and decade-to-decade variability within observations and models. Second, they are large because these differences are similar in size to the changes we are trying to predict. The target of the Paris agreement is to keep global warming below 2°C above pre-industrial times, so a bias of 1°C (or more) implies that the models are significantly adrift from reality in

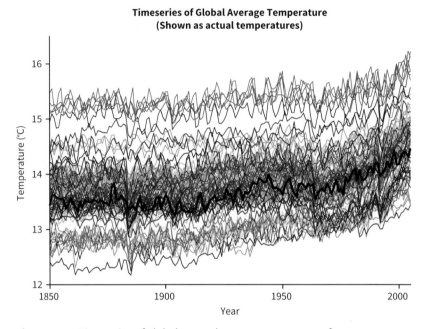

Figure 18.2 Timeseries of global, annual average temperature from observations (black line) and from the CMIP6 ensemble (colours). Temperatures are shown as the actual temperatures in degrees Celsius. Each line is a separate simulation. Lines of the same colour are from the same model, except dark blue, which is used for all models where only a single simulation was available.[e]

ways that are important for the things we care about; that's to say if 2°C is so big that the international community agrees we need to avoid it then a 1°C bias must surely be significant in terms of things that affect society. Finally, they are large because biases of this scale mean that if the model accurately represents the surface processes which govern the types of terrestrial biome (desert, savannah, forest, grassland, etc.) then the biomes will be in the wrong places because the temperatures (and probably rainfall, snowfall, etc., as well) will be inconsistent with reality. Alternatively the locations of the biomes may be consistent with observations, but if they are, then the way the surface processes are represented in the model must be unrealistic because it generates the right biomes from the wrong temperatures. In either case, it means we should not expect the models to capture the processes that govern the details of climate change in the future. Certainly not at regional scales, and potentially not at global scales either.

At this point, I could spend a lot of time giving you descriptions of other variables and other details but surface temperature is enough to be clear that climate models and the real climate system are very different things and hence a climate prediction made with a climate model cannot simply be taken as a prediction of future reality.

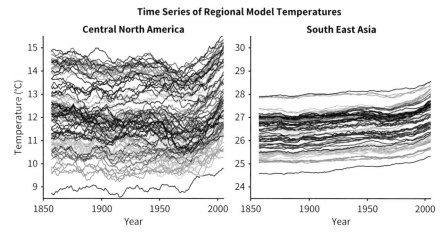

Figure 18.3 Timeseries of two regional average temperatures from the CMIP6 ensemble. The value for each year represents an average over 15 years. Otherwise they are the same as Figure 18.2.[f]

In the scientific modelling community, the limitations of today's climate models are widely acknowledged, but how to address them remains the subject of debate and they are nevertheless used as the basis for assessing climate impacts and supporting climate adaptation. Some put the models' failings down to them being insufficiently detailed. There are calls for ever greater investment in climate models. Higher resolution. Greater complexity. Make them bigger and better and eventually the problems will go away—that's the perspective of many climate modellers. Others, like myself, however, have concerns that we don't even have a framework for understanding when a climate model would be sufficiently realistic to answer a particular question about climate change—this might be described as the philosophical mathematician's perspective.[7] They might argue that research investment is better directed at studying how to use the models we have, and how to combine model results with theoretical understanding to provide useful and reliable information for society.

The former approach assumes technology will solve our problems, the latter that it's up to us to work out what our problems are first.

One of the key challenges brought to the fore by such debates is the question of how complex is complex enough. How many processes do we need to include? How realistic do they need to be? A higher-resolution grid provides more detail, but how much detail is necessary? How close to reality does a model have to be to provide reliable climate predictions?

---

[7] Although I'm neither a philosopher nor a mathematician.

18.4   From butterflies to hawkmoths: How realistic does a model need to be?

Imagine a band: a drummer, rhythm guitarist, lead guitarist, bass player, pianist, fiddle player, flautist, and singer. They're practising together but without the singer. They're playing a cover of some track and they want to play it in the same key as the original because that's what fits the singer's vocal range. In my peculiar imaginary world they can all play very well but they don't know very much about music, so they all start playing in different keys. It sounds awful, but they each listen to the other band members that are close to them and they adjust what they are doing. With a bit of practice they get to a point where all but the fiddle player are playing in the same key as each other. Unfortunately, that key is not the same as the original. The fiddle player is the odd one out. She is playing in the same key as the original. It actually sounds good most of the time because the fiddle player only plays during an instrumental in the middle, during which it sounds awful. Then the pianist has a quick listen to the original on his phone, realises his mistake and changes to the correct key. The piano and fiddle are now in harmony, both doing what they need to do, but the overall sound has got much worse. Now it sounds awful all the time even though it is actually closer to what they need to achieve for when the singer joins them.

There are a couple of messages from this analogy. The first is that improving one element can decrease the quality of the 'whole thing' even though it is taking the 'whole thing' in the right direction. We need to keep focused on what the goal of climate modelling is, just as the band needs to focus on producing a sound suitable for when the singer joins them, not just what sounds good in the practice session. If climate modelling is about understanding and predicting the future, then we should not expect our models to be on a steady process of getting more and more realistic; making progress with some elements will almost certainly degrade the collective results. The second message is simply that until everyone is doing the right thing, the collective piece may well be flawed. Something a little bit wrong in one part can make the whole thing horrendous. This applies to the band and also to climate models.

With this in mind, how realistic does a complex nonlinear model have to be to make trustworthy, detailed, extrapolatory predictions? Answering this question is one of the biggest challenges in climate prediction and, more broadly, in the use of computer simulations in academic research. It is a question we have hardly begun to address. It is, nevertheless, amenable to study both in principle and directly in climate models. A few tentative steps to exploring how we might study it in principle have been taken by researchers at the London School of Economics. Prof Roman Frigg and colleagues have looked at a simple nonlinear, chaotic system from the point of view of probabilistic predictions with an imperfect model. The system they used was the logistic map, one of the simplest chaotic systems: simpler than either Lorenz–Stommel or Lorenz '63, it has only one equation and one variable, but its behaviour is fascinating.

The logistic map became famous following a paper in Nature in 1976 by Robert May, later Baron May of Oxford. It originates in ecology as a description of population dynamics. We can think of it as describing how the population of some breeding animal varies from one season to the next. The population next season is taken to depend on the population this season as a result of two factors: reproduction and starvation. Higher populations lead to greater reproduction and therefore greater increases in population, but resources are limited, so higher populations, beyond a certain point, also lead to increased starvation—which decreases the population. The logistic map provides a description of how these two factors interact. It is a single equation that relates the population next season to the population this season.

Figure 18.4 shows an example of the relationship. It shows the population this season on the x-axis and the population next season on the y-axis. If this season the population is small then it grows, but if it is high then it will shrink. If it is middling, then next season it grows to be very high but subsequently collapses. Figure 18.5 shows an example of how the population can vary from one season to the next.

Figure 18.4 shows the simplicity of the relationship between one season and the next while Figure 18.5 shows how this can lead to complicated fluctuations over time. It turns out that the logistic map exhibits all sorts of wonderful features of chaos but that's not why we're looking at it. Neither are we interested in it as a description of population dynamics. For us we can simply view it as a model that takes us one step into the future—and repeating it many times takes us many steps into the future. You could even think of Figure 18.4 as representing a weather forecast: the x-axis is today's weather and the y-axis tomorrow's. Repeat the process and one gets further and further into the future. Of course, we never know perfectly the state of the

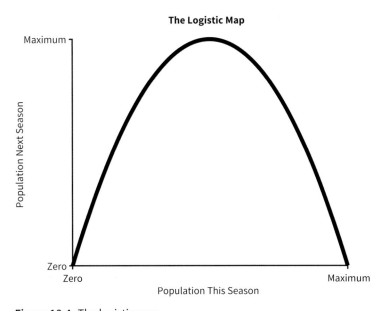

Figure 18.4 The logistic map.

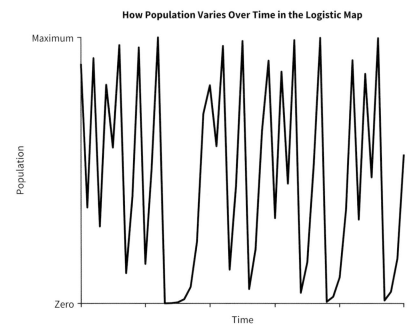

Figure 18.5 A time series of fifty points from the logistic map.

weather today so if we were trying to make a climate forecast, a forecast of a changing probability distribution, then we would use an initial-condition-ensemble: multiple examples of weather going forward with each one starting from approximately but not exactly the same initial conditions. The grey distribution in Figure 18.7 shows how this probability distribution can jump around.

What Frigg and colleagues did was to ask, what if the logistic map were our model but reality was slightly different, slightly more complicated. They constructed a new system where next season's population was 90% given by the logistic map and 10% by another nonlinear calculation based on this year's population. The new system gives a relationship between next season and this season which looks very, very similar to the original (Figure 18.6). If we describe the logistic map as our model and the new system as reality, then we would say that the short-term forecast with our model looks very good. We'd say our model was extremely good at making short-term weather forecasts, even though it doesn't capture all the processes that exist in reality.

So far, so good—but is it a good climate prediction model? The answer is yes and no. Figure 18.7 shows the climate of the new system (probabilities for reality in yellow) and the logistic map (probabilities for our model in grey). For a few steps ahead it is very good but then it gets worse and can give completely misleading probabilities.

What this demonstrates is that even with simple systems it is possible to have a model that gives very good short-term forecasts but is not at all reliable for longer-term climate forecasts because small errors in the model can lead to big errors in predictions. This result is of obvious relevance to climate predictions.

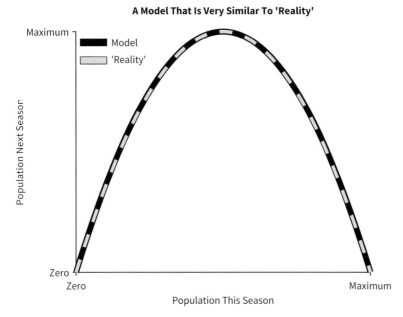

Figure 18.6 The logistic map (labelled as 'model') and some 'reality' that is similar to but not quite the same as the logistic map.

In her doctoral thesis in 2013, Dr Erica Thompson argued that small differences in the representation of a dynamical system like a climate model could potentially lead to it giving very different probabilistic predictions: different probabilities for future behaviour. She argued that even the smallest error in the formulation of the model could lead to different predictions for some point in the future. She gave the effect a name: the Hawkmoth effect. This lesser-known cousin of the butterfly effect captures the idea of sensitivity to model formulation in the same way that the butterfly effect captures the idea of sensitivity to initial conditions. The results of Prof Frigg and colleagues' work are arguably a demonstration of the Hawkmoth effect: if our model is slightly wrong, our probability predictions can be very wrong indeed, beyond some point in time.

The key question is whether the Hawkmoth effect is a problem for climate models and their climate predictions. It may be. Or it may not. It could be the case that with climate models the interactions of many geophysical processes leads to stability of outcomes rather than instability, so long as a certain degree of realism has been attained. Thinking again about the band, if a few of the piano keys are out of tune then the rendition may nevertheless sound largely okay so long as the relevant keys are not often played. On the other hand, if the rhythm guitar is slightly out of tune then the whole piece will sound awful: small errors can matter a lot or not very much. For climate models, this other hand is more complicated. If a few processes are missing or misrepresented, then model predictions might be fine for a period of time, say a year or a decade or so, but beyond that the nonlinear interactions might build

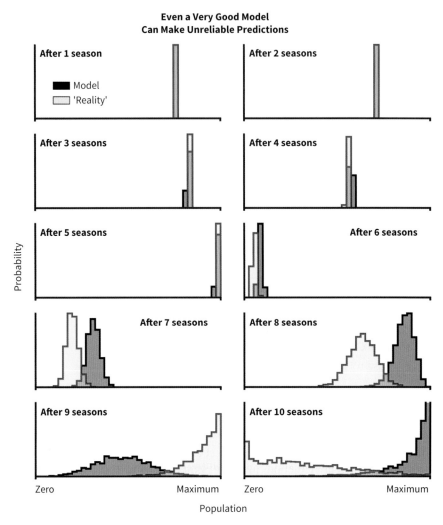

Figure 18.7 Probability distributions from an initial-condition ensemble using the logistic map (our model) and the extended logistic map (which I'm calling 'reality').

up and lead to the whole thing being completely different to what could happen in reality. Whether that actually happens might depend on the processes in the same way that the importance of a few keys being out of tune on the piano depends on which ones they are and whether they are central to the piece.

The Hawkmoth effect raises many questions for climate predictions, and extrapolatory predictions more generally. How can we go about studying whether and to what degree it limits model-based predictions of the details of future climate change? How can we tell which processes could lead to such sensitivity? These are questions that need to be studied and resolved before we should even consider pursuing the ever-bigger-models route to climate predictions. They are also critical to the modelling of ecosystems, river systems, star systems, brains, hearts, economies and much more.

19
More models, better information?

19.1 Are lots of models better than one?

The last chapter showed that today's climate models are a long way from being 'close to reality'. Yet there is an urgency to the climate change problem, so we very much want to know how we should interpret what we have as well as how we should move forward. Rather than aiming for one model that can make reliable climate predictions, is there perhaps a different route to confidence? Can we instead derive reliable probability predictions from multiple okay-ish climate models rather than one realistic one? A great deal of research investment is directed towards this approach. And yet stepping back from the practicalities reveals just another layer of conceptual, philosophical, and mathematical challenges—another front in our efforts to use computers to help us understand the real world. That's what this chapter is about.

It's difficult to be precise about the number of different global climate models there are because they are constantly under development. At any one time there are certainly several tens which might be considered the most up-to-date, and another similar size set which, while not absolutely cutting-edge, are nevertheless still considered very useful. These models are all used to simulate the twentieth and twenty-first century within model inter-comparison projects and Figures 18.1, 18.2, and 18.3 are examples of their results.

Each model is different and produces different simulations of climate change so how should we go about looking at them as a collection? One option is to calculate an average or create a distribution of outcomes from the different models. But is this a sensible thing to do? For the average to be meaningful, and for the distribution to represent probabilities, the models have to be independent of each other: like separate rolls of a dice. Another option is to ignore averages and probabilities and just take them as representing the scope of what could possibly happen. In this case—in fact, in both cases—we want the collection of models to broadly span the range of plausible behaviour given today's level of understanding. These two issues—model independence and whether they span the range of plausible behaviour—are central to what we can and can't get from climate models. You'll meet them a lot in multiple guises throughout the rest of this chapter.

Let me say a little more about independence. The distributions from micro-initial-condition ensembles give the probability of different outcomes in model world **because** the different simulations are independent. Changes to the initial conditions lead to different outcomes but there is no way[a] of choosing the initial states in order to achieve certain outcomes and there's no reason to expect that a particular group of simulations will be similar to each other. The ensemble, therefore, represents a random sample of possibilities—like multiple rolls of a dice. That's what I mean by independent.

When we have simulations from different models, this isn't the case. Modelling centres share ideas, approaches, priorities, and sometimes even code so none of them develop their models independently of the rest—they all share similarities. Indeed often multi-model ensembles—collections of simulations with somewhat different models—can include multiple versions of the same model. The distribution of outcomes from an ensemble like this doesn't represent the behaviour of any one model but neither does it represent some hypothetical random selection of models because they are all related to each other. Since they are not independent simulations, the distributions constructed from them can't be interpreted as probabilities of either reality or even of potential model behaviour. This is a pretty weird problem to have, but as usual, dice can help us get a handle on the issue.

If you play role-playing games such as Dungeons and Dragons, you'll be familiar with a set of dice with different numbers of sides: 4,6,8,10,12, and 20. In case you don't play, Figure 19.1 shows such dice. A set contains seven dice but two have ten sides; here consider just six, all with different numbers of sides. Imagine that they represent **all** possible types of dice and consider this a parallel with the set of all possible climate models.[1] Now consider throwing all six dice together and repeating that one hundred times. The six hundred numbers that you get could be presented as a probability distribution which would represent the probability of the outcome of throwing one of these dice, chosen at random, once.[2] If a multi-model ensemble of climate models were like this then the distributions they generate would represent a good starting point for understanding models and arguably a good line of evidence about how reality might behave. It would give the probability[3] of different outcomes on the basis that you don't have any reason to believe that any one is more likely to be informative about the real world than any other.

Figure 19.1 A collection of dice with different numbers of sides—as used in many role-playing games.

---

[1] If there were such a thing.
[2] Actually you would need to throw them many more than a hundred times to tie down the probabilities well but if you keep throwing them that's what you'll eventually get.
[3] If you had enough simulations with each of them.

The essence of the model-independence issue is that multi-model ensembles are not like this. Instead they're like having lots of some types of dice and none of the others; they're like throwing three 6-sided dice, two 8-sided dice, and one 4-sided dice. The resulting distribution doesn't represent the probability for one dice, or for the collection of all possible dice (i.e. for a dice chosen at random). In the same way, the distribution from a multi-model ensemble doesn't represent the probability of what will happen in any particular model world, but nor does it represent the probability of outcomes for a randomly chosen model out of all possible models—something which would represent the consequences of uncertainty in how to construct a model.

This sounds esoteric but it is actually tremendously important and relevant because the distribution of model outcomes from these ensembles is being interpreted as measuring our uncertainty in what could happen in reality.[4] If the selection of models available is limited and biased towards certain approaches, then this interpretation is ill-founded and is likely to mislead our efforts to tackle the consequences of climate change.

In the end, all a multi-model ensemble distribution tells us about is the behaviour of a particular set of broadly similar and interconnected models. It is no more and no less than a description of the outcome of the experiment which was run. It has no clear interpretation regarding how reality behaves, nor even how other models in the future may behave. If among thirty models, two models say New Hampshire will warm by 3°C, fifteen by 4°C, ten by 5°C, and three by 6°C, we cannot take this as an indication of a 10% chance of 6°C warming. The probabilities do not represent what we should expect to happen in reality or even a robust assessment of what models can tell us about this. The probabilities are arbitrary. The degree to which we should expect 6°C warming in New Hampshire—given today's level of understanding across diverse disciplines—could be much more or much less than 10%. A multi-model ensemble does not provide sound evidence regarding the probability of future real-world behaviour.

This is one problem for the interpretation of multi-model ensembles but there is also the second issue of whether they span the range of plausible behaviour—do they capture the range of climate responses consistent with today's best understanding? The answer is, unsurprisingly, no; they cluster round certain responses rather than representing the full range of uncertainty. Admittedly it's really hard to work out what the full range of uncertainty for future climate is, but it is certainly the case that today's models don't reflect it, as I'll show shortly. This is hugely problematic because climate models increasingly dominate the study of climate science, and as a consequence uncertainty across multiple climate models is increasingly, and mistakenly, seen as representing uncertainty in climate science, locking us into the historic scientific trap of over-confidence (see Chapter 12).

In terms of the dice, imagine that the different dice (4,6,8,10,12, and twenty-sided) are all equally likely. The first problem—lack of independence—is illustrated by having unequal numbers of each dice, such as three 6-sided dice, two 8-sided dice, and

---

[4] Sometimes directly, sometimes as the starting point for more complicated assessments.

one 4-sided dice. The second problem—lack of range—is illustrated by us failing to have any 10-, 12-, or 20-sided dice.[b]

19.2 Global climate models and climate sensitivity

One way to see how this plays out in climate models is to look at how sensitive they are to increasing atmospheric greenhouse gases. We can capture one aspect of this sensitivity by considering how much global warming would ultimately come about as a consequence of doubling atmospheric carbon dioxide concentrations. This warming is called the equilibrium climate sensitivity and it is a really big deal in climate science.

> Equilibrium climate sensitivity: The ultimate change in global average temperature as a consequence of doubling atmospheric carbon dioxide concentrations.

A major problem for climate predictions, and for responding to climate change, is that there is a lot of uncertainty in the value of the equilibrium climate sensitivity. There have been many assessments of it using many different types of analysis: models, observations, theory, and all sorts of combinations. Most present their results as a probability distribution: the probability of different values for equilibrium climate sensitivity. Often they come up with a distribution that looks somewhat like the curve in Figure 19.2: a pretty sharp cut-off at low values (usually between 1 and 2°C); a peak of probability at mid-range values (usually between 2 and 4°C); and a tail of small but non-negligible probability that stretches to high values. The details are often different but they usually have this characteristic shape. This has been the case since the late 1990s at least.

The lack-of-range problem for the interpretation of global climate models can be illustrated by the fact that the models we have don't span this range of uncertainty very well.

The roughly seven yearly assessment reports of the Intergovernmental Panel on Climate Change (IPCC) provide regular updates on their judgement of the value of equilibrium climate sensitivity. In 2013 they said it was likely (meaning more than 66% probability) to lie between 1.5°C and 4.5°C but they gave a probability of up to 10% that it could be above 6°C.[5] In 2021 they said it was likely to lie between 2.5°C and 4.0°C but with up to 10% probability that it could be either above 5°C or below 2°C.

The situation in 2013 reflected what it had been like for many years. Many studies had allowed for the possibility of high values of equilibrium climate sensitivity and this was reflected by the IPCC's up to 10% probability of it being above 6°C. However,

---

[5] The black line in Figure 19.2 has 66% probability between 1.5 and 4.5°C and slightly over 13% probability above 6°C.

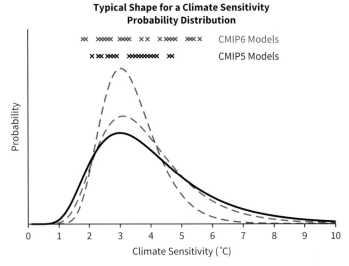

Figure 19.2 The black line is a typical shape for a climate sensitivity probability distribution; something like it is found in many studies.[c] This particular one reflects the 2013 IPCC assessment by having 66% probability in the 1.5–4.5°C range.[d]
The blue line reflects the 2021 assessment by having a 90% probability in the 2–5°C range. I would question whether this blue distribution should reflect our uncertainty because it is the result of four lines of evidence of which two suffer directly from the basic issues with the look-and-see approach and a third suffers from the limited diversity in our models. The final one—process understanding—on its own provides a 90% probability of 2.1–7.7°C,[e] which is captured by the red curve and which is closer to my own view of what the combined evidence can justify.
The crosses at the top show the climate sensitivities of the CMIP5 (black) and CMIP6 (blue) models: the main models used in the 2013 and 2021 IPCC assessments, respectively.[f]

none of the climate models at the time had values above 5°C. Now maybe 10% probability doesn't sound a lot to you, but in other walks of life—your health, your job, your house, your pension—a one in ten chance of severe damage or loss is far from negligible, so a 10% probability of much more severe global warming seems important to me. In climate modelling, however, all the models were clustered around the central, most-likely values, so the less likely, but arguably most relevant behaviours for assessing climate change risks, couldn't even be studied. In terms of guiding society regarding regional and local changes, the multi-model ensembles' ranges inevitably missed less probable situations even though they were still reasonably probable and would have high impact.

Has the situation improved since 2013? Maybe a bit, but not really. By 2021 the IPCC assessed that there was an up to 10% probability of equilibrium climate sensitivity being either above 5°C or below 2°C. From the shape of the typical probability curve (Figure 19.2) we should expect most of that probability to be above 5°C. At the same time, model development had led to models with a wider range of equilibrium climate sensitivities: a handful with values above 5°C but still none above 6°C. It's an improvement but it doesn't solve the problem. For a start, the upper tail of the probability distribution is very hard to constrain—it includes smallish probabilities which continue out to high values—so the lack of models stretching above 6°C or 7°C remains a problem. On top of that, we have to remember that equilibrium climate sensitivity is just one measure of the sensitivity of the climate system to increasing atmospheric greenhouse gases. In practice, we care about many more things, from changing rainfall patterns to changing frequency and locations of heatwaves and storms. When we talk about wanting models that span the range of plausible behaviour, we want them to do this in many dimensions of the climate's response. Equilibrium climate sensitivity is a blunt instrument that shows we currently don't do this, but it can't show us when we do.

Furthermore, it is worth noting that the 2021 IPCC assessment of climate sensitivity was based on four lines of evidence: one derived largely from multi-model ensembles, two from observations of the past and one based on understanding of the processes involved in global warming. As is no doubt clear to you by now, I would place the greatest weight by far on understanding the processes but this aspect of the assessment gave a 10% probability of equilibrium climate sensitivity being either above 7.7°C or below 2.1°C, not 5°C and 2°C as in the combined assessment (Figure 19.2). For me, this makes the limited range of model behaviours look even more severe and problematical.

These issues—of models having a limited range of behaviour and lacking independence—were first considered in the late 1990s. At that time a few researchers began thinking about how they might study the model error problem in a more systematic way. That's when I got involved in the study of climate prediction uncertainty. Rather than simply creating multi-model ensembles from the different models that were available, it was proposed that many versions of a model could be created from a single model simply by changing the values of parameters.

19.3   A mechanism for churning out more models

As described in Chapter 11, these models contain chunks of computer code designed to represent processes that take place on scales smaller than the resolution of the model grid: clouds, the land surface, convection, gravity waves, etc. These 'parameterizations' are usually written in terms of the physical processes expected to be taking place but they aren't actually representations of those processes—rather, they capture the statistical characteristics of their consequences on larger scales. And because they don't represent well-understood physical processes, they contain parameters which aren't constrained by physical understanding: there's no right or wrong for what values they can take. The values of these parameters therefore have to be chosen by a process of trial and error, by tuning. The new experiments proposed in the late 1990s were based on the idea that it would be interesting to change the values for these parameters and see what happens. This would be a way of generating lots of versions of a global climate model and maybe better spanning the range of plausible behaviours.

Ensembles created this way from a single model are called perturbed-parameter ensembles. One of the challenges in building such ensembles is our old friend, nonlinearity. You can change one parameter and see how that influences the model results, and you can change another parameter and see how that influences the model results, but if you add together the influence of each parameter change individually you don't get the same result as when you change both parameters at the same time. The combined effect can be very different indeed. As a result you can't just explore one parameter at a time—you have to try to explore all the combinations as well. And the consequence of this is that to investigate the possible behaviour of a climate model requires very, very large ensembles which are way beyond the computational capacity available in modelling centres. And that's why the climate*prediction*.net project, introduced in section 16.3, was created.

Remember, climate*prediction*.net asked volunteers to run climate models on their home computers so were able to run much bigger ensembles than modelling centres could. The first results were published in 2005 and involved 2578 simulations: 414 different combinations of parameter values, known as 'model versions', and a small micro-initial-condition ensemble of each. The project went on to generate a perturbed-parameter ensemble involving a few hundred thousand simulations. Nowadays the largest multi-model ensembles include around forty or fifty model versions, so even by today's standards the climate*prediction*.net experiment was an extremely large exploration of the uncertainty related to model formulation. And the results? Well the climate*prediction*.net ensemble succeeded in addressing, to a large extent, the problem of the limited range of responses in climate models as reflected by equilibrium climate sensitivity. Even in the first published results it generated model versions which had values ranging all the way from 1.9°C up to 11.5°C.

This was a great outcome and demonstrated that there were routes to exploring the potential behaviour of models. When it came to writing the results up, however, things got tricky. While writing the first climate*prediction*.net results paper, I spent many months debating with colleagues what the results of such a perturbed-parameter ensemble actually meant: what it told us about reality. It became clear that while it certainly provided more information than we previously had, it didn't represent an obvious route to solving the problems of climate prediction. It didn't resolve the independence problem or the question of how we would know when we have models that explore the range of plausible behaviour in reality. What it did, nevertheless, was provide a route to studying and thinking about these issues in a systematic way. It provided us with a structure to consider the conceptual questions of what we are trying to do when we use a computer model to predict the future extrapolatory behaviour of a complex nonlinear system. Reflecting on what we can and cannot deduce from the climate*prediction*.net perturbed-parameter ensemble forces us to consider what we can deduce from computer models more generally, and how we should go about identifying robust information.

The rest of this chapter is about reflecting on these points but first, let's have a quick recap of the issues so far regarding the relationship between climate models and reality. Today, we have around thirty to fifty cutting-edge, complicated global climate models that have been built in a handful of modelling centres around the world. On the one hand, they are fantastic achievements of modern science and fantastic tools with which to study the climate system: they generate somewhat realistic-looking very complex behaviour built up from the interactions of many diverse processes. On the other hand, **looking** realistic isn't enough to trust them as engines of prediction and when you start delving a bit, it becomes evident that they are still very significantly different from reality. We also have nascent work on the Hawkmoth effect which implies that even small differences between a model and reality could potentially matter a lot when trying to make extrapolatory predictions.

The issues around lack of realism have led to substantial effort being focused on extracting information from multiple models. That, however, leads us to statistical problems arising from the fact that the models are not random possibilities for how the real world behaves: rather, they are all very different to reality and all closely linked to each other.[6] On top of that, we don't have very many of them. To try to address these problems we have built perturbed-parameter ensembles—climate*prediction*.net was one of the first but there are now several worldwide—to systematically explore the potential behaviour of climate models. This gives us potentially hundreds of thousands of model versions, but we still have deep questions over what they tell us about reality.

Take a deep breath. The following sections are the pinnacle—or maybe the deepest pit—of this book. They deal with perhaps the most peculiar, conceptual limits of climate predictions. They may appear esoteric, philosophical, or just irrelevant. They are not. Assessing what perturbed-parameter ensembles can tell us provides the foundations for what multi-model ensembles can tell us. And multi-model ensembles are the foundations of modern climate science. They are widely used to provide detailed descriptions of the future under climate change—descriptions that are being used to guide society's response and direct hundreds of billions of dollars of investment. Unless we understand how to use them appropriately, we risk misleading efforts to prepare for and adapt to climate change in diverse domains ranging from finance to health to water provision to disaster risk, ecosystem management, and conflict. They also provide the foundations for how we assess the importance of climate change—both personally and at the level of national economies—the subject of Chapter 24. These issues of interpretation are therefore critical for society.

The following sections address the lack of independence problem and explain why there are no solutions on the horizon. The next chapter examines the challenges in achieving a collection of models that span the range of plausible behaviour. Beyond that things get less conceptual and more down to earth with the search for reliable ways to utilize computer models and identify the best, most robust, information.

---

[6] The lack of independence problem.

Before we get there though, we need to be clear why what might appear the most obvious interpretations aren't actually the best ways forward.

19.4 An opening gambit on interpreting multiple models

Climate*prediction*.net managed to generate model versions with equilibrium climate sensitivities up in the less-than-one-in-ten-chance region above 6°C. It demonstrated the possibility of achieving global climate models with that behaviour. It didn't generate any with equilibrium climate sensitivities less than 1.9°C. What should we take from such information? As an opening gambit for an answer, I'll propose that if we have models that span the range of equilibrium climate sensitivity from 1.9°C to 11.5°C, then we shouldn't rule out the possibility that the real-world sensitivity could be anywhere in that range.[7] We may be able to say much more, or we may conclude that the models don't even tell us this much, but let's take this as a starting position. Note, however, that I'm only talking about what we can take from a multi-model or perturbed-parameter ensemble. We may well have other types of information (physical understanding and observations) that make us think some values are more likely than others irrespective of the models, but here I'm only interested in what we can take from the models.

19.5 Beyond the opening gambit but held back by an anchor

To go beyond my opening gambit can't we just count the outcomes from the different model versions and build a probability distribution that tells us which values are more or less likely? This is a question I spent at least six months in 2004 debating most Wednesday afternoons with Prof Lenny Smith in a small but nice office above a great ice cream and sandwich shop in Oxford. Ultimately we concluded that we could not. We can, of course, count the outcomes and get a distribution. That's easy. The question is whether that distribution is a **probability** distribution, or, more saliently, whether it is a probability distribution for anything we're interested in. Does it have any meaning in terms of reality? The distribution from the first climate*prediction*.net results (Figure 19.3) looks very like Figure 19.2 and like many other probability distributions for climate sensitivity. Indeed it has been misinterpreted as a probability distribution for reality by subsequent research projects and even by a book on the public understanding of statistics[g] (about which there is more in section 21.2). So if it looks like a probability distribution and is being used as a probability distribution, why shouldn't we just accept that it is a probability distribution?

---

[7] We could make the same type of range-of-response statement about any other variable or quantity that the model simulates: regional precipitation, spring temperatures in Eastern China, etc.

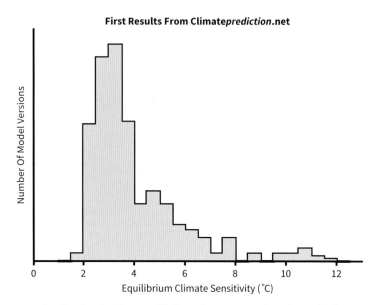

Figure 19.3 The distribution for equilibrium climate sensitivity from the first set of climate*prediction*.net results.[h]

Answer: Because it isn't a probability distribution. The probabilities don't reflect reality, or our understanding of reality, or any partial assessment of the behaviour of reality, or the results we should expect from ensembles with other models, or even the results of other ensembles with the same model.

Remember the multi-model ensembles. The distributions created from those ensembles don't represent the behaviour of any model or any random collection of models, because the models are not independent of each other. The probabilities don't represent anything apart from the very specific experiment that was performed. And since they don't represent any general result, even about models, there is no hope of relating them to reality.

It's tempting to think that the model versions in a perturbed-parameter ensemble are moderately independent and therefore somewhat solve this problem. Sure, they are all based on the same model but if that were the only problem then I could argue that the probabilities are a robust description of at least the behaviour of that model. We could then have a lively debate about whether the behaviour of that model is informative about reality but the probabilities would at least have some meaning: they would tell us about the potential behaviour of a model version with this model structure: they would tell us about a new model version created from the same model. Unfortunately we are not in such a rosy situation.

In climate*prediction*.net and other perturbed-parameter ensembles, the original model acts as an anchor for the results. The parameters whose values are changed are usually chosen with the hope of maximizing the diversity of outcomes, but thanks largely to the consequences of nonlinearity it is very difficult to know which parameters will have a significant impact and which will have almost none. Indeed, it is extremely likely that many parameters will have very little impact on any particular characteristic of interest—such as equilibrium climate sensitivity—simply because most parameters have only a limited impact on most aspects of the model. The aim of the ensemble is to search for the ones that do. Imagine, then, creating new model versions by varying one parameter at a time. If a significant fraction—say half—of those parameters don't make much difference to what we are interested in, then the distribution from the ensemble will have a strong peak at the value for the original model.

In practice, as mentioned already, perturbed-parameter ensembles don't always vary parameters one at a time; indeed it was one of the strengths of cli-mate*prediction*.net that it was able to explore many combinations of varied parameters. Even in this situation, however, the distribution of results is likely to be strongly anchored by the original model. In perturbed-parameter ensembles the parameters are usually varied around their original, tuned values: they are increased and decreased from that value. If some parameters have little effect, then there are many combinations which also have little effect and only a few combinations—perhaps where two or three parameters take outlier values—that do. This makes the behaviour found in the original model much more likely in the ensemble. We should, therefore, still expect a peak in the distribution near the behaviour of the original model. That's what we see in Figure 19.3 but it tells us nothing about the probability of real-world behaviour or even about what this particular design of model could produce. If, for instance, we took one of the model versions with a high equilibrium climate sensitivity and ran a new perturbed-physics ensemble where parameters were varied around the values in this high-sensitivity model, we should expect to get a distribution with a peak at high climate sensitivities. The point is that the shape of the distributions from these ensembles are themselves arbitrary: they are completely defined by the particular experimental setup, they are distributions but not probability distributions.

Just as in the multi-model ensembles, the model versions are not independent: they are anchored to the original model. Indeed it would be quite feasible to choose a host of parameters which we **expect** to have little effect, and in that way build an ensemble which would most likely show that the original value is by far the most likely: it would be very easy to cheat (but we didn't!).

19.6 Beyond the opening gambit but held back by the arbitrary shape of model space

Anchoring is not the only problem though. When interpreting perturbed-parameter ensembles we face two further, related issues. The first is that the size of the ensembles—the number of simulations—is tiny by comparison to the number of uncertain parameters and the number of different values we want to try for each of them. The second is that even if we were able to run vast ensembles and to get rid of the anchoring problem, the distributions would still be arbitrary because we've moved into a realm of dealing with a bizarre computational construct which is completely disconnected from reality. I'll do my best to explain why.

Remember we don't know which parameters matter or which interactions between parameters matter, due to nonlinearity. We therefore want to investigate as many parameters as we can. We also want to investigate as many values for each parameter as we can so that we can study just how the parameter value affects the response. By sampling lots of parameter values we can look for nonlinearities and oddities in behaviour that might tell us useful things about reality and enable us to look for signs of the Hawkmoth effect.

We can think about how well we've investigated a parameter in terms of how many values of it we've used in our ensemble. If a parameter has a default value of, say, 2.5 we might want to take values of 1.0, 1.1, 1.2, and so on up to 4.0. That seems like a reasonably high level of detail: 31 values. We can think of it as points on a line. Now imagine three parameters: if we want to explore all the nonlinear interactions and we have thirty-one values for each one then we'd need nearly 30,000 combinations—31×31×31. We can visualize this as 30,000 points in a three-dimensional space. With ten parameters it would be 800 trillion and we could imagine it as 800 trillion points in a ten-dimensional space—so long as you're now happy to imagine ten-dimensional spaces. On top of these numbers we would need initial-condition-ensembles so that we can measure climate change within each model version (that's the message from Chapters 14 to 16), so we actually want ensembles at least a hundred times bigger than these numbers. This requires infeasibly large ensembles.

Climate*prediction*.net explored more than twenty-one parameters but typically with only three values for each parameter and in any case it didn't get close to exploring all combinations; even it—by far the largest perturbed-parameter ensemble to date—only managed a few hundred thousand simulations. Most other similar ensembles are a few hundred or less. The distributions based on perturbed-parameter ensembles can't, therefore, be considered probabilities of the potential behaviour of a particular model simply because they don't have the numbers. Their sample of parameter combinations is just too small. It would be like thinking we had a probability distribution for the heights of the residents of New York after measuring only a handful of people. The first problem is therefore simply the size of the ensembles.

There is, however, a second, more subtle, important and general problem.

When we think about 30,000 parameter combinations we can visualize them as points in a three dimensional space like a sports hall. The 800 trillion combinations are points in a conceptual ten-dimensional space. This space is called 'parameter space'; every point within it represents a different model version. This space is a different space to the one in which the path moves[8] but they are similar in the sense that we're using a multi-dimensional space to picture the different combinations that variables or parameters can have.

Our aim with a perturbed-parameter ensemble is to study how the model's response to increasing greenhouse gas concentrations changes as we change parameter values and therefore move to different points in this space. Every point in parameter space has a value for the response we're looking at—let's say climate sensitivity—but we don't know what it is unless we run the model for that particular combination of parameter values. It's like having a map where each point on the map has an associated value for the height above sea level (the values are usually indicated by contour lines). In order to make a good map, one wants to measure the height at as many places as possible across the region. In a similar way, we want a perturbed-physics ensemble to explore different parameter combinations as densely as possible. We also want it to explore as much of parameter space as possible: we want the region covered by the map to be as big as possible.

If we don't have an ensemble which is close to 30,000 members (for three parameters) or 800 trillion members (for ten parameters), then what can we do? Well, we could call on statisticians to come up with clever ways to join the dots: to take what happens at the points where we've run the model and deduce what would have happened had we run the model at other points. This would be like deducing the height above sea level at places where it hasn't been measured by interpolating between points where it has been measured.

Historically, that is exactly what has been done. Indeed this method has been used to interpret another perturbed-parameter ensemble—not climate*prediction*.net—and generate climate change information which is provided to businesses, industry, and government[i] for use in support of climate adaptation; these issues have real-world consequences. If such techniques could be relied upon, then they would substantially reduce the size of ensembles required. They would give us the answer for our characteristic of interest all over parameter space. And from that we could build a probability distribution. Anchoring and the need for infeasibly large ensembles would no longer be issues. In practice, however, there are all sorts of complex challenges in actually doing this and many reasons to be sceptical about their ability to 'fill-in' parameter space and reduce the need for large ensembles. Fortunately, there is no need for me to tell you about them because even if they were successful, the probability distribution would still be meaningless.

The problem is that the shape of parameter space is arbitrary. Woah, let's roll back to see what that means and why it matters.

---

[8] The path moves in the space of model variables (whose values change during a model simulation) while the model versions are particular combinations of parameters which are fixed throughout a model simulation but can be represented by points in parameter space.

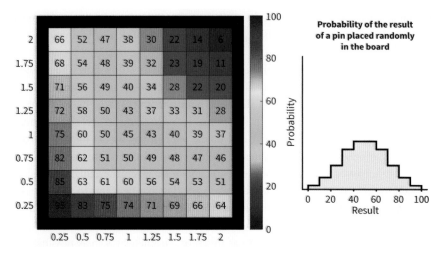

Figure 19.4 A conceptual board, like a chess board, with numbers on each square. One can imagine the numbers being the result of a model when two parameters in the model are given particular values—the parameter values are shown on the lower and left-hand side.[j] If a random position is chosen on the board then it will be associated with one of the numbers and the probability distribution for that number is a normal distribution—a bell curve—which is shown on the right.

Imagine a chessboard with a different number associated with each square (Figure 19.4). I can build a probability distribution from the sixty-four numbers on the sixty-four squares (Figure 19.4). If I am then blindfolded and I put a pin into the board—that's to say, I choose a random location on the board—then the pin will be in a particular square and I can associate it with that square's number. It's a way of selecting a number. The likelihood of getting different numbers is represented by the probability distribution, so I know what the chances of different outcomes are.

However, if I don't know the numbers on each square and I'm only able to look at, say, five squares beforehand, then I don't really have any idea of the probability distribution from which my random selection will be. This is the ensemble size problem. The chessboard is equivalent to parameter space. The numbers are equivalent to the response of a model version to increasing greenhouse gases.

If I could look at the numbers on all the squares, then I could create the probability distribution for the outcome of randomly putting a pin in the board (Figure 19.4). In the same way, if we could run enough model versions to densely explore the parameter space of a climate model then we would have a good measure of the probability of a randomly selected model version's response to greenhouse gases.

So far so good, but here comes the big problem. The way the parameters in the model are defined isn't set in stone; indeed it can be up to the whim of the programmer. There's nothing unusual or clandestine about this; there are often multiple ways of describing the same thing. We can describe a bus timetable in terms of their frequency (six buses per hour) or in terms of the time between buses (ten minutes per bus). We can describe the motion of a train in terms of its speed (speed) or its kinetic energy (mass x speed squared). We can describe the size of a cube by the length of one side (length) or by its volume (length cubed). When we're building a parameterization for a climate model we choose how to describe a quantity according to how we're thinking about the problem at the time or possibly according to what might be computationally most efficient.

On a computer, multiplication is faster than division so if we are going to repeatedly divide variables by some number it would be faster to initially calculate 'one divided by that number' and then multiply the variables by this new quantity. Doing so means we have many multiplications rather than many divisions and the code runs faster. Programmers may choose to define parameters with this in mind. Or they may choose parameters according to how they learnt about the science of the problem. Whatever the case there is rarely one and only one way to define a parameter.

One parameter which was explored in climate*prediction*.net is known as the ice-fall-rate, which conceptually represents how fast ice particles fall out of clouds. Of course, the actual processes which go on in clouds aren't represented in these models. The cloud parameterization gives the large-scale features of clouds in a grid box: things like the fraction of a grid box that contains cloud. The ice-fall-rate parameter isn't therefore the actual rate that ice falls in clouds but rather it represents the aggregated process of ice leaving the parameterized summary of clouds in a model grid box. This isn't something that is measurable in reality because it doesn't exist in reality—it is an entirely model-based quantity. The value of this parameter must therefore be chosen and tuned in order to get the overall representation of clouds looking reasonably realistic. It is not defined by physical understanding which is why it can be tuned and why it is reasonable and useful to run a perturbed-parameter ensemble to investigate the consequences of choosing different values.

However, the ice-fall-rate could just as well be defined as the 'ice residence time', describing how long ice stays in clouds rather than how fast it falls out of them. This is just like the situation for buses. The number of buses per hour (six in the example above) is the number of minutes in an hour (sixty, in case you've forgotten) divided by the number of minutes between buses (ten in the example above). So we can describe the situation equally well by buses per hour or minutes between buses. It's the same for ice in clouds. The ice residence time is how far a typical ice particle falls through a typical cloud, divided by the ice-fall-rate. In both cases, one description is equal to some factor divided by the other.[9]

In other parameterizations, one representation of a parameter could be related to the square or the cube of another, or some more complicated mathematical function of another. The point is the same—there is no preordained definition for these parameters.

That there are different ways of expressing the same thing has a big impact on the interpretation of perturbed-parameter ensembles. Let's go back to the chessboard. Imagine taking the chessboard in Figure 19.4 and adding axes to represent two

---

[9] They are said to be inversely proportional with each other.

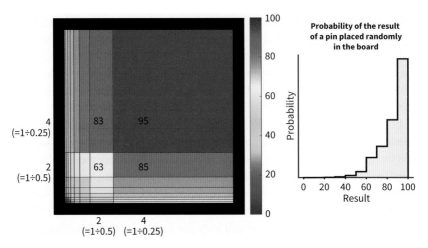

Figure 19.5 As Figure 19.4, except that the 'parameters'—the axes—are redefined as one divided by the values shown in the previous figure. (This is called their inverse.) The boundaries of the boxes are at the equivalent of the boundaries in Figure 19.4. The numbers associated with each box are the same as with the original parameter values in Figure 19.4. However, the size of the boxes has changed. The probability distribution representing a randomly chosen position on this board—shown on the right—is completely different to the previous one. The point is that changing how a parameter is defined can be done without changing the result from a model but sampling the new parameter definition gives a different distribution.

parameters like the ice-fall-rate. The numbers on the squares represent the model's response for the combination of parameter values shown on the axes. The distribution of values across all squares represents the distribution of our model output across parameter space (Figure 19.4). But now imagine creating a new board by redefining the parameters as their inverse, as 1 divided by the original parameter. The squares on the original board have equivalents on the new board, but they have all changed size (Figure 19.5). If I put a pin randomly into the new board then the probability of getting the various numbers has completely changed (Figure 19.5 vs. Figure 19.4).

When I say the shape of the parameter space is arbitrary, I mean that it can be transformed in the way the chessboard is transformed and still represent the same perturbed-parameter ensemble, the same exploration of model uncertainty. The way the model works doesn't change at all when you make this transformation. If we redefine ice-fall-rate as ice-residence-time, the outputs of the model stay the same; they can be identical. There is no change to the model, so the response for particular ice-fall-rates is the same, but the shape of the parameter space is entirely different. When we create a distribution by uniformly sampling parameter space, we are assuming that any particular area, volume, or perhaps ten-dimensional equivalent of volume, is as likely as any other area or volume.[10] On the chessboard we are assuming that

---

[10] This is known as the principal of indifference.

one square is as likely as another (Figure 19.4). But if we redefine the parameters on the board, then the sizes of the squares change, they are no longer equally likely, and the probability distribution is very, very different (Figure 19.5). In the same way, imagine a model where the original developers had chosen to define their parameters differently. The shape of the parameter space would therefore be different, and hence the probability distribution from a perturbed-parameter ensemble would be entirely different. Uniformly sampling this new parameter space gives us a very different distribution even if we're using the same experiment with the same model.

The upshot of all this is that however big our perturbed-parameter ensemble, however well we sample our uncertain parameters, and however well we might fill-in parameter space with clever statistical methods, we still don't have a probability distribution. The resulting distribution is arbitrary. We can't therefore have probabilities that represent what we expect in reality because the distributions themselves depend substantially on arbitrary decisions in the formulation of the model.

The consequence is that even a complete exploration of all parameter values in all combinations doesn't solve the independence problem that we had with multi-model ensembles (section 19.1). We can't therefore get a probability distribution directly from multi-model or perturbed-parameter ensembles.

We also need to be very careful about letting these ensembles influence our experts' opinions. It is very easy to see a distribution from ensembles of this nature and take it as evidence that certain behaviour is more likely and other behaviour less likely or even impossible. Without careful analysis of these seemingly esoteric but actually highly relevant issues, such conclusions will mislead climate science and consequently mislead society.[11]

These issues raise questions for whether we can get probabilities from multiple computer models in many fields, from pandemic behaviour to galaxy formation,[12] but there is also a philosophical question to consider. Why would we think that a distribution across multiple models, or parameter space, should represent our uncertainty about the behaviour of reality in the first place? Computer models are a new and different source of evidence for scientific understanding and there are significant conceptual challenges regarding their role in understanding the natural and social world. In extrapolatory situations—where we don't have observations to keep us on the straight and narrow—it's tempting to assume that behaviour which is common across multiple models is more likely to reflect what will actually happen. It is understandable and perhaps intuitive to take this perspective, but these are models not random samples of reality. The diversity in our computer models shouldn't be taken to represent our best understanding, or our best expectations, of the behaviour of reality.

---

[11] For those of you familiar with the terms, what I am saying here is that the ensemble distributions do not provide frequentists probabilities but neither should they be treated as evidence to support expert opinion and hence Bayesian probabilities. This definitely **is** an esoteric statement.

[12] See 'Escape from Model Land' by Erica Thompson for much more on the wider issues in model interpretation.

People often fail to consider these challenges, so they often over-interpret climate models. The dash for answers undermines consideration of what we can and can't get from them. This does not, however, imply in any way that we can get nothing from them. Remember my opening gambit was that the climate*prediction*.net results represented a range of behaviour which shouldn't be ruled out. That particular perturbed-parameter ensemble found climate sensitivities ranging from 1.9°C to 11.5°C. We may not have probabilities but we do have information. The question then is whether we can do better than this range. Can we narrow it? Do we need to broaden it?

20
How bad is too bad?

20.1 Targeting the informative, not the perfect

It doesn't look hopeful for getting probabilities from multi-model or perturbed-parameter ensembles but can we perhaps rule out some models and thereby reduce our uncertainty about what will happen in the future—reduce the range for what is considered plausible? The answer turns out to be: not easily. There are many pitfalls in attempting to do this. At the very least we're going to have to think about it very hard.

If our yardstick for model assessment were simply an ability to produce simulations consistent with past observations, then we could rule them all out; they are all inconsistent with reality to a significant degree. But ruling them all out isn't very helpful because they all capture many difficult-to-study processes and interactions and are valuable tools to help us understand climate change. We don't want to rule them all out; we want to rule out some while keeping others. The question is, therefore—how unrealistic does a model have to be to be discounted as a tool for providing information about reality? How bad is too bad? What is the appropriate yardstick? There is also a subsidiary question: when does rejecting a model lead to rejection of the behaviour predicted by that model? These are the issues for this chapter.

All of today's models are very different from reality, as illustrated by Figure 18.2 and Figure 18.3. Statisticians have tried and tested methods for evaluating the likelihood that one sample of data is from the same underlying probability distribution as another, so we can compare observations and model output and put a number on the probability that a model could have generated those observations. In practice this is not often done because the models and the observations are quite obviously and substantially different: we don't need detailed statistical tests to tell us that the model worlds are very different from the real world. Nevertheless, it's worth saying that by one measure their probability of being consistent with observations is less than 0.00003%.[a]

I repeat though that this doesn't mean that the models are bad or un-useful. It simply means that they are different to reality. They may be realistic in the way a painting of a landscape or a photograph of a cake is realistic. They capture certain aspects very well and they are useful in many ways, but they don't let you explore the hidden details of the landscape or taste the cake.

If all models are substantially different to reality, then the question of how bad is too bad becomes one of whether a model can realistically represent the processes that are important for climate prediction. So which ones are those? Unfortunately, the nonlinear interactions between processes in our climate models means it is difficult to say that any of the processes are unimportant for climate change predictions (extrapolation). It is easy to identify some that are particularly important (for instance, clouds), but it is difficult to say that any are unimportant. Furthermore, the non-linear interactions between the processes in our models mean that if there are errors in some processes, they can show up as differences from reality in any or all of the model variables. As a consequence, it could be that a particular model simulates the key processes for extrapolatory prediction in a highly informative way but that due to interactions with errors elsewhere in the model, the historic simulations are particularly unrealistic. That's to say it might get key aspects that govern climate change right—for instance, the way land biomes respond to temperature and rainfall—but get other aspects that affect weather patterns and variability wrong and hence not simulate present or historic climate at all well. This gives us a serious problem if we want to use observations to assess when a model is too bad to be taken into consideration. A cricket ball can help illustrate the issue.

Judging models is like trying to judge how high a cricket ball will bounce if I drop it on the floor, without actually dropping it. In this situation, I'm not allowed to drop the cricket ball (which here represents the real world) but instead I can drop a tennis ball, a super bouncy ball, a table-tennis ball, a golf ball, and a dice (which represent a multi-model ensemble—Figure 20.1). I can also compare each object with the cricket ball by looking at them. So I drop each of the objects from head height and see how high they bounce. Then I compare each object with the cricket ball. The four balls are moderately 'realistic' versions of the cricket ball—most are the wrong size, all the wrong colour and texture, but they are all the right shape. The dice is one of the least realistic objects in terms of size and colour and it is also completely the wrong shape. On the basis of this, it is tempting to dismiss or rule out the dice because it is a substantially worse representation of the cricket ball than the other objects in terms of the things I can assess (a parallel with historic observations). In terms of bounce, however, it turns out that it is arguably the most realistic.[1] The point is simply that when all our models are substantially

Figure 20.1 Bouncing balls and dice.

---

[1] I did some not very scientific ad hoc tests of this.

different from reality it is very difficult to know which ones are the most informative. In such situations we need to be extremely careful about deciding to dismiss some, or even consider some as less credible than others. This is where the extrapolatory characteristic really hits home because it means we have no observations of the aspect we're interested in; we have no observations of dropping the cricket ball—not even once.

In this illustration we don't want something that looks like a cricket ball; we want something that bounces like a cricket ball. We therefore need to be very careful about how we judge our models. We also, though, need to think carefully about how we build our models. Are we designing them for the purposes which they get used for, and are the targets of model design achievable?

Climate model development takes place in various large modelling centres around the world but what are they aiming for? They're certainly aiming for something bigger and better than has gone before. Something with more detail and more physical processes. Maybe something that compares more favourably with observations—not just recent observations but climate from long ago: palaeo climates such as ice ages and interglacial periods. They're essentially aiming for the perfect model of the real-world climate system: something that can be considered equivalent to reality. Yet this is a very different target to other great scientific endeavours.

Putting a human on the moon? We expected we could build the rockets and we knew when we had succeeded. Finding the Higgs boson? If it existed in the form proposed, we knew what was required of a particle accelerator to detect it and we saw when it was successful. But climate modellers don't expect to ever actually achieve a perfect climate model—no one believes it is an attainable goal. Even if they did they wouldn't be able to tell that they had done so due to the one-shot-bet nature of the climate prediction problem. So we don't expect to achieve a perfect model, but neither have we laid down an alternative, achievable goal, still less a yardstick by which we could judge progress towards it. Instead, there is an ill-defined assumption that any step closer to realism will necessarily improve the reliability of the information that the models provide. At the end of this chapter I'll argue that this is fundamentally the wrong approach to model development. What is needed for climate science and for climate predictions is greater diversity in our climate models, not greater realism. First, however, some words about why it is so unhelpful to target model improvement without a sound grasp of what is potentially attainable.

The key characteristic of climate change that is important when thinking about model development is, yet again, nonlinearity. If the errors in our predictions were related to the flaws in our model in a linear way, then each improvement in the model would lead to a better prediction and the process of model development would be a steady trudge towards reliability. Unfortunately, that's not the case. The models are nonlinear, the climate system is nonlinear, and we should expect that improvements in our model might be related to prediction reliability in a nonlinear way. Improvements in the way the model represents reality could either improve the overall prediction or degrade it. Remember the band—when the pianist joined the fiddler in playing in the right key, the overall performance became much worse even though the change was in the right direction.

Errors in some aspects of today's models are likely to be balancing out different errors in other aspects: something known as 'compensation of errors'. This means that improvements in one part can actually degrade the model as a whole because they take the model out of balance. Unfortunately, there is always a pressure to ensure that simulations look good **with the current model,** so with each iteration a model is retuned to look as realistic as possible. But that's not what we want. We don't want them to look realistic, we want them to provide useful predictions. The unattainable target of the perfect create a driver to make them look as good as possible now rather than focusing on what's necessary to answer particular questions for science and society. If model developers were in my peculiar band, they'd tell the pianist and fiddler to change key to be consistent with the rest rather than the other way round, because that would be the quickest way to get things sounding good now even though the goal is to sound good with the singer. If they were assessing the behaviour of the cricket ball, they'd focus on something that looks like a cricket ball so they might well throw away the dice and focus on the tennis ball—exactly the wrong thing to do. Climate prediction is a problem of extrapolation, so getting close to being able to represent historic observations may be desirable but is certainly not sufficient to give us confidence in a model's predictive ability. It is plausible that a model could be very close to representing aspects of the real world historically and still be unable to make good climate predictions. And the opposite is also true: the worst models could possibly be the best route forward.

These concerns apply even before considering the possibility that the Hawkmoth effect could undermine even the possibility of ever building a good predictive model for climate change at some levels of detail and for some timescales.

Without a good understanding of what an achievable goal is for a modelling system, there is a significant risk that modelling centres take whatever funding is available to make their models 'better' without actually increasing the reliability of the predictions or providing better information to society.

There are modellers who argue that the basic problem in climate modelling is that the models need to be higher resolution and more complex. Doing so certainly has benefits in terms of being able to study more interactions, but without a robust approach to evaluating reliability for prediction, we will never know to what extent we can use them to predict the future or guide society. Until climate science addresses the challenges of reliability assessment for the extrapolatory problem of climate change, the nebulous target of the perfect will lead to models which are constantly getting 'better' but which may never be sufficient for the tasks we set them. Continually making a 'better' model will absorb vast research funding and provide little in return. We need to change our approach and focus on understanding both the potential and the limits of the modelling process, while at the same time seeking models with a greater range of behaviour so that we can study and explore a wider range of possibilities.

20.2 The urge to over-constrain

In responding to climate change our societies could benefit from a better under-standing of the risks it poses. Greater diversity in our models would enable us to study plausible, if less-likely, responses, and provide society with descriptions of their potential consequences in support of risk-based planning.

However, while diversity in model behaviour may be the best thing for scientific understanding and for guiding society, the structural aspects of science work against it. Nobody wants an extreme model—a model that is considered less credible than the rest. That's understandable. It's human. It's a consequence of competing organizations and simply organizational pride. Everyone is aiming for the best model, the perfect model. That's why perturbed-parameter ensembles are so important and useful: they can create the diversity of models that we need to build pictures of possible futures which span the range of the credible.

This sounds great but do the perturbed-parameter ensembles actually go too far: beyond the credible? Can we rule out some of those model versions and hence con-strain the future? Are the model versions with very high sensitivities, for instance, so bad that they should be dismissed?

We're back to the question of how bad is too bad, only now the in-sample/out-of-sample problem from section 18.2 raises its head again. Earlier I discussed this problem it in terms of how assessments of models based on their predictions (out-of-sample) are different and better than assessments based on their ability to simulate historic behaviour (in-sample). The same problem arises if we decide what consti-tutes 'acceptable' in a model after having looked at the model results (in-sample) rather than deciding what would be acceptable beforehand (out-of-sample). When we assess models based on historic behaviour, there is a route for them to be tuned, or even designed, to achieve relatively realistic historic simulations, without necessarily providing similarly realistic future simulations. When we decide what is accept-able after having seen how a model behaves, there is a route for our assessments to rule out things that we already believe to be unlikely, thereby reinforcing our opin-ions by design rather than allowing them to be supported or undermined by new evidence.

Consider, for instance, examining the climate*prediction*.net model versions, select-ing those with high climate sensitivities and seeing if there are problems with their representation of climate. If there were, then we might conclude that these high sen-sitivities are unrealistic and dismiss them. It sounds like a perfectly reasonable way to proceed but it is not. By choosing to look at the high-sensitivity model versions, we are biasing our assessment of what is acceptable. In practice, all of the current models and all of the climate*prediction*.net model versions would fail significantly any robust comparison with observations (just look at Figures 18.2 and 18.3), so it would be easy to dismiss any and all of them for different reasons. If we pick any subset of them, we will find flaws which might tempt us to dismiss that subset, and with them a whole collection of outcomes. By picking the high-sensitivity ones we would essentially be choosing the ones we want to dismiss and then finding a reason to dismiss them:

deciding what's unacceptable by choosing to look only at those that show behaviour that is uncommon and thus maintaining the status quo by design. A better and more robust approach would require us to decide—and ideally publish—what we will consider acceptable or unacceptable **before** running the experiment. Or at least before seeing the results. Doing so would, however, require a wholesale change to the way model development and assessment is carried out. Now that's a challenge.

The in-sample/out-of-sample issue is important for forecasting but it doesn't necessarily apply when simply studying climate science. For climate science, the most interesting model versions in climate*prediction*.net are the extreme ones. A physicist is most likely to look at those simulations first, look at what processes are driving the extreme responses, and perhaps consider how those processes differ from observations. This is in-sample analysis, but it is a reasonable thing to do because it illuminates the processes that lead to extreme responses and helps us understand the climate system better. Unfortunately, if such analysis shows us a lack of realism in these extreme model versions we can easily fall into the trap of thinking that therefore this extreme behaviour can be ruled out. The in-sample nature of such studies implies we should be extremely cautious about such conclusions. It might be that they are not worse than other model versions, only differently bad; or that they are indeed worse but since all models have such a substantial lack of realism it's not clear what we should conclude from that. We're back to needing to define when a model is too bad for consideration.

On top of these concerns, even if these particular model versions have fatal flaws, that doesn't mean that we can't find alternative models with similar high sensitivities but without such flaws. Any conclusions need to be put in the context of how much we've looked for such models. Ruling out particular models is not the same as ruling out all the behaviour shown by such models.

In practice, when climate*prediction*.net first published its results, studies did indeed follow that looked at the extreme sensitivity model versions.[2] Sure enough, they found flaws and raised concerns. Sure enough, climate*prediction*.net later found other model versions with different perturbations which gave similarly high climate sensitivities. This simply highlights two related challenges. The first, which I've already mentioned, is the need to define what model behaviour should be considered too unrealistic to be taken into consideration, without ruling out potentially valuable and informative models. Remember the bouncing dice—models can be substantially unrealistic and still valuable and informative. The second challenge is to effectively link the forecasting demands on modellers (the need for out-of-sample analysis) with their interests in the physical system (the need for in-sample analysis). How do we explore the most interesting behaviour while ensuring that the mere choice of what to investigate is not misinterpreted as implying that particular behaviours are considered unlikely and thus biasing our predictions?

As it turns out I, together with climate*prediction*.net co-authors, did rule out a small number of simulations. In these simulations ice formed in the tropical pacific

---

[2] And other models perturbed in a similar way to the extreme model versions.

ocean and this led to widespread and dramatic cooling in the model, but it wasn't directly due to either of these factors that we discounted them. No, the reason to rule them out was that the dominant response of the model, the widespread cooling, came about because of limitations in the way physical processes were represented in the model. If you remember from Chapter 16 these experiments in climate*prediction*.net used a model without a full representation of the ocean. As a consequence the model itself was founded on the premise that there wouldn't be significant changes in ocean circulation patterns. The formation of ice in the tropical pacific could only happen in a model like this because in reality, or in a more complex model with a full representation of the ocean, ocean currents would have changed and carried heat from elsewhere and warmed that region before ice could ever have formed. Thus the dominant behaviour in these simulations resulted from processes that broke the assumptions on which the model was designed.

You might think of this as like throwing a dice which falls apart when it lands such that two sides are face up and you get a total of say nine (a six and a three) from a standard six-sided dice. That behaviour breaks the assumptions on which we are using the dice: that only one of the six sides can be face up. When we talk about the behaviour of a standard dice, we are interested in the behaviour of a dice that generates the numbers one to six, so we wouldn't want to accept a situation where two sides were showing. This suggests a route forward for specifying when a model is too bad to be included in perturbed-parameter or multi-model ensemble results. If the results of a simulation break the assumptions inherent in the design of the model, then that model version—or maybe just that simulation—is not acceptable. We nevertheless need to be aware of the in-sample problem—ideally, we shouldn't make that judgement after having seen the results. Rather we should invest significant effort in describing what those inherent assumptions are, and how they may be broken, **before** we see what results the model produces.

20.3   Going beyond what we've found so far

The discussion so far has been all about taking the ensembles and trying to reduce the uncertainty they represent. Reducing uncertainty is one of the drivers of climate science: funders request it and many scientists believe it is what decision-makers in society need or want. This framing is deeply detrimental to climate science in the early twenty-first century. We have actually only begun to scratch the surface of understanding uncertainty in climate predictions. For most variables we have no robust assessments of uncertainty in future climatic behaviour. We know to expect warming and many forms of associated climatic disruption, but quite what form that can take, particularly at local scales, is a field that is still wide open. Targeting a reduction in uncertainty at this stage is premature and represents an unhelpful motivation for research.

Instead, we really need to explore uncertainty more thoroughly. More perturbed-parameter ensembles would be helpful, as would extending them to explore not just parameters but the way physical processes are represented within models: that's to say, structural aspects of model designs. We might call these perturbed-**physics** ensembles, a term that has sometimes been used interchangeably with perturbed-parameter ensembles. These experiments should explore the sensitivity of models to both variations in parameters as well as different structural possibilities for representing climatic processes. In using models, we want to explore how far we can push a model and still get something that has a reasonable level of credibility.

One question this raises is how we should choose the range of parameter values to explore.

The relationship between model parameters and real-world quantities is unclear at best. To date, their exploration has been guided largely by expert judgement regarding what values they could take. This, by design, limits the values studied. The complexity of the models and the nonlinear interactions between their diverse elements implies that expert judgement is likely to be of limited value. Experts often know a lot about their particular aspect of the model but much less about its interactions with other aspects and how that plays out on timescales of *decades* or *centuries*.[3] Only running a perturbed-physics ensemble can tell us about these interactions.

So what values can parameters potentially take? What actually constrains them? This is something that itself requires study and exploration. Some may have limits beyond which they make no sense within the design of the parameterization scheme: for instance, a negative value for the ice-fall-rate implies that ice tends to go upwards which is inconsistent with the concepts behind the representation of clouds in the model. There will also be limits beyond which the computer code simply crashes. Apart from such absolute limits (the latter) or somewhat arguable absolute limits (the former), it is simply a matter of investigation. One of the challenges for climate modelling therefore is to explore the behaviour of models across as wide a range of parameter values as possible. Doing so would focus us on understanding the behaviour of the model itself rather than treating it as almost equivalent to the real world, which it is not. Understanding the whole range of potential model behaviour would provide a much better foundation for relating it to reality. A very wide exploration of parameter values will likely lead to some very unrealistic model versions, but even those are useful in understanding the model system. They have a valuable role in guiding the debate over when a model is too bad to be considered for the study of climate science and/or climate predictions. Such a model-focused approach would also be much less susceptible to bias toward the original model's behaviour—to anchoring (section 19.5).

Of course, experiments of this nature would require very substantial computing power. There are, however, already calls for vast computing power to be invested in climate modelling.[b] Existing calls though are focused on the development of more complex, higher-resolution models. They are founded on the assumption that bigger is better but they have no strong basis for saying how big is big enough. A more useful application of such resources would be to demonstrate how to explore the

---

[3] Remember italics means time within a model.

full potential behaviour of a model and to study how to relate it to reality. Experiments of this nature would provide a solid foundation for the interpretation of ensembles and the design of subsequent experiments. Higher resolution and more complex models will no doubt be developed, but the top priority at the moment should be to understand how we would use them. What is the minimum ensemble size needed to extract robust information? How should experiments explore parametric and structural uncertainty? How should they balance this with the exploration of initial condition uncertainty?

These questions need to be addressed with current models so we don't waste effort running potentially uninformative or misleading experiments with new and bigger models. Climate science can learn from other areas of science here: the Large Hadron Collider was designed to be big enough to find the Higgs boson—smaller would have been a largely pointless waste of time and this was known at the start. In climate modelling, building and running more complex models has little value until we, first, understand what ensemble designs are required to study the behaviour we're interested in, and second, have the computing capacity to run such ensembles with these new and better models. That computing capacity is likely to be some orders of magnitude bigger than what is required to simply run the model.

Fortunately, demonstrating how to build and design these mega-ensembles is itself open to study using smaller systems such as extensions of the Lorenz–Stommel model from Chapter 15. Research of this nature is therefore possible with much smaller computing resources. This, however, is not the only way to reduce the demand for computing resources. Earlier I mentioned the concept of 'filling-in' parameter space using statistical methods. These techniques are far from 100% reliable, particularly where there are substantial nonlinear interactions between the parameters. Nevertheless, they point the way to reducing the size of ensembles needed to usefully explore a model's behaviour. They could be used to direct us to interesting combinations of parameters which would be worth testing by running a model version. This approach represents an integration of statistics and physical modelling and provides a way of optimizing ensemble designs so as to reduce the demand on computational capacity. It's a way forward.

Of course, using statistical techniques to point to interesting parameter combinations raises the question of what constitutes 'interesting'. And so we come back to my opening gambit regarding the interpretation of multi-model and perturbed-physics ensembles: they represent a range of behaviour for reality that we can't currently rule out. Interesting parameter combinations are therefore those that extend that behaviour. The aim should be to understand just how big our uncertainty is in any particular quantity. This information is useful for society and also puts us in a better position to apply whatever techniques we might develop later for ruling out models.

A key goal is therefore to find out how wide a range of behaviour our models can show given their basic structure. This could be done for aspects of relevance to society as well as aspects of interest to climate scientists. It would be interesting, for instance, to see just how different the change can be in the probability of extreme storms or heatwaves in some location. This is something that could be achieved by using statistical approaches to guide our choice of parameter variations and help push out the range of such behaviour within a perturbed-physics ensemble.

Climate modelling and climate prediction would benefit from replacing the target of approximately perfect models with a target of broad diversity in a collection of models. Such an approach would represent a new way to interpret models as evidence about future behaviour. If despite thoroughly exploring a model's parameter space, we can't find any model versions that simulate changes[4] outside a certain range (for a particular variable), then this arguably provides evidence that more extreme changes may not be possible. It's not completely solid and reliable evidence because other models—available now or in the future—may show such behaviour, but it's much better than simply taking today's models as representing the most likely outcome in reality. It provides a path to a different source of confidence in climate predictions that doesn't rely on having a perfect, or almost perfect, model.

In this way the combination of physically based computer models and statistical representations of the behaviour of such models in perturbed-physics ensembles holds out the potential for more reliable, user-focused information about future climate. Not yet the probability distributions we might desire, but at least much better sources of information than we have today.

20.4 Looking forward optimistically

The structure of the relationship between science funders and climate scientists creates a pressure to over-interpret climate models. This is a relatively new phenomenon. Models of this nature were not originally built as engines of prediction but as engines of understanding. They are now so heavily ingrained in the structure of addressing climate change that it has become difficult to step back and question what the fundamental constraints on using climate models for climate predictions actually are—what their limitations are and how we should use them effectively. These questions however should be at the forefront of academic research. Currently they are not.

There is a dissonance in the research community. On the one hand the fallibility and limitations of climate models are widely acknowledged. On the other the outputs of these models are discussed and presented as if they are representative of reality. The detail they provide undermines the development of alternative methods for understanding future climate and supporting societal decisions in the context of future climate change. A reliable climate model prediction would always be better than any other method because it would provide the best conceivable information: probability distributions for every possible variable or vulnerability of interest, locally and regionally, in 2050, 2100, 2200, and any point in time you wish. If we had

---

[4] In response to assumed changes in greenhouse gas concentrations.

these distributions, then it would simply[5] be a matter of using them to optimize our planning decisions. The current models appear to provide this holy grail of reliable climate prediction and the vision of the grail hampers the development of alternative methods which would be much more helpful, if not so comprehensive.

Challenge 5—the last three chapters—has been about the many challenges associated with how we explore and interpret complex simulation models. The challenges are fascinating and intriguing, but overall these chapters are a bit of a downer because they highlight just how difficult it is to know very much in detail about future climate under climate change. We are not, however, without information and considering these challenges will help us produce better information in the future. It's not just about doing better in the future though. We have lots of information now, or at least lots of information that we could easily get with today's models and data. If we manage to avoid being distracted by the grail of the perfect, then climate science, and even climate modelling, has many ways to support society now in responding to climate change. We need to focus on them.

---

[5] My policy and social science colleagues will rightly laugh at the idea of describing this as simple. It in fact just takes us to a new and different form of complexity and a whole new set of challenges but it would at least end the role of the physical scientists.

Challenge 6: How can we use today's climate science well?

The model experiments we have today provide potentially valuable information but not climate predictions of reality. Even those we could possibly have in the near future are inevitably going to be substantially less than ideal. Nevertheless, society is basing actions and investments on climate science and on the output of models: actions and investments with expensive, long-term consequences. How can we support them with the model simulations and the observational data we have? How can we identify robust information and avoid over-interpretation of models? How can we encourage the use of reliable but potentially vague information, when detailed but less reliable information has the appearance of providing everything that decision-makers want? How can we provide information that leads to decisions that are more robust to climate change, without aiming for unachievable perfection? How can we use today's climate science well?

21
What we do with what we've got

21.1 Avoiding the lure of the perfect

'Il meglio è l'inimico del bene', Voltaire, 1770[a]
'The best is the enemy of the good'

Back in the late eighteenth century, in his *Dictionnaire Philosophique*, Voltaire quoted an Italian proverb: 'the best is the enemy of the good'. The phrase can be interpreted in the context of climate predictions, as meaning we should not hold out for perfect information in the future when we already have good information today. Perfect information may come too late to be useful or it may simply not be worth the effort to seek it out. The message is that it is better to use imperfect information now than to wait for perfect information later. The phrase is an appropriate aphorism for climate change information: we should definitely act on the information we have now rather than waiting for better information which may or may not come along in the future. After all, time is of the essence here.

This simplistic interpretation of Voltaire's proverb is not, however, the most apposite or interesting interpretation of the phrase in relation to climate information. The real problem for climate information is that it is all too easy to believe that we already have 'the best'. That's what high-resolution computer models appear to provide. We have data from multi-model ensembles which can be processed and presented as what look like probability distributions for future climate at any location, at any time, and for any variable you might be interested in.[b] Climate modellers seem to be providing almost perfect information, so there's no need to settle for only 'the good'. To many users of climate information, therefore, Voltaire's proverb doesn't feel like it has any great relevance; we seem to be 99% of the way to 'the best' anyway.

And yet, as has been described in previous chapters, these predictions are not actually as perfect as they look. Far from it: they may be detailed and generate realistic-looking pictures but this masks the fact that they aren't reliable predictions of reality. The biggest challenge—for both academics and those making practical decisions in our societies—is to see through this veil of perfection and accept that although they look like what we want, actually they are not. The challenge is to recognize that we are a long way from having 'the best', so we need to put in a lot of effort searching for 'the good': searching for new ways of interpreting the information we have, even if they provide less comprehensive, less detailed, and less precise information than the illusory 'best'. We need to seek out information that is more reliable, more relevant to real-world decisions, and less likely to encourage over-optimization, than our current presentations of multi-model ensembles. A central challenge in responding to climate change is giving space for 'the good' to be utilized in place of the oh-so-tempting, mirage of 'the best'. With climate change, it is the misleading apparition of the best that is the enemy of the good. *Il miraggio del meglio è l'inimico del bene.*

Searching for the alternative 'good' is made particularly difficult because conventional approaches to climate prediction are neat and tidy. If you believe that the models, or the distributions from multi-model ensembles, are actually predictions of the real-world climate, then you are sorted. Finished. As a scientist or a modeller, your job is done. A single approach has neatly solved all our prediction problems. All the information anyone could possibly require is available. The only thing you need to do is make the data available to the rest of society and they can pick out whatever it is they want. It's up to them to work out how to make good decisions with it. It's up to them to work out what it means for food production, land stability, vector-borne disease risks, and so on.

In a similar vein, if you believe that the current models don't quite provide such information but that the next generation of models will, then all you need to do is focus your effort on improving the models. In that case, scientific effort and modelling should be concentrated almost entirely on building something bigger and better.

In both cases what is needed is investment in computers, modelling, and software systems for making data available. There is little need for scientists to work with business or policymakers, specialists in climate change impacts, or those who study the theoretical foundations of forecasting. There is little value in multi-disciplinarity at all.

The alternatives are messy—messy in the sense that they don't provide a one-stop shop for climate predictions, and they do require complex multi-disciplinary perspectives. They often involve integrating diverse strands of information in the context of specific decisions or questions. They lose the beautiful generality of pure science—an aspect of physics that has always appealed strongly to me—but they gain practical relevance and reliability. In the alternative approaches discussed in this chapter, science and modelling cannot be separated from how the information will be used; the science and the application of the science are completely and inextricably intertwined. Unfortunately, this means they often need to be repeated or reworked to support different societal decisions or answer different questions about climate impacts. They present a much less tidy approach than the wholesale processing of multi-model outputs into a single climate prediction product that appears to answer everything.

These different approaches also require a substantial shift of emphasis regarding what a government, business, or research funder might want to spend their money on. In the perfect world of models and perfectly processed ensembles, you would want to spend a lot of your money on computer hardware. By contrast, with the alternatives you would want to invest in people, expertise, and understanding of complex systems; you would want to bring together the physical and social sciences; you would want to create a range of methods that are targeted at a variety of different situations;

and you would want to support people in gaining the skills and knowledge to apply them.

In an attempt to be clear and avoid criticism (unlikely given the subject of this book) I should say that no one I know in the climate science or climate modelling community claims that the models are 'almost perfect', or that the probability distributions derived from multi-model and perturbed-parameter ensembles are 'almost perfect'.[1] To note this, however, is missing the point. The way the model outputs are made available and presented for wider use, invites other research disciplines and other users of climate information to treat them as if they were reliable predictions. This makes it appear as if reliable predictions are available and as a result it's unlikely that either funders or users would want anything else. If model results appear to be comprehensive, detailed, and easily available, why would you seek anything different? The basic answer, of course, is that the alternatives might be more reliable: they might help you make decisions that are more robust to what will actually happen as a consequence of climate change and avoid potentially huge costs associated with maladaptation. It's a simple answer but in the shadow of such detail from the models, it's not often considered. In any case, there is also another answer, and it's to do with eggs and baskets.

If we adopt a single approach to the provision of climate information to support climate-sensitive decisions then we are gambling the whole of society on that approach being right. Preparing our societies for the future state of climate is not about protecting this bridge, that coastal community, this ecosystem, or that coral reef—it's about doing our best to protect everything: making our societal systems as a whole resilient to the changing risks from climate. If all our approaches to preparing and protecting what we value use the same underlying sources of information, then we risk either protecting everything or making everything vulnerable to the flaws in that information. We end up putting all our eggs in one basket but due to the complexity of the issues it will not be obvious that that is what we are doing.

Consider, for instance, a situation where a country provides a standard set of climate predictions[2] based on one methodological approach, and requires or encourages business, industry, and civil society to prepare and adapt to climate change based on that information. The whole of that country then becomes vulnerable to the information being incorrect or misleading. Everything is vulnerable in the same way. Everything: systems to prevent flooding, supply water, resist storm damage, maintain energy and communication services during extreme events, ensure tolerable internal environments, and even achieve financial stability during multiple correlated climate shocks. If the information is flawed or misleading then whole sectors of the country would end up sharing vulnerabilities by not preparing for threats that turn out to be highly likely and by optimizing for threats that aren't.

---

[1] No one claims this, but the implication is often that the next generation could be.
[2] Predictions based on a variety of assumptions regarding international efforts to reduce greenhouse gas emissions and atmospheric concentrations.

In this situation the risk of the country succumbing to unmanageable impacts is high. The way to avoid this risk is to encourage diverse approaches to the provision of climate predictions. We need a marketplace of ideas and approaches, and users of climate information need the skills to assess the pedigree and suitability of information for the particular issues they face. It's certainly not a situation where anything goes, but more things go than we have access to at the moment.

This is an exciting time for research in practically useful climate predictions but only if the research community can escape the lure of perfect, model-based predictions. The aim should be to encapsulate the best possible integrated understanding of what we can know today about the consequences of climate change, and to do so in a way that is useful for those who can act on such information. A number of alternatives to simply trusting model-based information are available and the rest of this chapter presents three of them. They have all cropped up in earlier chapters in one way or another so little here is new. Earlier, however, these ideas were embedded in the complexities and difficulties of other methods. Now we are past all that I can give them a positive spin. It would help science and society a lot if we could develop them into practical, decision-focused methods. All require models, theory, and observations but the first is led by model information, the second by theory, and the third by observations. Remember, though, that they are all about addressing user-driven questions; the alternatives to models are messy and specific, they aren't global and one-size-fits-all.

21.2 Can I have a bigger envelope, please?

To start a discussion of user-focused methods for climate predictions, let's reflect on what we can say from multi-model and perturbed-physics ensembles.

First, a reminder of why this is difficult. For multi-model ensembles the lack of independence between the different models means that the average of their future climate simulations is largely meaningless. There is no reason to expect any version of the average (mean, median, or mode) to reflect reality in any way. Consider the model simulated change in summer temperatures in the Western USA (Figure 21.1). The median across the models tells us that 50% of the models show a change above some value—5.7°C. It tells us about the models in this particular multi-model ensemble but it does not tell us that there is a 50:50 chance of reality being above or below this value. Similarly the shape of a distribution built from the different model simulations (Figure 21.1) isn't a probability distribution that reflects the probability of different real-world outcomes. The twenty-fifth and seventy-fifth percentiles of the distribution don't represent a range (4.7–6.6°C for the Western USA) in which reality is 50% probable to lie; there isn't a 90% probability that reality will be between the fifth and ninety-fifth percentiles (4.0–8.1°C for the Western USA). All these numbers are no more and no less than a description of the particular models in this ensemble.[3]

---

[3] This is the type of information which is currently made easily available.

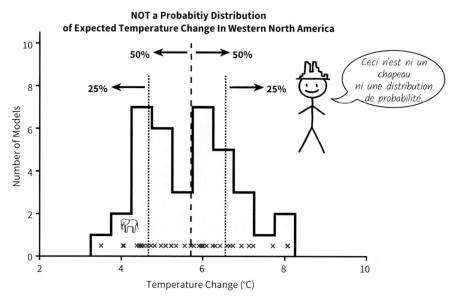

Figure 21.1 No more and no less than model output. Temperature change in Western North America by the end of the twenty-first century (2070–2100) compared with pre-industrial times (1850–1900) in thirty-seven climate models under a high scenario for greenhouse gas emissions.[c] The actual values are the blue crosses, while the distribution of results is the histogram. Acknowledgements are due to the artist Rene Magritte[d] and the writer Antoine de Saint-Exupéry[e] for highlighting the difficulties we have interpreting images.

The distributions from multi-model and perturbed-physics ensembles may have no meaning in themselves but the models nevertheless provide information about the different ways in which the interacting physical processes of the climate system could respond to climate change. They contain information that we should not ignore. My opening gambit in Chapter 19 was that they represent a range of behaviour for reality that we can't currently rule out. By the end of the last chapter I hadn't made much progress beyond this minimalist interpretation but one crucial jump to be made in climate modelling is to realize that even this is potentially extremely valuable. A second crucial jump is to see that perturbed-physics ensembles could be targeted, with this basic interpretation in mind, to be much more relevant to other disciplines and to societal decisions.

To begin to see how to do that we need to acknowledge that in most cases vulnerabilities to climate change are not related to a single variable, such as maximum daily temperature, but rather to combinations of variables such as monthly average temperature and rainfall, or daily rainfall and a combination of temperature, rainfall, and humidity over the previous month, or other more exotic combinations of variables and time periods. Large perturbed-physics ensembles would enable us to build a picture of the range of plausible responses across such combinations of variables.

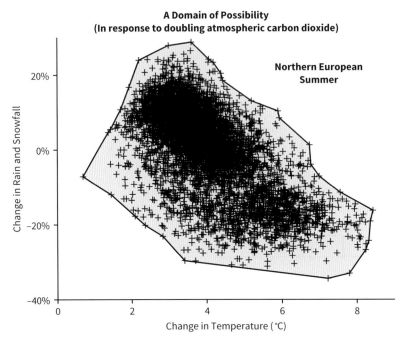

Figure 21.2 A 'domain of possibility' or a 'non-discountable envelope' for northern European summer temperature and precipitation from the climate*prediction*.net perturbed-parameter ensemble.

Figure 21.2 illustrates this type of multiple variable response using the idealized double atmospheric carbon dioxide experiment of climate*prediction*.net. For summer in Northern Europe, there is a wide range in the response of both temperature and rainfall. We can draw a line around the domain of response. I call this a 'non-discountable envelope' of climate change behaviour[f]: it represents the range of responses which we can't currently rule out, for multiple variables. It provides a 'domain of possibility', and it shows how the variables are related. It may be a simple interpretation of the ensembles but it is still valuable for assessing the potential impacts of climate change and for supporting decisions across society.

A non-discountable envelope like this doesn't say that the real world couldn't be outside the domain; the real world is different from our models, and in any case a future model might be outside the envelope. It also doesn't say that any value within the envelope is more likely than any other, even if some areas are populated by many model versions and others by only a few.[4] The envelope simply represents a domain of behaviour which we can't currently discount (rule out) as a possible consequence

---

[4] This is because the likelihood of finding a model with certain behaviour in parameter space doesn't represent likelihood in reality—it's simply a consequence of the model structure, the somewhat arbitrary decisions in the model development process, and the design of the perturbed-physics ensemble. (See section 19.6.)

of the changes in atmospheric greenhouse gases. It is 'non-discountable' in the sense that nothing within it can be ruled out.[5]

For many people this interpretation is deeply counterintuitive. We all have a feeling for likelihood and when we see lots of models doing the same thing, we want to believe that that thing is most likely. But computer models are a new source of evidence in scientific understanding and we haven't yet developed intuition for how to interpret them. It looks like a statistical problem but is actually a philosophy of science problem. It's a question of what has meaning, not the practicalities of processing numbers. This friction between perspectives creates a substantial problem in communicating what we know.

The BBC has a radio programme, *More or Less*, which investigates the accuracy of statistics in the media and the public domain. It is often excellent but in this case it stands to illustrate the problem of limited perspectives. Shortly after the publication of the climate*prediction*.net range of climate sensitivities, the creator and the presenter of the programme published a book about the illusions in numbers called *The Tiger That Isn't*.[g] In the book they interpret the climate*prediction*.net results as a probability distribution, arguing that it is alarmist to pick out the extreme values: the range. They describe it as '*akin to a golfing experiment: you see where 2000 balls land, all hit slightly differently, and arrive at a sense of what is most likely or typical; except that climate*prediction.net *chose to publicise a shot that landed in the car park*'. I sympathize with their frustration but their interpretation was completely wrong. They were producing a programme about statistics and they could only see the data through the lens of the statistical tools they were familiar with. In the climate*prediction*.net paper that they were discussing was the following statement:

> The 'lack of an observational constraint, combined with the sensitivity of the results to the way in which parameters are perturbed, means that we cannot provide an objective probability density function for simulated climate sensitivity. Nevertheless, our results demonstrate the wide range of behaviour possible within a GCM and show that high sensitivities cannot yet be neglected as they were in the headline uncertainty ranges of the IPCC Third Assessment Report'.[h]

The interpretation in *The Tiger That Isn't* would have been spot-on for a micro-initial-condition ensemble with a perfect model, but is completely wrong for a perturbed-parameter ensemble.

In a different but not entirely unrelated context, the philosopher Ludwig Wittgenstein illustrated the problem by considering someone buying 'several copies of the morning paper to assure himself that what it said was true'[i]. A more suitable analogy than golf would have been someone buying 1999 copies of today's edition of a newspaper with, say, a right-wing bias and one of a newspaper with a left-wing bias. That

---

[5] Of course you could rule out all the models as they are all inconsistent with reality but if you choose not to do this because the models nevertheless contain useful information, then you have only a non-discountable envelope.

1999 out of 2000 copies present one interpretation of the effectiveness of the national government's policies doesn't make that interpretation more likely to be accurate. Rather what we should take from the 2000 copies of newspapers is that there are differing views. Without any prior understanding or belief as to which newspaper is likely to be correct, the sensible approach is to look at the extremes of the range. Of course, further analysis and assessment may lead you to believe one perspective more than another, but that is not related to the number of copies of the paper you bought! Similarly the number of models that show some sort of behaviour is not an indication of the likelihood of that behaviour.

This tale is simply meant to illustrate the difficulties of communication across disciplinary boundaries. Identifying when a climate prediction question is one of statistics, philosophy, physics, computing, economics, or some combination of them all, is difficult and requires a much higher level of cross-disciplinary expertise than is encouraged within research institutions today, or within the world of public communication of science and statistics.

The concept of a non-discountable envelope of possibility may be alien to many but it does point to a way forward for designing perturbed-parameter ensembles that provide more and better information about future climate. As discussed in the last chapter, if our interest is in the range of behaviour, or the size of a non-discountable envelope for multiple variables, then the aim of our ensembles should be to attempt to increase the size of that envelope. Trying to find models with behaviour beyond that seen in current models provides us with more information and if, despite looking, we find that some types of response are unachievable with our models then this might, perhaps, be taken as evidence that such a response is implausible or less likely in reality.

'Despite looking' is the crucial point here. Fortunately, the statisticians can help us look. The statistical tools mentioned in the previous chapter could take the results from a perturbed-parameter ensemble and predict what the model would simulate with different values of the parameters. They aren't necessarily reliable, particularly where the interactions between the parameters are complicated and nonlinear, but they could provide a starting point. They could suggest parameter combinations where a model version is more likely to produce different outcomes from those we have come across before. The model could then be run with those parameter values to see if that is actually the case. In this way we could design experiments to push out the bounds of the envelope without requiring vast computing power. This approach would by design increase our estimates of uncertainty but would ultimately help us gain confidence in our predictions.

What makes this a messy, user-focused approach is that the statistical tools need to focus on particular variables or combinations of variables; there are too many to imagine doing it for all them. Indeed it is difficult to imagine doing it for more than a handful at once. We have to choose what we're interested in: what the particular envelope is that we are trying to expand. A climate physicist might choose equilibrium climate sensitivity but a national bank, water planner, hydrologist, or health researcher would have their own variables of interest. The ensemble could be—should be—tailored to their needs. This approach is a route to designing climate model ensembles to inform specific questions.

If we endeavour to find models with more and more diverse responses to increasing greenhouse gases and then find that there are bounds beyond which it becomes impossible for a model of a given structure to generate such behaviour, well that is a new type of evidence. It begins to give us confidence that behaviour beyond that bound may actually be impossible in any world which is vaguely similar to our own. If we could do that with a perturbed-parameter ensemble then that would be a good beginning. The evidence would be stronger still, however, if we were able to repeat it with perturbed-parameter ensembles based on a variety of model structures. And of course, if we can find a physical explanation to justify the result then the conclusion becomes even stronger still. This is the ladder we want to climb to achieve increasing levels of confidence. It uses computer models to generate a new type of evidence to help us understand the limits of future behaviour.

Of course these issues of model interpretation don't just apply to the climate system. Where any natural or social system is changing, or where we have no observations for the state of the system that we are trying to predict, computer models can end up being a major source of information and insight. Exploring the domain of behaviour which is achievable within a certain type of model provides better evidence about reality than can a single model. The extent to which it does so comes down to a judgement regarding the value of the particular type of model and the extent to which the range of model behaviour has been explored.

The limits beyond which no model can go represent some sort of maximum range on uncertainty. Our current ensembles don't get close to providing these ranges because we haven't designed them to push out the limits of behaviour and they are of very limited size, so they only give us a lower bound on what they may be. The 'non-discountable envelopes' from current ensembles must therefore be described as providing a 'lower bound on the maximum range of uncertainty'.[j] I like both terms. Feel free to use whichever one you feel trips off the tongue most easily when you're discussing them with friends on the bus.

21.3 Maybe I can use a smaller envelope?

The last chapter went into the details of why it's difficult to reduce the range, or the size of the non-discountable envelope, by ruling out models. To simply say that a model is not realistic is not enough because all models are unrealistic. Nonetheless, we can imagine a more nuanced interpretation to the non-discountable envelope of Figure 21.2. If some regions of the envelope were only populated by models with high climate sensitivity, for instance, then you might consider those regions to be less likely to reflect reality. Why? Because there are a number of different sources of information that provide strong evidence that climate sensitivity could take some values but less evidence that it could take other values. There is deep uncertainty—that's to say, we don't know the probabilities that we should associate with different values of climate sensitivity—but there is nevertheless general agreement on the shape (Figure 19.2) of the probability distribution. Values around 2.5 to 4°C are considered most likely—or at least best understood—while values above, say, 5°C[6] are considered less likely but still plausible.

The information we have on climate sensitivity suggests we shouldn't consider all parts of a 'non-discountable envelope' as equally likely. If some parts can only come about from models with very high or very low climate sensitivity then we might reasonably consider them to be less likely to represent reality than those with more central values. We can therefore use climate sensitivity to designate some regions of the envelope as less likely, though not to rule them out (see Figure 21.3).

This is not the same as simply comparing models with observations because judgements about equilibrium climate sensitivity use a diverse range of information sources, including theoretical understanding which can reflect on plausible future behaviour, not just what has been seen in the past.

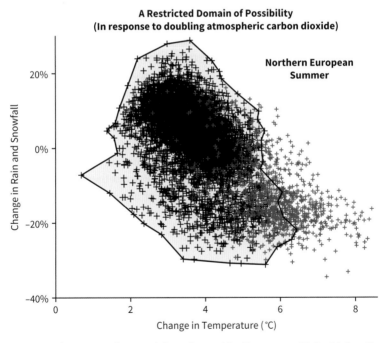

Figure 21.3 As Figure 21.2, but model versions with climate sensitivity higher than 5°C are coloured red and the domain of possibility is only drawn round those with climate sensitivity less than 5°C.

---

[6] Have a look back at section 19.2 for more about what a reasonable value to choose would be.

Setting these considerations of interpretation aside though, the aim of future modelling activities should be to make the envelopes bigger and encompass as wide a range of behaviour as possible—we need to be confident how big our uncertainty is before trying to reduce it.

21.4 Telling tales

An alternative to starting from the climate models is to start from physical understanding of the climate system. We know a lot about how the climate system behaves. We understand many of the processes that drive global warming and we also understand the character of the uncertainties we have in those processes. This is why there is a level of agreement in the scientific community regarding the shape of the probability distribution for climate sensitivity and to all intents and purposes consensus about the reality of recent and expected warming.

Given all this understanding, one option we have is to formulate physically consistent tales or narratives of how the physical system's response to increasing atmospheric greenhouse gases could play out. Such tales would be very general. We could talk about how a low, medium, or high level of climate sensitivity would translate into levels of global warming over time in a particular scenario for greenhouse gas emissions. For some level of warming we could provide possibilities for how that might change ocean circulation patterns, ocean temperatures, the cryosphere, and land cover, on large scales. Changes in these aspects of the climate system are particularly pertinent because they tend to vary more slowly than atmospheric conditions. We could hypothesize changes in other aspects too: ecosystems, river flows, circulation patterns in the atmosphere, and atmospheric/oceanic behaviour such as the El Niño/La Niña oscillation. In this way we could build up plausible, physically consistent pictures of the future, all accompanied by explanations of the physical processes that would lead to such a future. The explanations leading to these pictures would represent self-consistent, scientifically founded tales of what might happen.

Once we have such pictures, we could return to the models. Remember, one of the big problems with the reliability of climate models is that they are being used to extrapolate into a never-before-experienced state of the world. However, very similar models are used to make weather and seasonal predictions, and for these purposes the slowly changing aspects of the system and their response to increasing greenhouse gases are unimportant because they don't change during the period of the forecast. Furthermore, weather models have evolved their ability to make predictions by doing so repeatedly—they have many forecast-outcome pairs—so we know that on weather timescales they are now pretty good. This opens up an opportunity for reliable—if conditional—climate predictions. Once we have a theory-based narrative for how the general picture of climate change might evolve, we can fill in the details using weather models. In this way we could create credible tales not just about large scale climate features but about future weather and climate on local scales; even, perhaps, to build up probability distributions for climatic behaviour consistent with such a tale.[7]

An approach like this doesn't provide probabilities for different tales but it could provide probabilities of behaviour within any particular tale. More importantly, a tale can explore a much wider range of behaviour than seen in current climate models. This means we are not constrained by the models but rather liberated by them to use our intellectual expertise to explore and consider what is and isn't plausible. The models become tools for filling in the details rather than predicting the entirety of the future.[8]

The tales approach allows us to focus on the local and regional aspects of future climate that are often most important for policy- and decision-makers. It allows us to tailor our assessments to particular demands from society. Indeed, modern weather modelling techniques provide tools to push a model to simulate a particular type of weather system using a technique called 'data assimilation'. By doing this in a model which is already setup to represent some hypothesized future state of climate, we could present a picture of how a particular type of weather could look different in the future. The first proposal of this nature came from work led by the Dutch Meteorological Office[k] (Koninklijk Nederlands Meteorologisch Instituut—KNMI). It illustrated the tales concept by considering a particular weather system that had previously threatened severe flooding in the Netherlands and looked at how this weather system could be different in the future. Such methods would allow us to address questions framed along the lines of 'what would the 2010 floods in Pakistan look like in 2070?', or 'what would the 2003 European heatwave look like in 2050?'.

The strength of this approach is that it maintains focus on the subject of interest to the ultimate users while enabling the assumptions and judgements of the specialists who developed the tale to be made clear and therefore open to questioning and debate.

The most important aspect of the tales approach though is that it is driven by physical understanding and hypotheses—not by models. Unfortunately the terminology has been somewhat adopted by the modelling community to simply represent descriptions of what happens in a model—the antithesis of what this approach is actually about; users of climate information need to beware.

A good example of a true tales approach[9] examined the consequences of climate change for water resources in southern India. In this case experts on the Indian Summer Monsoon were gathered and encouraged to discuss the different ways in which climate change could affect the character and intensity of the monsoon. In initial discussions, their responses were anchored to the behaviour seen in models, but once

---

[7] Just 'perhaps'. There's a lot more work needed to demonstrate that these models could provide such conditional probabilities.

[8] There are still some conceptual issues to consider with this approach. For instance, weather models have been built and tested to simulate weather under (roughly) today's climate, so we don't know how reliable they'd be with different levels of greenhouse gases and different levels of global warming. However their ability to simulate and predict weather in many different regions of the world suggests that they remain quite reliable under a wide range of climatic conditions so we have reasons for confidence.

[9] On this occasion it was described as a 'narrative'.

the discussion got going, a much wider variety of potential responses was considered possible, though not all were considered equally likely.[1] The study took these responses and used them to create multiple tales for potential changes in rainfall in the region. These were then input into a model of the hydrology of southern India which was used to explore options for water resources management in the region. This is an example of using theoretical understanding to create tales and then using a model to fill in the details and make it relevant to practical decision-making.

To date, studies of this nature have been limited, but there are vast opportunities to use them to assess climate change impacts and to provide information which is directly relevant to society and which is not constrained by the limited range of climate models that we happen to have.

21.5 Observations through the lens of a decision maker

So far in this chapter the methods have been driven by how-it-functions approaches: first through how we interpret models, then through how we use theoretical understanding directly with additional information from models. The look-and-see perspective is nevertheless an extremely valuable additional source of information. It may not be able to tell us how things will change in the future but being clear about what climate is now and how it has changed over the last 50–100 years is an essential baseline for understanding society's vulnerabilities.

We can get lots of practically useful information from observations but it's not a simple or generic task to find it. Chapter 17 described the challenges in measuring 'climate' and 'climate change' in observations but we also face challenges regarding what we decide to look at. Typically climate scientists decide what to look at, driven by what is interesting in the physical system and/or likely to have the clearest signal and therefore most likely to give a publishable result. But this is not just an academic exercise—decisions responding to climate change are being made right now and the drivers of scientists—the need to publish—shouldn't be the drivers of what climate information is available to society. Just as with models and with theory, the most useful information can only be extracted from observations if the data are examined through the lens of particular impacts or planning decisions. What is useful depends on what you are vulnerable to. The relevant variables, locations, seasons, types of variability—and the relationships between them—will vary according to your—or your organization's—role and aims. There is no one-size-fits-all; the practical application of the data is messy.

Unfortunately the prevailing culture in organizations working to provide climate information to support society encourages the provision of generic datasets rather than working together to find the most relevant data and the most suitable interpretations. The separation between science and social-science, as well as between climate science and its practical applications in society, is holding back our ability to identify the best possible information for climate change planning. For observations the issue is about choosing what lenses or filters to use: choosing where in the data to look.

In Chapter 17 the discussion was about the shape of climate distributions and how they were changing. The methods of Chapter 17, though, can be easily refocused on particular variables and particular parts of the climatic distribution: the parts that are most relevant for some decision or subject of interest.

First, a quick reminder of the issues from Chapter 17. Imagine that we are interested in a location where we have very good historic observations going back seventy years. In that case, we can construct the probability distributions for our chosen variables. These distributions reflect the probability of, say, low temperatures or intense rainfall or some combination of temperature, rainfall, and humidity, over the last seventy years. This is a good starting point but this distribution doesn't represent the probability of these variables today because we know that climate change has been taking place. We don't know what the probabilities are today. However, we can use our time series of observations to get an idea of whether—and potentially how— these distributions have been changing shape over recent times. Figures 17.3 and 17.4 provide examples of this type of information.

The small jump we need to make is to select the variables and thresholds of vulnerability through the lens of a specific decision or area of interest. We can use the same methods and concepts from Chapter 17, but focus on the parts of the distributions that are important to us. For instance, if we're interested in building design, overheating risks, or the ability of humans to work effectively, we might want to look at how very specific thresholds of temperature are changing across a region: perhaps the probability of exceeding 28°C or 33°C on summer days. As shown in Figure 17.3, we can use distributions constructed over short periods to give us an indication of how the probability of exceeding a **particular threshold** is changing locally, as well as the robustness of our assessment and how these things vary across regions and nations (Figure 21.4). We can supplement that information with distributions constructed over longer periods, say thirty or seventy years (Figure 21.5). Together these give an indication of how **the relevant aspect** of climate today might differ from the longer term, historical value: together they provide a more complete picture of what we can say about the aspect of the climate distribution we're interested in.

Information from these short period assessments also tells us about the variability and reliability of the probability assessments **for the threshold of interest**. This is potentially valuable in assessing risk and supporting practical decisions.

Of course these assessments rely on having local time series of climate variables so we are limited by the availability of historic observations. There's not much we can do if the data doesn't exist. It's worth noting however that research on climate proxies—and even citizen science projects to collate and digitize data held on paper[10]—are continually expanding our historic datasets. Nevertheless, there are only a limited number of locations at which we have detailed data going back

---

[10] For instance, the Rainfall Rescue project.[m]

Observed Changes in the Probability of High Temperatures

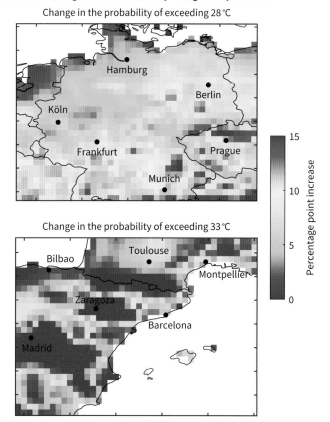

Figure 21.4 Maps of the 'at least' change in probability of exceeding 28°C (top) and 33°C (bottom) in summer in two regions of Europe. For each point on the map these are the equivalent of the green vertical line in Figure 17.3—the smallest change in probability over multiple samples. They show the change between the 1950s/1960s and the 2000s/2010s.[n]

to at least the mid-twentieth century. This highlights the importance of continuing to gather observations and to expand our observing systems. It is crucial that we continue to gather the data that will help us understand how local climate has changed—and is changing—so that we can know more in the future.

Any lack of reliable and geographically-dense observing systems will limit our future ability to understand local climate distributions, evaluate how they are changing, and assess the reliability of our models. This may seem obvious but these systems are not always being maintained, let alone expanded. In Africa, for instance, the number of meteorological observing stations has substantially decreased since the 1980s.[o] Where observing stations are sufficiently spatially dense, there are good statistical methods to fill in the gaps and provide local details, but where they are scarce such methods become unreliable. The emergence of satellite observing systems over recent

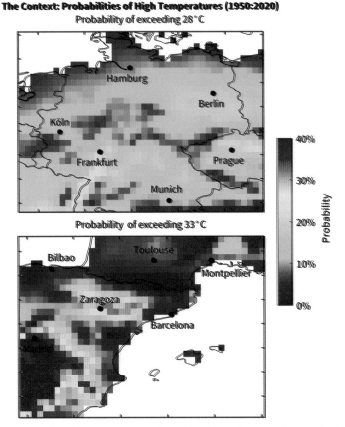

Figure 21.5 The context for Figure 21.4. The probability of exceeding 28°C (top) and 33°C (bottom) in summer in the same two regions of Europe, calculated over the whole 1950–2020 period. If the change (Figure 21.4) is large compared to the probability over the whole period (Figure 21.5), then it's clear that climate change is a big deal. In Germany, for instance, the probabilities of exceeding 28°C are typically around 15% and the 'at least' change at 28°C is typically 5–10 percentage points—that's pretty big.

decades has massively increased the amount of data available but it would be wrong to see these satellites as a panacea to the observing problem; there are many aspects of the climate system which are not easily observed from space, including precipitation and oceans—well, the oceans beneath their surface.

21.6 Digging deep and maintaining focus

We know that greenhouse gas concentrations in the atmosphere have increased substantially and will continue to change climate in a substantial way, so we know we need to prepare our societies for these changes. We don't know what these changes will be in any detailed, quantitative, and local way: we face deep uncertainty. So what do we do? The way to cut this particular Gordian knot is to focus not on what will happen to climate but on what we're trying to achieve while nevertheless making the best possible use of the information we have however qualitative or vague it might be.

In this chapter I've talked about various routes to more robust information but they all embrace two key features. First is that they aren't in any way generic or global but rather they are local and specific to how the information will be used. Second they aren't focused on finding optimal information but rather about seeking out the best we can achieve.

This avoidance of the optimal leaves the physicist in me feeling a little uncomfortable. Physical scientists are often looking for 'the answer': the correct way of describing a system or of responding to a problem—the best possible solution. If we had perfect probability predictions, we could spend ages optimizing our decisions and perfectly addressing society's problems. But we don't. So instead our aim should be to accept the messiness of real-world systems and look for solutions which achieve what we want to achieve, even if they aren't the perfect solutions.

Perhaps all this feels rather unsatisfactory to you. Climate change is a huge global issue so a few highly specific assessments of, for instance, northern European temperatures and rainfall, or of storms in the Netherlands or of Southern Indian rainfall, don't seem to grasp the scope of the issue. Yet that is the point. The consequences of climate change are in some senses always local—global consequences are the result of local consequences and we will feel the impact—of both local and global consequences—locally.

The examples herein are rather small and limited because the mirage of the best—all the investment in climate modelling—undermines the possibility of seeking alternatives. It completely skews the climate research and climate information landscapes. The point of this chapter is simply to show that alternative routes to societally relevant information do exist, should we choose to invest in them.

And can we use this less-than-perfect information? Yes. There are whole domains of expertise focused on decision-making under deep uncertainty. Some examples are: Robust Decision-Making, Dynamic Adaptive Pathways, Info-Gap Theory, Engineering Options Analysis.[P]

Robust Decision-Making is about identifying approaches that are robust to the envelopes of plausible future behaviour. Dynamic Adaptive Pathways place the emphasis on maintaining flexibility to adjust practical and policy decisions as more information becomes available about how climate is changing. Info-Gap Theory focuses on prioritizing between alternatives given what is known, in the context of what we'd ideally want to know to make a decision. Engineering Options Analysis is about maintaining flexibility in the design of infrastructure and systems.

How we approach a decision influences what information is desired or required from climate science but these approaches can all make use of the various types of information discussed in this chapter. Nevertheless, the most relevant approach to climate science, climate modelling and the extraction of information about future climate will depend on the specific decisions being considered and the way those decisions are being addressed. At the risk of sounding like a broken record—or an mp3 on repeat—multidisciplinarity is central to achieving robust decisions for society. Those who study and use decision theory need to do so in the context of plausibly-achievable information about current and future climate, and those who study future climate and its impacts need to do so in the context of the types of decisions faced by different parts of society.

There are challenges in building these robust decision methods into real-world decision systems: regulators and drivers of industrial strategy are not well aligned with responding to deep uncertainty in climate information. In principle, however, there is no reason to think that building climate-resilient societies in any way requires climate predictions with high levels of detail and precise probabilities. The question is whether we can accept a focus on less-than-perfect information when our models give the illusion that the perfect is available. Will we allow this illusion of perfection to undermine and distract us from investing in how to make good decisions in the context of climate change now? As I write[q] my fear is that we will. I fear that the illusion of the best will indeed be the enemy of the good.

Challenge 7: Getting a grip on the scale of future changes in climate

The impacts of climate change are felt locally but the context for them is global. Constructing a coherent response requires a sense of the size of the problem at the global scale. But how should we go about understanding and constraining the processes that control climate change at the global scale? What future changes are consistent with our current understanding of the physical system? Are the physical consequences of greenhouse gas emissions going to be small, medium, large, huge, vast, colossal, or catastrophic? It's essential to address these questions if we are to engage with the risks facing society and constructively debate what action is appropriate.

22

Stuff of the Genesis myth

22.1 How human beings handle knowledge

Today,[a] two decades into the twenty-first century, concerns about climate change are widely acknowledged politically, in the media, in many business sectors, and in grassroots movements. It is difficult to imagine that they'll go away. Yet I'm cautious. My own interest in climate and climate change began a few years before the Earth Summit in Rio de Janeiro in 1992. It was at that meeting that the United Nations Framework Convention on Climate Change (UNFCCC) was negotiated and signed. The convention agreed aims to stabilize greenhouse gas concentrations at 'a level that would prevent dangerous anthropogenic (human-induced) interference with the climate system'.[b] My recollection is that in the UK at least, there was substantial news coverage of the convention and a general sense that this was an issue that was going to be taken very seriously. It wasn't going away.

Yet over the intervening years there have been many swings in the public's interest in climate change. Political efforts have also fluctuated back and forth between serious concern and substantial reticence to engage, or simply dismissal of it as an issue. The negotiation of the Kyoto Protocol in 1997 seemed like a significant achievement, but it was more than seven years before it came into force. Indeed it often looked like it wouldn't make it because there appeared to be insufficient countries willing to ratify. The Copenhagen Summit in 2009 was the next big opportunity for coordinated international action, but the resulting accord was weak. Then came the Paris Agreement in 2015. This is viewed by many as much more successful, although President Trump's early commitment to withdraw seemed to limit its potential. In the end, the USA did withdraw, in 2020, but President Biden brought the United States back in, in 2021.

Throughout the political tos and fros there have been debates over the reality and seriousness of human-induced climate change. There have been questions about the reliability of the science[c]—and the scientists.[d] The lack of significant increases in global average temperature in the first decade of the twenty first century was presented by some as an indication that climate change had stopped and that we no longer needed to worry.[e] Such arguments never held water because global average temperature varies a lot naturally and it was always likely that we would see a decade or more with little or no increase—even within a bigger picture of significant global warming over multiple decades and centuries. It would have been helpful though, if scientists had made this clear to the media before it actually happened. But then it's easy to be wise with hindsight. In any case, from the mid-2010s the steady rise was clear again.

At the moment[f] we're in a phase of widespread acknowledged concern about the issue. Even in the midst of a global pandemic, it is a central element of international politics and of the news cycle. Or so it seems to me, living in the UK. My expectation is that it will remain this way going forward but then I remember that I expected the same thing in 1992. And 1997. And 2009. I am minded to be cautious.

If climate change is going to maintain political attention, we, and our leaders,[1] need to grasp the scale of the issue and of the consequences. This is not a trivial task. You've made it this far through this book, so you'll understand that climate change and climate predictions are complicated subjects with many uncertainties. In any case, politicians are driven by aims and aspirations that are usually unrelated to climate change. It is understandably difficult for them to put in the necessary time and effort to get a grip on the wide-ranging implications and inter-related issues of the topic. For that matter, this is similarly difficult for many experts working in the field.

Through the swings in public interest levels in the 1990s and 2000s, there were two Members of parliament in the United Kingdom who had responsibility for the natural environment for substantial periods of time. One was from the right of the political spectrum and the other from the left. John Gummer MP[2] held the role from 1993 to 1997, Michael Meacher MP held the role from 1997 to 2003. Both went on to be strong advocates of climate change action. In 2006 John Gummer MP said:

This is 'the biggest threat that human beings have had. This is the stuff of the Genesis myth. This is about how human beings handle knowledge'.[g]

In 2003 Michael Meacher MP said:

'The lesson is that if we continue with activities which destroy our environment and undermine the conditions for our own survival, we are the virus. Making the change needed to avoid that fate is perhaps the greatest challenge we have ever faced.'[h]

Governmental responsibility for the environment is of course a much wider remit than just climate change and in the 1990s that was substantially more so than today. Nevertheless, by the nature of their job these particular politicians had to frequently engage with the subject and discuss it repeatedly with scientist and other experts over a number of years. It appears they came out of that process having judged the scale of the issue to be extremely large.

Of course, most politicians don't get that immersive experience and four years as environment minister should not be a prerequisite for a politician to grasp the importance and implications of climate change. To ensure this is not necessary, we need to overcome the challenges in characterizing and communicating the scale of the issue. Doing so is not simply about presenting the latest predictions—it's about putting across the existence of myriad connections between impacts, and grasping the basis

---

[1] Political leaders, business leaders, community leaders, religious leaders, etc.
[2] Now Lord Deben.

for confidence in the scale of the threat. It's also very much about representing how uncertainties in our knowledge translate into risks for society.

As John Gummer said, '*This is about how human beings handle knowledge*'. Responding to climate change is not about responding to what we see around us or what we may personally have experienced in our lives, but about how we handle diverse sources of knowledge involving a variety of disciplines and disparate sources of expertise.

22.2 What's a lot?

Climate change is full of complexity but we do need to be able to talk about it as one thing. We need a way of honing down the knowledge and capturing it in a simple form. We need a metric that cuts through the complexity and gives us an idea of the scale of the issue, the scale of the uncertainties, and the scale of benefits related to any actions we take. What should that metric be?

The answer depends on your perspective. From the perspective of social science one might reasonably argue that the appropriate metric should be a measure of the consequences for human welfare: wealth, quality of life, our ability to access what we need, our ability to protect what we value. From a physical science perspective, one might reasonably argue that the appropriate metric is the heat content of the climate system because greenhouse gases trap heat so heat content is a direct measure of the consequences of emissions. In practice, neither of these is a very suitable metric. This is partly because they aren't easy to relate to each other (they keep disciplines apart rather than helping them come together) and partly because they aren't very familiar to most people (they're a bit rubbish in terms of wider communication). So instead of these, we use a compromise. We use annually and globally averaged surface temperature: the temperature averaged over the surface of the planet and averaged again over a year. I've been calling it simply 'global average temperature' although the scientific literature refers to it as global mean (surface) temperature, abbreviated to GMT or GMST.

We aren't directly impacted by global average temperature but most of us at least have a sense of what 'temperature' is. It's a familiar concept and therefore better for communication than heat or human welfare. Furthermore, global average temperature strikes a balance between the needs of the physical and social sciences. Surface temperature is a variable that affects many societal and physical systems, so it's not a huge conceptual leap to relate changes in global average temperature to changes in human welfare (much more on this in Chapter 24). At the same time it can be related to changing heat content and greenhouse gas concentrations through fundamental principles of physics. It's also easier—if not easy—to measure. And on top of these benefits, people are used to it: the goals of the 2015 Paris Agreement are framed in terms of change in global average temperature; regular updates on the level of global warming are presented as changes in it; and many of the headline predictions of the Intergovernmental Panel on Climate Change (IPCC) are presented in terms of it.

Nevertheless it is worth being aware of its flaws as a metric. We may talk about targets for limiting the change in global average temperature to, say, 2°C, but that doesn't tell us about the social consequences or even the physical impacts of that target. At the same time the link between atmospheric greenhouse gas concentrations and global average temperature is less clear than the one with global heat content: unlike the latter the former suffers from substantial natural variability, which is why we might see many years in a row without apparent global warming despite there being a clear underlying trend. Despite its flaws, though, global average temperature is really the only metric in town for climate change.

And despite its disadvantages, it's actually a pretty good tool for describing the scale of the issue. During the last interglacial period[3]—that's to say, just before the last ice age[4]—the planet was about one degree **warmer** than during the pre-industrial period.[i,5] That makes it similar to the present day in terms of global average temperature but unlike the present day, it had a long time to get used to those conditions. And what was the world like? Polar temperatures were 2–4°C higher, sea level was at least 6 m higher, Scandinavia was an island, forests stretched further north, and there were hippos in England and Germany[j]. Bones of elephants, hippos, rhinos, and hyenas have been found in Northern England and dated to this time[k]. The world was very different and many of those changes are easily related to the sustained higher temperatures. At the height of the last ice age,[6] on the other hand, global average temperature was about 6°C **below** pre-industrial levels,[l] sea level was around 120 m lower,[m] and 32% of land was covered in ice all year round as compared to about 10% today[n].

The past provides no direct equivalents to the present day, but information of this kind gives us a handle on what size of change in global average temperature should be considered a big deal. A degree or so is a lot—a huge change. It's important to say, however, that for sea level to rise or fall by several metres can take a long time, depending on how it comes about. The paleo-climate comparisons above are not meant to imply that we'll see many metres of sea level rise in the next century,[7] only that today's global average temperature should be seen as a big change because it has set us on a path in that direction. It is not plausible to think that that path won't involve substantial, much more immediate changes to those parts of the system that can change more rapidly, such as weather, ecosystems, or the likelihood of floods and heatwaves. The historic observations simply tell us that even a degree or two's change at the global scale leads to very significant changes in the state of the climate system. They give us a sense of what is big as regards change in global average temperature.

With this in mind, we can return to the problem of prediction. How will global average temperature change in the future if mankind continues to emit greenhouse

---

[3] Interglacial periods are the warm periods between ice ages.
[4] About 130,000 to 115,000 years ago.
[5] That's the eighteenth century (and earlier), but it's sometimes approximated by the nineteenth century.
[6] Around 20,000 years ago.
[7] The IPCC's current assessment projects sea level rise in the twenty-first century to be most likely in the 0.3–1 m range, although values over 1.6 m are considered possible. Under a high greenhouse gas emissions scenario the projected change by 2300 is assessed to be most likely in the 2–7 m range but a rise of over 15 m on that timescale is not possible to rule out.

gases at the current rate? How will it change if we achieve NetZero by 2050, or under any number of other scenarios? Are the changes we expect in the future big or not so big? This chapter is about answering these questions: it's about the relationship between global average temperature and greenhouse gas concentrations and emissions. Chapter 8 provided an eleven-paragraph summary of why the seriousness of climate change is not in doubt. We need to revisit the physics parts of those paragraphs here because they provide the foundation for how we can go about prediction at the global scale.

22.3 The knowns, the fundamentals, and the confirmed expectations

There is a chain of connected pieces of knowledge that link mankind's emissions of greenhouse gases with increasing global average temperature. This chain provides the basis for us to be extremely confident that human emissions are leading to global warming. It also provides the foundations for predicting future changes.

Each link or section of the chain comes with a level of confidence: (i) **known beyond reasonable doubt (known BRD)**, (ii) **fundamental expectation**, or (iii) **confirmed expectation**. Most are **known BRD**: they are well understood, they have been widely studied and they represent our best understanding of how the universe and its contents work. It is completely implausible that these are misleading in any significant way because they have been shown to be consistent with physical behaviour in many, many different types of situation. Next come aspects which are not quite so absolutely solid but are pretty close. These are **fundamental expectations** which it would be crazy to doubt: expectations like expecting a ball to come down when you throw it up in the air. For these you can construct circumstances where they would be misleading but such situations require a very particular set of conditions to be true. If we have no evidence to suggest that these conditions apply to our current situation, then we should take fundamental expectations to be reliable. Lastly, if fundamental expectations are backed up with evidence from observed behaviour in the climate system then we can upgrade a fundamental expectation to a **confirmed expectation.**

The first section of the chain revisits the explanation in Chapter 8 for why the earth's temperature is what it is.

22.4 Why is the temperature of the world about 14°C?

The chain begins with a couple of definitions: definitions for 'shortwave radiation' and 'longwave radiation'. Everything in the subject of climate change is ultimately related to how these types of radiation differ. To understand what they refer to, we need to talk about light.

Light comes in many different types: different colours of visible light—red, green, blue, yellow, etc.—but also ultraviolet and infrared, which we can't detect with our eyes. They're all part of a continuous spectrum—the spectrum of electromagnetic radiation which ranges from gamma rays to ultraviolet to visible light to infrared to radio waves (see Figure 22.1). Shortwave radiation refers to a range of that spectrum that includes visible light along with some ultraviolet and a little infrared. Longwave radiation consists entirely of infrared radiation.

With these definitions in our pocket, we can start making our chain.

> **Known BRD 1**: Greenhouse gases absorb longwave radiation but they let most shortwave radiation pass straight through.

We know this from laboratory experiments and from observations of the atmosphere. We also have good theoretical understanding of why it is the case from our understanding of atomic and molecular behaviour. There is really no doubt about this.

> **Known BRD 2:** The sun emits shortwave radiation and the earth emits longwave radiation.

All objects give off energy in some form of light, but the hotter the object the more of that energy is in the form of shortwave radiation (mainly visible & ultraviolet) as opposed to longwave radiation (infrared). Fundamental, well-tested physics tells us how the spectrum of light emitted by an object is related to its temperature. The sun is hot, so it radiates shortwave radiation. In fact, by studying the distribution of solar radiation—the amount of light of different colours in sunlight—we can work out that the surface temperature of the sun is about 5770°C. The earth is much, much

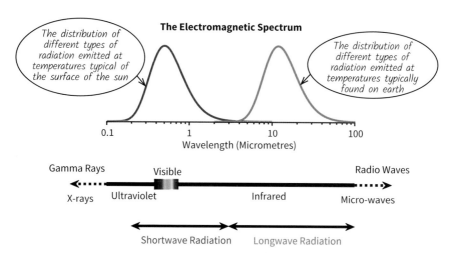

Figure 22.1 The electromagnetic spectrum and the distribution of radiation emitted by objects at temperatures representative of the surface of the sun and of the surface of earth.°

cooler: the global average surface temperature is roughly 14°C at the surface.[8] The local temperature varies of course—from below −50°C to above +50°C—but the temperature anywhere on earth is much much cooler than the surface of the sun. Our physical understanding of how objects radiate therefore tells us that we expect objects on earth to emit longwave radiation. Observations of those objects confirm that this is the case. So we know that the sun emits shortwave radiation and the earth emits longwave radiation—Figure 22.1.

Known BRD 3: Energy conservation.

The law of conservation of energy is a deep and fundamental principle of our physical understanding of the universe. It says that energy cannot be created or destroyed and it applies at scales ranging all the way from subatomic particles to frying pans to galaxies. Temperature is simply a measure of the energy contained in an object due to the jiggling of the particles—the atoms and molecules—that make it up. For an object such as a frying pan or the earth's oceans, if the rate of energy entering it is the same as that leaving it, then energy conservation tells us that there can be no change in the amount of energy it contains and consequently it will stay at the same temperature. For a frying pan on a hob, when you set it just right to make a batch of pancakes, the energy heating up the pan equals the energy it loses through infrared radiation and transfer of heat to the surrounding air or batter, so it stays at just the right temperature. Turning the hob up changes the energy transferred to the pan so the amount of energy in the pan increases and it heats up until the amount of energy entering and leaving balances again; it then stays at this new temperature. Getting this set just right is a key element of making good pancakes. The physical principle of energy conservation is familiar and applies to atoms, frying pans, and planetary atmospheres (Figure 22.2).

Known BRD 4: The amount of energy emitted by an object is related to its temperature.

Figure 22.2 Energy conservation applies to everything from frying pans to planetary atmospheres.

---

[8] Or was, before the global warming of the last couple of centuries.

An extension of known BRD 2 is that we don't just have a good understanding of how the types of light emitted by an object vary with temperature; we also have a good understanding how much energy it emits in total. This too depends on temperature and there is a well-tested formula for this relationship.[P] Since the surface of the sun is about 5770°C, we can use this formula to get a good estimate of how much energy is in the light from the sun that makes it to the earth. Together with an estimate for how much sunlight is reflected straight back out to space—the earth's albedo, which is currently about 30%—we can calculate how much energy from the sun is absorbed by the earth. This tells us how much energy is coming into the climate system—like the heat setting of the hob under the frying pan—from which we can calculate what temperature the earth would have to be to radiate the same amount of energy back out to space and therefore to not heat up or cool down. And the answer is −18°C. We should expect the average temperature of the planet to be about −18°C for it to be in a stable state.

Known BRD 5: The greenhouse effect.

Known BRDs 2, 3, and 4 are based on reliable, demonstrable, widely studied physics but they lead to the wrong answer. Observations tell us that the surface temperature of the earth is,[9] on average, about 14°C, not −18°C. Something is missing. This is where known BRD 1 comes in. The earth emits longwave radiation (known BRD 2) and greenhouse gases absorb longwave radiation (known BRD 1), so the energy emitted from the surface doesn't make it out to space. Rather it is absorbed by the greenhouse gases in the atmosphere. It is then remitted, but not out towards space: not in the same direction it was already going. Rather it is emitted in all directions. Some of it is directed back down to the surface where it is reabsorbed, some makes it out to space, and some is reabsorbed by greenhouse gases elsewhere in the atmosphere, where the process repeats itself. None of this happens to the shortwave radiation coming to earth from the sun because that radiation is not absorbed in the atmosphere.[10] The upshot is that when we think about conservation of energy in the calculation of the **surface** temperature, we need to think about incoming energy not just from the sun but also from the greenhouse gases remitting radiation back to the surface. When you include that in the calculation, the answer comes out at about 14°C, which matches observations. This is the greenhouse effect (Figure 8.1). It arises from and is deducible from the four previous known BRDs. It's completely consistent with physical understanding so it deserves to be called known BRD 5.

---

[9] Or was, before the global warming of the last couple of centuries.
[10] Although a different process does scatter some of the solar shortwave radiation, which is why the sky is blue. But that's a different story.

22.5 What do you expect to happen when you turn on a kettle?

Fundamental expectation 1: Increasing atmospheric greenhouse gases should lead to surface warming.

When you turn on a kettle you expect the water to warm up. At a similarly basic level if the concentrations of greenhouse gases in the atmosphere increase significantly we expect the greenhouse effect to strengthen and the surface to warm. We expect the extra greenhouse gases to trap more energy in the lower atmosphere, the surface and the oceans, and we therefore expect the temperature of these components to increase. It is a fundamental expectation.

Yet there are ways in which these expectations could fail. In the case of a kettle it may be broken, there may be a power cut, or the water may already be at boiling point. But these possibilities don't influence what I expect, by which I mean what I consider to be the likely outcome. I do not have peer-reviewed, well-quantified, statistically tested information about kettles but even without such data I confidently expect that by far the most likely outcome of switching on a kettle is warming water because the alternatives are clearly and substantially less likely. It's just obvious.

In the case of increasing atmospheric concentrations of greenhouse gases, there could be some complex interactions whereby the increasing greenhouse gases lead to decreasing clouds in such a way that the clouds' contribution to the greenhouse effect is reduced and hence the temperature doesn't change much at all. It could be that the earth is at some sweet spot in terms of the greenhouse effect so that changing atmospheric greenhouse gas concentrations have little effect—at least for a while. An effect of this nature has been proposed and evidence for it has been looked for but not found.[9] Even if it hadn't been looked for though, assuming that it's the case would be like assuming that the kettle you're about to switch on is broken. It's not an impossible outcome, but it'd be bizarre to assume it's a likely outcome without supporting evidence. The fundamental behaviour of greenhouse gases and radiation provides a confident expectation that you would want solid contradictory evidence before doubting. The fundamental expectation of increasing atmospheric greenhouse gases is surface warming.

22.6 Did we really do that?

Known BRD 6: Atmospheric concentrations of carbon dioxide have increased a lot in the recent past.

If increasing atmospheric greenhouse gases would lead to warming then our next question must surely be, are atmospheric greenhouse gases increasing? This one is easy to answer thanks to monitoring stations around the world. In particular, it is worth highlighting the name Charlies Keeling, and what is now referred to as the Keeling curve. Charles Keeling of the Scripps Institute for Oceanography in San Diego set up a monitoring station for carbon dioxide in Mauna Loa in Hawaii in 1958 and we have observations of atmospheric carbon dioxide concentrations from that location continually from that point to the present day. The data from Mauna Loa shows that atmospheric carbon dioxide concentrations vary between the seasons but that these variations are small compared with the steady year-on-year increase through the late twentieth century (Figure 22.3). It's not just Hawaii, though. There are monitoring stations around the world and they all show a steady year-on-year increase in atmospheric concentrations of carbon dioxide. Since 1958 concentrations have increased from about 315[r] parts per million (ppm) to over 414[s] ppm in 2021.

We can go further back than 1958 though. By examining carbon dioxide contained in air bubbles trapped in ice sheets, we can get data for atmospheric carbon dioxide concentrations going much further back. Using these methods, we know that when the industrial revolution was beginning in the middle of the eighteenth century, carbon dioxide concentrations were about 280 ppm, and that they had been around that level for the previous 2000 years. In fact since the end of the last ice age, about eleven thousand years ago, atmospheric carbon dioxide concentrations have been between about 260 ppm and 280 ppm.[t] We can be confident therefore that atmospheric concentrations of carbon dioxide have increased very substantially over the last couple of hundred years.

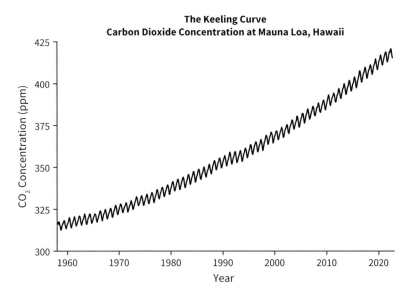

Figure 22.3 The Keeling curve. Observations of atmospheric carbon dioxide concentrations at the Mauna Loa Observatory in Hawaii.[u]

Using the same bubbles from ice sheets we can go further back too, back through previous ice ages and the interglacial periods between them. These bubbles tell us that until the industrial revolution, atmospheric concentrations had not been above 280 ppm for more than 800,000 years[v], although they had been much lower—sometimes down to about 180 ppm during ice ages. These details are interesting and important for understanding how the earth's climate behaves, but for humans the context is simply that until the last two hundred years, all human civilisations had lived in a world where atmospheric carbon dioxide levels were in the range of 260 ppm to 280 ppm. Now we're living in one where the levels are 410 ppm and most of the increase has been in the last seventy years. They are expected to go substantially higher still.

It's worth repeating this: atmospheric concentrations of carbon dioxide are now nearly 50% higher than at any point during the history of human civilisations up until the industrial revolution. They are substantially higher not just than the time of the industrial revolution but than the time when humans were hunter-gatherers, before even the development of agriculture. Mankind has never before experienced a world with such levels. If we're interested in the consequences for humans and human society, this is the context that matters.

> **Fundamental expectation 2:** In our current circumstances we should expect potentially significant warming.

Carbon dioxide is not the only greenhouse gas, of course—there are also many others. The most important ones in terms of their contribution to the total greenhouse effect (that is, the effect that takes us from minus 18°C to plus 14°C and makes the planet a pleasant place to live) are, however, water vapour and carbon dioxide. Clouds also have a substantial contribution to the total greenhouse effect:[w] their constituent water droplets and ice particles also absorb and emit longwave radiation. It's been estimated however that carbon dioxide contributes about 20% to the total greenhouse effect[x]. The exact fraction doesn't matter here, though. So long as it is significant, then if concentrations are increasing substantially we should expect noticeable and potentially important levels of warming. The contribution of carbon dioxide to the greenhouse effect is significant, and its concentrations in the atmosphere are increasing substantially, so fundamental expectation 1 (that increasing greenhouse gases leads to warming) takes us to a fundamental expectation of potentially significant warming in our current circumstances.

There are also other greenhouse gases which are less significant for the total greenhouse effect but are nevertheless important in the context of how it is changing. Some of them, notably methane and nitrous oxide, have also increased substantially in the atmosphere since the industrial revolution and they too contribute to fundamental expectation 2. When changes in these other gases are taken into account, the equivalent level of atmospheric carbon dioxide was, in 2019, over 460 ppm,[y] more than 60% above pre-industrial values. These other gases are extremely important in climate change negotiations and debates—when we're considering mitigation responsibilities and actions—but for the conceptual chain of connections regarding the seriousness of the issue, it is enough for us to consider carbon dioxide alone.

Known BRD 7: Mankind is responsible for the change in atmospheric carbon dioxide concentrations.

Carbon dioxide levels are going up, but where is it all coming from? Those of you already deeply embedded in climate change issues might say, 'duh, from us, of course'. That's actually a pretty reasonable response but its's still worth asking the question because it's an important link in the chain. There are various reasons why we can be confident that mankind's activities are the source of increasing atmospheric carbon dioxide. There are technical lines of evidence to do with the amount of oxygen in the atmosphere and the type of carbon in atmospheric carbon dioxide, but I'm going to stick with the non-technical explanation which sits in the 'duh it's obvious' camp.

The observations in Hawaii and elsewhere give us data from which we can calculate how much extra carbon dioxide is being added to the atmosphere each year. Meanwhile, inventories of fossil fuel use[11] tell us how much carbon dioxide mankind's activities are emitting. It turns out that over the last 70 years, the amount we emit each year is very approximately twice the increase that we see in the atmosphere (Table 22.1).

The point here is simply that carbon dioxide concentrations were roughly level for eleven thousand years and they started to increase just at the point when mankind started emitting large quantities. Those emissions are of quantities which are large in the context of the atmospheric concentration changes, so it would be pretty bizarre to think they were unconnected. It is fairly obvious therefore that mankind is the main source. In fact the real question is not whether the growth in atmospheric concentrations is due to mankind's activities, but where has all the rest gone? It turns out that the rest has been absorbed by the oceans and by the land surface—vegetation and soil[aa]—but in any case there is no reasonable doubt that mankind is responsible for the change in atmospheric carbon dioxide concentrations.

Table 22.1 Emissions of carbon dioxide from fossil fuels along with growth in carbon dioxide in the atmosphere.[z] All numbers are in gigatonnes of carbon per year (GtC/yr)

Values in GtC/yr	1960 to 1969	1970 to 1979	1980 to 1989	1990 to 1999	2000 to 2009	2010 to 2019
Fossil CO_2 emissions	3	4.7	5.4	6.3	7.7	9.4
Growth in atmospheric CO_2	1.8	2.8	3.4	3.2	4.1	5.1

---

[11] And other activities such as land use change.

22.7 Water magnifies

There's one more important component of the chain linking mankind's emissions and global warming. That's water vapour.

Observations show that atmospheric concentrations of carbon dioxide, methane, nitrous oxide, and a number of other extra-strong greenhouse gases such as certain halocarbons are all increasing. Discussion of climate change tends to focus on these gases, but in terms of the total greenhouse effect, it's water vapour and carbon dioxide that have by far the greatest effect—so why do we rarely hear about water vapour?

The thing is that water vapour behaves very differently to all the other greenhouse gases because it's part of the hydrological cycle. If we decided to put lots of water vapour into the atmosphere in an attempt to warm the planet, it wouldn't work because the water would simply condense, form clouds, and then rain and snow out; it would simply fall to earth. The typical period that water remains in the atmosphere is only about ten days.[bb] Similarly if we wanted to cool the planet's surface by taking water vapour out of the atmosphere, it would simply be replaced by evaporation from the surface, mainly from the oceans. Water vapour in the atmosphere is not something we can control. Well, not directly. Time for another known BRD.

> **Known BRD 8:** The amount of water vapour that air can hold without it condensing increases with temperature.

This is another of those scientifically well-studied properties of the universe; in this case it's a property of water. The amount of water vapour that a chunk (meteorologists tend to say 'parcel') of air can hold without it condensing goes up by about 7% per degree Celsius.[cc] The important relationship here[12] dates back to the mid-nineteenth century and has been widely studied and tested. This known BRD leads to another fundamental expectation.

> **Fundamental expectation 3:** If the temperature of the surface and the lower atmosphere increases, then we should expect it to contain more water vapour. This simply follows from known BRD 8.

In terms of the total greenhouse effect, water vapour and water in clouds (ice and water droplets) are thought to have the biggest effect: about 75% of the total. Carbon dioxide is nevertheless very significant with about a 20% direct contribution. The important point here though, is that the warming generated by carbon dioxide enables the atmosphere to hold more water vapour, which itself leads to more warming. Carbon dioxide therefore is responsible for much more than its direct contribution to the total greenhouse effect. Water vapour is the greenhouse gas that has the largest contribution to trapping heat in the lower atmosphere, but it responds to any warming generated by the other gases rather than being an independent agent in

---

[12] The Clausius–Clapeyron relation.

itself. The term for this is a 'climate feedback'. Understanding this, and other, climate feedbacks is tremendously important for climate prediction—more on this in a few pages time. Water is a positive feedback: it magnifies the effect of the other green-house gases but it can't drive change on its own because, unlike the others, it doesn't hang around for long enough. In terms of mankind's impact on the planet, the concentration of carbon dioxide is something we can influence directly, while we can only influence water vapour indirectly.

22.8 Does it all add up?

We now have all the pieces in place to understand the principles of twentieth and early twenty-first century climate change. Fundamental expectation 2 is that carbon dioxide increases should be leading to warming while fundamental expectation 3 is that that warming will lead to increased water vapour in the atmosphere, which itself will lead to more warming. The final links in the chain are to do a quick sanity check—ask whether these expectations of warming are consistent with what we see.

Pause a moment though to consider what the purpose of this sanity check is. It's only to see whether observations of temperature provide any indication that our basic expectations are flawed. It's not to prove the link between greenhouse gases and warming; that has been dealt with in the last few pages. Still less is it to provide an exact figure for how much warming is due to mankind's activities. We should not expect a simple relationship, with a high correlation, between greenhouse gases and surface temperature over the last fifty or a hundred years. Such a relationship is likely to be tricky to find because increasing greenhouse gases trap heat in the lower atmosphere but that heat can go to heating or changing many different bits of the system: for instance, the atmosphere, the land, the surface oceans, the deep oceans, or melting ice. The extra heat can end up in many different places, so surface temperature can only ever be an approximate measure of change: global average temperature is a flawed metric. Wherever the heat ends up though, it is changing the earth's climate and ultimately, as the system comes into balance with the new levels of greenhouse gases, the surface will warm.

Furthermore, what warming there is on the surface will not be uniform. Some places will heat up more than others, and most places will warm up more during some periods but less during others. The fundamental expectation is not that everywhere will always be warming or that everywhere will warm to a similar extent. Indeed it would be entirely consistent with our expectations of warming and climate change that some places might not have warmed much or at all in the last century. That's one reason why we focus on global average temperature. Global average temperature is a flawed metric but it's better than focusing on a specific location if we want to get an idea of the scale of the issue.

Flawed as it may be then, our sanity check involves looking at how global average temperature has changed. Figure 22.4 shows the global average temperature change on an annual basis over the last 172 years. There is a large degree of variability—from

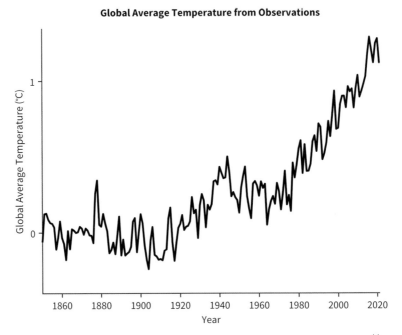

Figure 22.4 Global average temperature time series from observations.[dd]

year to year and from decade to decade—but even so, picking out the consequences of increasing atmospheric greenhouse gases isn't difficult. The temperature is going up. I mean, honestly, it's just obvious. You can draw straight lines to fit different periods and conclude different rates of rise but it's pretty obvious that the temperature is going up. So fundamental expectation 2 is actually a confirmed expectation.

> **Fundamental expectation 2 becomes confirmed expectation 2:** In our current circumstances, we should expect and we are seeing potentially significant warming.

And it's not just the change in global average temperature that supports our conclusions based on well-understood physical processes. Other observations also demonstrate consistency with theory in the same way: these include the heat content of the upper oceans, sea level rise, sea ice extent, and the profile of temperature change through the atmosphere. These observations may not be necessary to expect greenhouse gases to lead to climate change but they do provide a wide range of additional supporting evidence that backs up our expectations. And what about water vapour—is that doing what we expect? Yes.

> **Fundamental expectation 3 becomes confirmed expectation 3:** Water vapour concentrations are expected to increase and are increasing.

Observations are tricky for water vapour because it varies so much by location and season, but recent satellite observations show an increase in water content in the lower atmosphere on average across the globe.[ee] So fundamental expectation 3 has also been confirmed by observations.

So far so good. We have a chain of knowledge and observations that link increasing atmospheric greenhouse gases with heat and temperature. But what about prediction? The really important question is how much hotter it will get in the future? And how quickly? How do we use the knowledge in the chain to talk about the future?

23
Things . . . can only get hotter

23.1 Controlling factors

There are five main factors that control how global average temperature will change in the future. First is the way societies around the world evolve and the emissions they make. Second is how much of mankind's emitted carbon dioxide remains in the atmosphere. Third is how effectively heat is carried down into the depths of the oceans. Fourth are the knock-on effects of warming which can lead to further warming: the feedbacks over and above the direct heat-trapping behaviour of emitted greenhouse gases. And fifth is the unavoidable variability that arises from heat moving around the various components of the climate system.

These factors can be explored together in a very simple representation of the climate system but before we get to that, let's have a look at each of them in a bit more detail.

Any prediction of future physical climate is conditional on assumptions about how global society is going to change and adjust, and what action we take to mitigate climate change. The most significant assumption relates to mankind's future use of fossil fuels but there are others that relate to land-use change (including deforestation), emissions of other greenhouse gases such as methane and nitrous oxide, and emissions of a separate set of gases and aerosols which unlike the main greenhouse gases only stay in the atmosphere for days to years, rather than decades to centuries. Many of these so-called 'short-lived' emissions, including aerosols and nitrogen oxides, turn out to have a cooling effect and are currently partly offsetting the warming due to the more familiar, 'long-lived' greenhouse gases: carbon dioxide, methane, nitrous oxide, etc.[a]

Global society's behaviour influences what will happen in the physical climate system, so any prediction of future climate has to either predict what future society will do, or be conditional on assumptions about it. Predicting future societal behaviour, and the politics which influence it, is even more difficult than predicting the behaviour of physical climate, so here I'm going to stick to predictions based on assumptions about future societal behaviour.

How much of the emitted carbon dioxide remains in the atmosphere controls how much extra heat is trapped and therefore how much warming occurs. This is governed by the carbon cycle and how it may change as the planet warms.

The third controlling factor relates to how much energy it takes to increase the surface temperature of the planet. When you turn on a kettle, you don't immediately have boiling water: it takes time to heat up because a certain amount of energy is needed to heat the water to boiling point. The heating element only provides energy at a particular rate, so there's a delay between switching on the kettle and getting water hot enough to make a cup of tea or coffee. In the same way, it takes a certain amount of energy to increase the surface temperature of the planet by, say, one degree Celsius and the increase in atmospheric greenhouse gases is only trapping energy at a certain rate, so the surface temperature increases only gradually rather than changing suddenly. How fast it increases depends on both the rate at which energy is being trapped—like the power rating of the kettle—and how much energy is needed to warm the surface by a fixed amount. The rate at which energy is being trapped depends on the change in greenhouse gas concentrations. How much energy is needed to warm the surface of the planet by one degree depends, more than anything else, on how much energy is taken up by the oceans and transferred down into the depths. You can think of it in terms of how much of the oceans we need to warm in order to increase the temperature at the surface. Is it just the top metre, or the top 30 m or the top 100 m or the entire depth going down to more than 4 km?[1] It's like needing to know how much water there is in the kettle to say how long it will take to boil.

The fourth controlling factor relates to how sensitive the climate is: how much warming is generated by a particular increase in atmospheric greenhouse gases resulting directly from human activities. The third factor was about how fast it warms, while this is about how much it warms. For this one we need to throw away the kettle and get out the frying pan—the one we used to think about energy conservation. Turn the hob to a heat setting of 1 and let the pan heat up. It warms to a certain level and then stops warming when the energy input from the hob equals the energy lost, mainly by radiation and convection. Now turn the hob up to setting 4 and the pan will heat up some more until it stabilizes at a new temperature. But how much hotter will it be? How sensitive is the temperature to the change in heat coming in? How much does it need to warm so that it is again gaining energy at the same rate it is losing it? That's the question for climate: if we increase atmospheric greenhouse gases such as carbon dioxide by some amount, how much warming will we see before it stabilizes again? This question is addressed conceptually by the idea of climate sensitivity which was introduced in Chapter 19: how much warming do we expect to ultimately see if carbon dioxide levels in the atmosphere were doubled? But we're not interested in the conceptual now—we want the result. How big is it? Unfortunately that's difficult to answer. Increasing temperatures lead to increased loss of energy through radiation, just like in the frying pan, but this is not the only effect that controls the ultimate temperature. Another is the resulting change in water vapour: increasing carbon dioxide leads to warming, which leads to increasing water vapour, which leads to more warming. And it turns out there are a host of other mechanisms like this. They are referred to as climate feedbacks. Getting a grasp on the scale of these climate feedbacks is at the core of understanding how much global warming to expect.

The final factor is simply that there is a lot of natural variability in the climate system, as we've already seen in global climate models and in observations (Chapters 16

---

[1] The average depth of the ocean is about 3.7 km but at its deepest it is about 11 km.

and 17). An important question is therefore how much this could affect the change we actually see in global average temperature and in the consequences for our societies. It's related to the one-shot bet nature of the problem. I'll come back to this later when thinking about the economics of climate change in the next chapter. For the moment though, I'm going to ignore it in order to get a handle on how fast the surface of the planet might warm and at what level it might ultimately stabilize.

23.2   The only equation in this book (and it's hardly even an equation, more a relationship)

Up to this point, I have avoided equations and formulae so as to maintain a broad focus on the whole, messy, complex problem of climate prediction. I'm breaking with that principle here because saying anything solid about future climate involves thinking about the relationship between the various factors that affect the scale of future changes at the global level. That relationship can be encapsulated in a simple equation and for those of you happy to think in such terms, it can be handy to see it that way. The relationship, or equation, is at the heart of understanding climate change and linking the physical science of climate change to the social science of climate change. It provides a tool for considering the sources of uncertainty in climate predictions at the global scale and thus for identifying the types of research that matter most for grasping the issues. We can deduce the relationship by starting from known BRD 3: energy conservation.

Thanks to the law of energy conservation, we know that the energy content of the lower atmosphere, surface and oceans is increasing because greenhouse gases in the atmosphere are trapping more energy in this part of the climate system. We can use energy conservation to make this precise. It's a simple matter of balancing content and inputs. Consider a toy box. If we put a toy in a toy box every ten minutes, then the contents of the box—the 'toy-content'—increases. It increases at a rate of one toy per ten minutes: six toys per hour. Similarly if we add one chunk of energy to the climate system every second then the 'energy content' increases at a rate of one chunk of energy per second.

Imagine that you were sitting in a pre-industrial world, say in 1700, and all of a sudden the amount of carbon dioxide in the atmosphere doubled. Before the doubling, the energy coming into the lower atmosphere equalled that leaving, so the energy content was not changing. After the doubling, the extra carbon dioxide traps some extra amount of energy (per second) so the energy content starts increasing by that amount (per second). This is the first step—the first aspect of the relationship. It's shown as step 1 in Table 23.1.

Of course, we really want to know about temperature, not energy. Fortunately, temperature is a measure of energy content. It's not a good measure when things are freezing, melting, evaporating, or condensing, but for the vast majority of the climate system that isn't happening, so temperature is a good approximate measure of energy content. The question then is how much energy does it take to increase the surface temperature of the planet by some amount? A similar question has been asked about all sorts of materials. For instance, how much heat does it take to increase a fixed amount of water by one degree? This is one those really well-studied and widely known quantities. It's called the heat capacity of water and is approximately 4200 Joules[2] per degree per kilogram. This tells us that if we had one kilogram of water in a kettle then it would take 4200 Joules to warm it by one degree, so if the power rating of the kettle were 3 kW (which means 3000 Joules of energy per second) it would warm by about 0.7°C every second and hence take about 2 minutes to boil.[3] The heat capacity gives us a way of converting from heat to temperature: the change in the heat content of the water in the kettle is the change in temperature times the heat capacity.

If instead of the temperature of water in a kettle we consider the global average surface temperature of the planet, we can ask how much energy is needed to raise it by one degree. Put another way, we want to know the heat capacity of the lower atmosphere, land surface and surface oceans.[4] We'll call it the 'effective heat capacity' because it's not something that has a precise meaning the way the heat capacity of water does; nevertheless, it captures our conceptual ideas about how climate change fits together. Using it, we can translate between the changing heat content of the climate system and the changing global average surface temperature. With it, we can upgrade the relationship in step one into a relationship between energy trapped by greenhouse gases such as carbon dioxide, and the changing surface temperature of the planet (Table 23.1—step 2).

So far, so good, but these greenhouse gases are going to continue pumping in extra energy every second, so if the relationship is simply as we currently have it, then the surface temperature is going to rise and rise forever. In practice, it is not going to rise and rise forever because, just like a frying pan, as it warms up, it increases the rate at which it loses heat. Known BRD 4—the amount of energy emitted by an object is related to its temperature—tells us that as the surface of the planet warms, it radiates more energy, and some of this energy will make it out to space so the greater the warming, the greater the rate of heat loss to space. This works to counteract the effect of the extra greenhouse gases. After some amount of warming there is no extra heat trapped because the climate is losing heat again as fast as it's gaining it: the climate is back in balance, and the surface stops warming. This process is a feedback which keeps the planet relatively stable. Step 3 is simply to add this extra heat loss to the relationship (Table 23.1—step 3).

We expect the extra heat lost to space to increase as the temperature rises, so it's helpful express the balancing heat loss as some multiple of the number of degrees of warming. The multi-trillion dollar question (literally, as will become evident in the coming chapters) is how much extra energy does the planet lose (per second) for

---

[2] Joules is the standard unit for measuring energy.
[3] Assuming the original temperature was about 15°C.
[4] The oceans are the most important component by far.

Table 23.1 Deducing the most important equation in climate change science. Note: I say that many things are 'per second' because I'm a physicist and that's what we're like, but it could be written in terms of 'per hour' or 'per year'—it's actually just 'per unit of time'

The Only Equation In This Book

	Factor 3: Ocean heat uptake		Factors 1 and 2: Global society and the carbon cycle	Factor 4: Climate feedbacks
Step 1 (Energy conservation)	Change in energy content (per second)	=	Extra energy trapped by human-emitted greenhouse gases[a] (per second)	
Step 2 (translate heat into temperature)	Effective heat capacity x Change in global average temperature (per second)	=	Extra energy trapped by human-emitted greenhouse gases (per second)	
Step 3 (Account for extra heat lost because the surface is hotter)	Effective heat capacity x Change in global average temperature (per second)	=	Extra energy trapped by human-emitted greenhouse gases (per second)	− Extra energy lost to space (per second) because the surface is warmer
Step 4 (Represent extra heat loss in terms of total warming.)	Effective heat capacity × Change in global average temperature (per second)	=	Extra energy trapped by human-emitted greenhouse gases (per second)	− The feedback parameter × change in global average temperature from pre-industrial levels

Step 5 (Write it in a mathsy way with Greek letters and everything.)

$$C_{eff} \times \frac{d(\Delta GMT)}{dt} = F - \lambda \Delta GMT$$

where:

C_{eff} = Effective heat capacity of the lower climate system.

ΔGMT = Total change in global average surface temperature from pre-industrial levels but using the more common term: Global Mean Temperature (GMT). Pronounced 'Delta GMT'.

F = Extra energy trapped by greenhouse gases per second; this is known as radiative forcing.

λ (lambda) = The feedback parameter (the extra energy lost to space (per second) per degree of warming).

t = Time.

d means 'a change in something'

so 'dt' is a change in time and d(ΔGMT) is a change in ΔGMT,

and d(ΔGMT)/dt is a change in ΔGMT divided by a change in time (i.e. the time it takes

for ΔGMT to make the change), which means d(ΔGMT)/dt is the rate at which ΔGMT is changing.[b]

---

[a] Or the consequence of human activities such as land-use change.

[b] This is the essence of mathematical calculus and mathematicians would want to add extra details to this description but what I've presented here is enough for this book.

each degree that it warms? We don't know the answer but it has got a name: it's called the feedback parameter. The total extra energy lost (per second) is the total change in global average temperature from pre-industrial levels multiplied by the feedback parameter. Expressing it this way gives us a complete relationship in terms of the change in global average temperature, which is what we want to know (Table 23.1—step 4).

And there we have it: the most important relationship in climate change science. The final step is just a matter of making it look mathsy by using maths notation and adding some Greek letters to give it credibility and keep it short (Table 23.1—step 5). Whether you see it as a relationship or an equation, it encapsulates the key factors that influence the scale of climate change and where to look to understand the uncertainties in that scale.[5] It encapsulates and allows us to quantify—or at least helps us think about—the balances between the first four controlling factors for how global average temperature will change: society, carbon cycle, ocean heat uptake, and the consequential effects of warming: climate feedbacks (Table 23.1).

This equation (Table 23.1, step 5) describes how global average temperature changes over time so taken at face value we should be able to use it to predict global average temperature in the future. Only, of course, we can't—well, not reliably. The problem is that to use it to predict the future, we need to know F (the extra energy trapped by greenhouse gases per second), C_{eff} (the effective heat capacity of the climate system), and λ (lambda), the feedback parameter (the extra energy lost to space (per second) per degree of warming). We don't know any of these things very well. More than that, we don't expect these things to have fixed values: they're all likely to change over time. We expect their values in the future to be different to those representative of the past, so we don't expect the look-and-see approach to be able to step in and solve our problems. So this isn't an equation that tells us the future—it's a relationship that tells us what it's important to consider when thinking about the future. It's a tool to help us play around with ideas and understanding, and to explore what is plausible or implausible, likely or unlikely. For this purpose it is extremely valuable, so it's worth taking a few pages to examine what we can take from it and how the uncertainties arise.

23.3 Knowing emissions doesn't tell us how much heat is trapped

The first thing we don't know is the extra energy trapped by greenhouse gases and how this will change through the twenty-first century; that's to say, we don't know F and how it will change. We don't know this because we don't know how human society will develop and how it will respond to the issue of climate change. Will greenhouse gas emissions reduce? Will they reduce rapidly? If so, how rapidly? What are

---

[5] Well, it doesn't include natural variability but I'll deal with that in the Chapter 25.

the consequences of societal development for the emissions of short-lived gases that counteract the warming effect of long-lived greenhouse gases (like carbon dioxide) and reduce F from what it would otherwise be? What will be the consequences of human development for land management practices? There are many issues tied up in these questions but they aren't a big problem for climate predictions because we're only really interested in conditional predictions: what will happen **if** society behaves in this way as opposed to that way, what will the future look like **if** we follow some particular scenario for future development? This conditionality seems to get us of the hook. But it doesn't.

Even if we knew what human emissions of greenhouse gases would be for each year in the next century, that wouldn't tell us how much energy would be trapped in each of those years—it wouldn't tell us F. The problem is that emissions don't tell us concentrations. When we emit greenhouse gases, they don't simply stay in the atmosphere. There are chemical and biological processes by which they are converted into other gases or compounds or taken up by the oceans or the land surface. The biggest question, of course, is what happens to carbon dioxide. At the moment about half of carbon dioxide emissions remain in the atmosphere, with the rest absorbed by the oceans and the land surface.[b] But the processes of the carbon cycle which lead to this removal from the atmosphere are not fully understood and, most importantly, may well change as carbon dioxide concentrations increase and the planet warms.[c] All this means that converting emissions—and other human driven changes in the climate system—into changing concentrations of greenhouse gases, and thus estimates of extra energy trapped, is not simple. Even if we assume we know future emissions, we only partly know the consequences for trapping energy.

23.4 How much water needs heating?

Then there is the question of how much heat it takes to warm the surface of the planet: the effective heat capacity. It's not well quantified or understood how rapidly heat will get taken up by the oceans and transferred to the subsurface. Huge ocean currents carry water and heat around the world, and between the surface and the depths, but it isn't clear quite how they may change in response to global warming.

Some of these circulations have already changed substantially over the last century,[d] which gives us a basis for look-and-see approaches to speculating on how they may change in the future. We also have ocean models and substantial understanding of ocean processes which can support informed speculations. Together these provide ways for better understanding what might happen in the future but at the moment the link between flows of heat in the ocean and the consequences for changing global average temperature is not well made. As a consequence, we not only don't know what the representative value of the effective heat capacity should be for the past, we don't know how it might change in the future as circulation patterns change. This is of great importance because the higher the effective heat capacity, the slower the planet will warm. Unfortunately, even the high-end estimates based on historic observations are quite enough to imply massively concerning rates of warming.

As an aside, it's worth noting that a large contribution to sea level rise is the fact that water expands when it warms.[e6] Even sub-surface heating leads to water expanding, so although heat being carried down into the deep oceans leads to less surface warming, it has little effect on this component of sea level rise. It could however affect contributions to sea level rise from other sources including glaciers and ice sheets which are impacted by surface temperatures.

23.5 When will the warming stop?

Finally there is the question of how much extra energy is lost to space for each degree of global warming: the feedback parameter. This parameter governs when the warming will stop for any given level of greenhouse gas concentrations in the atmosphere. The equation shows this because when the feedback parameter (λ) times the change in the global average temperature (ΔGMT) equals the extra energy trapped by increasing greenhouse gases (F), the right-hand side of the equation becomes zero which means there is no extra energy entering the climate system and therefore there is no further warming.

The feedback parameter can be broken down into a number of 'feedback processes'. Getting a grip on these feedback processes and how they may change in the future is the most critical physical science research challenge because it is at the core of judging just how big a deal climate change is, what the social impacts will be and how we should view the whole issue. So I'm going to take a few pages outlining what these feedback processes are and how they are related to climate sensitivity.

The first feedback process is warming itself. A warmer surface emits more energy in the form of long wave radiation and some of this makes it out to space. This is called the temperature feedback. For a doubling of carbon dioxide it would lead to about 1.2°C warming.[f]

Next comes the water vapour feedback which I've already mentioned. A warmer atmosphere can hold more water vapour and since water vapour is a strong greenhouse gas, this leads to significantly more warming than would result from greenhouse gases alone. Since this leads to increased warming, it's described as a positive feedback.

Another feedback is the surface albedo feedback. Remember, about 30%, of incoming solar radiation is reflected straight back out to space without ever being absorbed by the earth's climate system. Much of this takes place from clouds but some occurs at the earth's surface. As the planet warms, the character of the planet's surface changes, altering the amount of incoming shortwave radiation that is reflected back out to space. The most obvious illustration of the surface albedo feedback comes from

---

[6] So long as it is above 4°C.

considering ice and snow. Generally speaking, ice and snow are relatively bright—they reflect a large fraction of shortwave solar radiation. That means much of the energy from the sun that hits them is reflected back out to space, contributing to the planetary albedo. But as the surface warms, the amount of snow and ice decreases—including sea ice on the surface of the oceans. This leads to more radiation being absorbed by the surface, which leads to more warming. In this case, the albedo is decreasing, so more energy is being absorbed, which makes it—like water vapour—a positive feedback. There are other parts of the earth's surface which absorb a very large fraction of solar radiation—forests, for instance. If such ecosystems change and, for instance, are replaced by savannah, then these parts of the planet could end up having a higher albedo than before and somewhat reducing the ultimate level of warming: their contribution would be a negative feedback. When all types of surface behaviour are considered the surface albedo is expected to be, overall, a positive feedback: it increases the level warming above what we would get due to greenhouse gases alone. However, there is a lot of uncertainty regarding just how big a positive feedback it is.

The surface albedo feedback is uncertain but even more uncertain is the cloud feedback. Clouds play a crucial role in controlling the temperature of the surface of the planet in a variety of ways. Most important for climate predictions are, first, that they reflect shortwave solar radiation, contributing to the albedo of the planet, and, second, that they absorb longwave radiation in their water droplets, contributing to the greenhouse effect. As the surface and the atmosphere warm, the amount, type, and location of clouds will change. How they change will affect the ultimate level of warming. Changes in low clouds have little impact on the strength of the greenhouse effect but a significant impact on the albedo, whereas changes in high-level clouds can strongly affect both the greenhouse effect and the albedo.[g] Which effect dominates depends on the type of cloud; are they thick high clouds or thin cirrus clouds? Furthermore if high clouds move upwards in the atmosphere but don't change their character or quantity, the shift in height itself intensifies the greenhouse effect but with little effect on the albedo, meaning that simply a change in height represents a positive feedback. There is evidence that this shift is to be expected in a warmer world.[h]

The upshot is that how clouds change—the cloud feedback—is tremendously important for predicting the change in global average temperature. It is affected, however, by many different processes: the processes within clouds themselves of course, but also circulation patterns in the atmosphere and even the behaviour of the oceans because that influences how and where clouds form above them. Overall the changes in clouds are expected to be a positive feedback, leading to more warming than would otherwise occur, but by how much, well that we don't yet know. It is arguably the most significant uncertainty in climate predictions at the global scale.

There is though, one more feedback that's important to mention but difficult to describe. It's the lapse rate feedback. As you go up through the atmosphere the air gets colder; you can feel it just by climbing a reasonably high hill or small mountain. This steady decrease in temperature is a characteristic of the lower atmosphere—the troposphere. The rate at which the temperature decreases as you rise is called the 'lapse rate'. The question is whether with global warming that rate gets bigger or smaller.

To see why this is important, imagine that the lower atmosphere is made up of many layers and in each layer a certain fraction of the longwave radiation from the surface (and elsewhere in the atmosphere) is absorbed and then re-emitted. The higher layers can emit energy to space more easily than the surface because there's less atmosphere above them and it's less dense. Now consider two climate change situations in which the surface warming is the same. In the first, all layers from the surface upwards warm by the same amount: the lapse rate stays the same. In the second, the upper layers warm more than the lower layers and the surface: the lapse rate decreases. We know (known BRD 4) that warmer things emit more energy, so in the second situation (where the higher layers warm more than the surface), the amount of energy they emit increases **more** than in the first situation (where they only warm by the same amount as the surface). In that case more energy makes it out to space **for the same level of surface warming**. That means that the surface doesn't have to warm as much to bring the climate system back into energy balance with increased levels of greenhouse gases.

This is quite a tricky concept. The result though is that if a consequence of climate change is that temperature decreases more slowly as you move up through the atmosphere, then there will be less surface warming than would otherwise be the case: a negative feedback. If, on the other hand, the temperature decreases faster than is currently the case, then there will be more warming: a positive feedback. This is the lapse rate feedback. But is it positive or negative? Unfortunately, just like with clouds, there are counteracting effects. In the tropics there are well-understood physical reasons to expect this feedback to be negative. In higher latitudes it's more likely to be positive. Overall it's expected to be negative and to lead to less warming than would otherwise be the case, but there is large uncertainty in the scale of the effect, just as there is with the cloud feedback. Having said that, the uncertainties regarding lapse rate aren't thought to be as significant as with clouds because what happens to the lapse rate is closely linked to what happens with the water vapour feedback, so when you look at them together, the uncertainties are smaller than when they are viewed separately.

These feedback processes—water vapour, clouds, albedo, lapse rate, etc.—control the feedback parameter (λ) and, therefore, how much warming we will get in response to a particular increase in atmospheric greenhouse gas concentrations. They impact how the amount of extra energy trapped in the lower atmosphere changes over time and as the surface warms. Better understanding them is crucial to judging the risks to our societies from climate change. Nevertheless, they are rather peculiar beasts, so it's worth connecting them to the easier-to-grasp concept of climate sensitivity.

23.6 Climate sensitivity and climate feedbacks

The idea behind climate sensitivity is to give us some sort of aggregated measure of how strongly changes in climate are related to changes in greenhouse gas concentrations. If climate sensitivity is large we should expect big changes in climate for even fairly small changes in greenhouse gas concentrations. If it's small then not so much.

This concept could be applied to anything from rainfall, to global economic output, to ecosystem loss, but in practice the term is applied to global average temperature. It's defined as the change in globally averaged, annually averaged surface temperature following a doubling of atmospheric carbon dioxide concentrations, measured after a long enough period for the climate system to have stopped warming. This is known as the 'equilibrium climate sensitivity' because it is measured after the climate system has returned to equilibrium. There are other definitions but when used by itself the term climate sensitivity usually refers to this equilibrium climate sensitivity.

Uncertainty in climate sensitivity is large and was discussed in Chapter 19. It arises principally from uncertainty in the feedback processes. In fact for those of you happy to play around with it, the one equation of Table 23.1 tells us that climate sensitivity equals the extra energy trapped by doubling carbon dioxide levels (F for doubling CO_2) divided by the feedback parameter (λ), with the feedback parameter being by far the most significant source of uncertainty. Here, the feedback parameter, λ, describes the feedback processes which are taking place when the climate system has returned to equilibrium so it is actually 'λ a long time after doubled CO_2 has been reached'. More about that in a moment.

The latest Intergovernmental Panel on Climate Change (IPCC) report gave their best estimate for climate sensitivity as 3°C. It may seem odd that we are talking about numbers around 3°C for a change from 280 ppm of carbon dioxide in the pre-industrial era to 560 ppm in the future, when the mere existence of the greenhouse effect at 280 ppm leads to approximately 32°C of warming. If this seems odd to you, it's because of our tendency to assume things are linear. If 280 ppm gives 32°C of warming, our instinct is that if we add another 280 ppm we will get another 32°C. But the relationship between carbon dioxide concentrations and warming isn't linear. It's nonlinear. Indeed for a wide range of concentrations it's what's called logarithmic, which means that each doubling has the same effect, so a change from a very low level has a much greater impact than the same change starting from a high level. This breaks down at very low levels but it gives a sense of why these numbers are so different.

Climate sensitivity is very uncertain. Indeed it is **deeply uncertain** which means we don't just not know what its value is, we don't know what probability we should give to different values.[7] Different studies come up with quite different probability distributions for what it could be (see section 19.2), particularly as regards the relatively low probabilities associated with high values (see Chapter 25). These uncertainties

---

[7] Deep uncertainty is also sometimes called Knightian uncertainty after the American economist Frank Knight.

arise mostly from uncertainty in how feedback processes will behave in the future. Nevertheless, many studies in climate physics, and most in climate economics, utilize climate sensitivity rather than the climate feedback parameter. This is perhaps understandable because it is more easily communicated—it is presented in terms of the familiar 'temperature' rather than the cumbersome 'energy lost to space per second per degree of warming'. However, it is also unfortunate because it obscures the source of the uncertainty and tends to lead studies to ignore the deep uncertainty and instead simply adopt some particular climate sensitivity probability distribution. In this way it works as a barrier to communication between disciplines because it is a simplification too far.

There is, though, another reason to prefer feedbacks to sensitivity when we're thinking about climate predictions. Climate sensitivity is, by definition, the response at equilibrium when the climate has reached a new steady state. It's related to feedbacks but only as regards what they will be when the climate has reached equilibrium. In practice the feedback processes will change over time. If we're interested in what happens in the next century while the level of warming is changing then we need to look at how the feedback processes could change. Indeed there are indications in models that the feedback parameter could decrease as we move through the twenty-first century,[i] suggesting that deductions of climate sensitivity using observations of the twentieth century may underestimate the ultimate level of warming—and also the warming in the twenty-first century. The possibility of such behaviour is an example of where the look-and-see approach can mislead us. In any case, if we are to create credible narratives which explore the range of plausible futures, we have to focus on the processes taking place, which means concentrating on feedbacks not sensitivity.

This may seem like an unimportant distraction in understanding climate change but in practice climate sensitivity is widely used as a tool for communicating uncertainty between disciplines—particularly between physics and economics. The lack of climate science's ability to recognize the limitations of that tool undermines our ability to distil the most relevant knowledge from diverse disciplines; it obscures the existence of a chain of understanding from, say, climatic circulations to cloud responses to global average temperature response to societal impacts.

23.7 Enough with the uncertainty, just tell me how much is it going to warm, please?

I said that the one equation of the book (in Table 23.1) can't tell us what the future will be, but it is nevertheless a tremendous tool for exploring possible futures. By making some assumptions we can use it to tell us how global average temperature might change, and thereby get a handle on the scale of the issue. It provides a tool for debate because it focuses attention on the credibility of the assumptions.

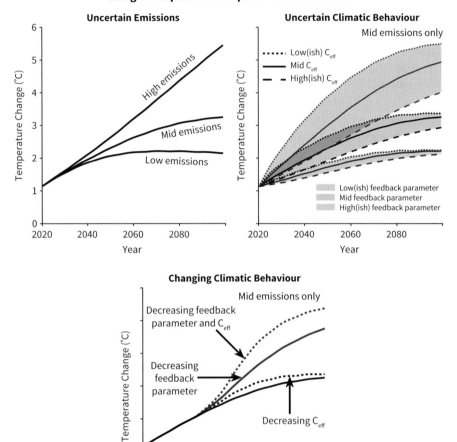

Figure 23.1 Using the one equation to explore possible futures. Change in global average temperature above 1850–1900 average. The low(ish) values of the feedback parameter are roughly equivalent to a 5°C climate sensitivity, mid to a 3°C climate sensitivity, and high(ish) to a 2°C climate sensitivity.[j]

If we have a scenario for future concentrations of greenhouse gases in the atmosphere and how much energy they trap (F) and choose a value for the effective heat capacity of the system (C_{eff}) and the feedback parameter (lambda), AND assume that they don't change as the planet warms, then the equation neatly tells us what will happen to global average temperature (Figure 23.1— top, left panel)

We can then go further and use it as a tool to investigate the impact of what we don't know. If we still assume we have a scenario for future concentrations of greenhouse gases, we can use the equation to show the impact of not knowing the heat capacity or the feedback parameter very well. We can use a range of values for them—values that we think are credible—and see how that changes the outcome (Figure 23.1—top, right). Or we could go further still and consider how the heat capacity and the feedback parameter could change as the planet warms and see how that influences future temperatures (Figure 23.1—bottom). The equation gives us a tool for exploring the consequences of physically plausible narratives of climatic behaviour without relying on the most complex computer models.

The value of the equation and the relationship it represents is not in predicting the future but in exploring the consequences of assumptions. We are nevertheless left with the question of how to come up with credible assumptions for the effective heat capacity, for the feedback parameter and for the translation of future potential greenhouse gas emissions into atmospheric concentrations and heat trapped (per second)—something which involves speculation about the carbon cycle.

For effective heat capacity we can get some estimates of how big this might be from observations of historic warming and historic levels of greenhouse gases. The uncertainty is large. For the future, how much heat is taken up by the oceans globally will depend on how the circulation patterns of the oceans change both in response to climate change and as a result of natural variations from year to year and decade to decade. This means answering questions such as: how do changes in rainfall and ice melting in the north Atlantic affect the global circulation and how do changes in the southern ocean circulations affect the transport of heat from the surface to the depths and from the tropics to the polar regions? Understanding the global effective heat capacity means understanding how oceans respond at sub-global scales.

The feedback parameter can be broken down into components representing the different physical feedback processes. This helps prioritize research but each one depends on the complexities of the way the atmosphere circulates and carries moisture and heat between different locations. The biggest uncertainty—and the focus of much recent attention—is cloud feedbacks. To reduce this uncertainty it might feel sensible to focus on understanding clouds better and representing them better in climate models, but is that really the case? If you were a research funder that certainly seems like a sensible place to invest your money and indeed such work is important. Yet clouds respond to atmospheric circulations which are driven by the whole range of atmospheric processes: heating and cooling at the earth's surface; convection and storms; evaporation and rainfall and how they transfer heat through the atmosphere; the factors that control how storms move around the planet (storm tracks); how heat is carried from the tropics to the poles; the relative changes in the temperatures of the poles versus the tropics, etc. You can't get a good representation of the impact of clouds just by studying clouds and representing cloud processes well in models. Rather it requires a good understanding of the whole atmospheric system and how the many different aspects interact.

And you can't stop there. The feedback processes aren't just affected by what happens in the atmosphere. Even those that take place in the atmosphere (clouds, water vapour, lapse rate) are substantially affected by the surface and particularly the temperatures at the surface of the oceans. Yet the surface temperature patterns of the oceans are affected by the circulation of water throughout the oceans of the world. So to get feedbacks right, including getting the behaviour of clouds right, requires that we have a realistic representation of what will happen in the oceans.[k]

In a similar vein, the greenhouse gas concentrations in the atmosphere are dependent on emissions but also on details of the global carbon cycle which involves physical processes that take place at regional scales on the land surface and in the oceans. So to some extent we also need to get all the ocean and land surface processes right, or reasonably realistic, to get the concentrations right, in order to predict changes in global average temperature under some emissions scenario.

Just like in the relatively simple nonlinear systems of Chapters 14 and 15, everything can affect everything else. It's tempting to think that if one aspect of the system—for instance, clouds—is the source of the greatest uncertainty, then that should be the focus of our attention, but the real challenges are understanding how we study and extract confident information about the whole messy, complicated, nonlinear, interacting, physical, chemical, biological system that is the earth's climate system. Even though we can construct a simple equation—a simple relationship—for the global temperature response, trying to use it rapidly takes us back into the mire of uncertainties related to regional behaviour and the complexity of the details of the system.

Nevertheless, this isn't a counsel of despair. It may be that the response to increasing greenhouse gases is little affected by some detailed aspects of the climate system. There may be sensitivities to the initial conditions and to the structure of our models (the butterfly and hawkmoth effects) but that doesn't mean our predictions are automatically sensitive to **all** of the finest details. It seems unlikely, for instance, that we would need to differentiate between the consequences of climate change for birch trees and oak trees separately in order to get a reliable prediction of future global average temperature. On the other hand soil respiration—respiration of a whole range of organisms found in soil—could be a very important component of the carbon cycle[l] that may change as the surface warms and which needs to be understood to a high level of detail if we are to make confident predictions of even global average temperature. What level of climate complexity matters for global scale predictions—and for regional and local predictions—is something we don't yet understand; the question itself has been little studied. We can't assume that processes not included in our models—or not included realistically—don't matter at the global scale without having asked the questions and come up with justifying arguments, without having thought carefully about each process in the context of the rest of the system.

This chapter started with an attempt to get a handle on the scale of the climate change issue. To do that we can focus on a number of global-scale factors such as energy trapped by greenhouse gases, energy being transferred down into the deep ocean, and feedbacks which increase or decrease the overall long-term level of warming. It is certainly useful to consider these issues on the global scale but it turns out that to understand the uncertainties—the range of plausible global responses—we have to drill down to the sub-global, the continental, the regional, sometimes perhaps even the local level.

The interconnections between all the relevant processes might lead us to think that the only way to address the problem of climate prediction is by developing more and more complex climate models. And yet the limited ability of these models to represent reality (Chapter 18)—particularly at the regional scales—and the lack of any substantial studies into what needs to be included and what can be left out means we can't trust them to provide the answers we're looking for. Not today, anyway. Maybe in the future, but not without substantial effort to understand the nature and requirements of the climate prediction task: what needs to be included and to what degree of realism.

Even only considering the global scale—the change in global average temperature—leaves us needing to integrate messy information from different sources. The challenges are about integrating information from computer models, observations, and theoretical understanding of diverse processes. The research challenges in the coming decades are not about making good predictions but about capturing as comprehensively as possible what we know across diverse disciplines and grasping the complexity of the climate prediction challenge. It's not about producing probability distributions for the future, because they will not be accurate. It's about identifying which changes are plausible and which implausible, which changes are more likely and which less likely—but not necessarily by exactly how much. It's about embracing the vague—not something scientists are very good at.

The biggest physical science challenge is integrating knowledge from different disciplines while at the same time capturing the consequences of the nonlinear interactions between them. The separation between disciplines might work fine for the haphazard pursuance of scientific knowledge but for climate science to be useful and relevant to society, the walls between them must be quickly and completely demolished.

The last chapter was about what we know well. In the light of all this uncertainty that information is still rock solid. We know to expect warming. There is an abundance of credible evidence to show it will be of an important scale for our societies and for what many people care about. The feedbacks affecting the change in global average temperature are uncertain but that is not a source of comfort. The assessments of feedbacks and effective heat capacity[8] show that the warming in the twenty-first century is very unlikely to be small and we have already seen more than one degree of warming since pre-industrial times.[m] Indeed, the message is one of risks of substantially more damaging consequences than are currently considered. The reasons for concern are robust.

The Paris Agreement has a goal of keeping global average temperature to well below 2.0°C—preferably 1.5°C—above preindustrial values.[n] If we want to achieve

---

[8] In the climate physics and oceanography worlds this is more usually discussed in terms of 'ocean heat uptake'.

these targets then our policies need to reflect not just our best estimates of what will happen but our best estimates of the uncertainties in what will happen. In 2021 the IPCC concluded that the response to doubling carbon dioxide levels—that is, climate sensitivity —was likely (66% probability) to be in the range 2.5–4°C. They said it was very likely (90% probability) to be in the range of 2–5°C but that they had less confidence in the upper bound than the lower bound. The critical message here is actually that in their assessment, there is up to 10% probability that it is outside those bounds and if so, it is much more likely to be above the range than below it. When translated into a scenario which might achieve the Paris target (Figure 23.2) it is clear that the uncertainties are large and are skewed towards worse outcomes. They don't include any significant likelihood of little change in global average temperature. But then

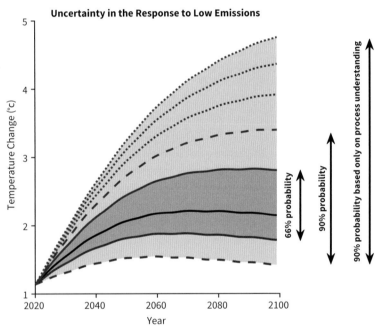

Figure 23.2 Even a low emissions scenario, designed to limit global warming to 2°C, has high uncertainty. Using the one equation, the IPCC uncertainty estimates for climate sensitivity can be translated into uncertainty in future temperatures. All lines use a low emissions scenario and mid-range estimate for effective heat capacity. Solid lines use feedback parameters equivalent to climate sensitivities of 3°C (black), 4°C (red), and 2.5°C (blue)—the IPCC 'likely' (i.e. at least 66%) range. Dashed lines have climate sensitivities of 2°C (blue) and 5°C (red)—the IPCC 'very likely' (i.e. at least 90% probability) range. Dotted lines use climate sensitivities of 6°C, 7°C, and 8°C—these illustrate some of the remaining upto 10% probability of the main IPCC assessment but are within the 90% probability of the IPCC assessment based only on process understanding.°

the observed changes are already significant in the context of the last few hundred thousand years so that, of course, is what we should expect.

Of course we aren't actually interested in the change in global average temperature; that's just a metric to help us get a handle on the scale of the issue. We are actually interested in the impacts on the things we care about: our societies, the world's ecosystems, our welfare, other people's welfare. That's where we need to go next if we are going to evaluate how we should utilize our uncertain scientific knowledge to make real-world decisions which support our aims and balance climate change with other social issues. The connection between physical climate and the social system is usually mediated by global average temperature change, so understanding where the uncertainties in its predictions come from is important for us all. As Lord Deben said, 'this is about how human beings handle knowledge'. I agree but would add that in particular it is about how we handle knowledge and uncertainties in that knowledge. It's about our ability to respond to risk—not just risk at a personal or business level but at national and global scales.

Challenge 8: How can we use the information we have, or could have, to design a future that is better than it would otherwise be?

How should we evaluate the consequences of climate change for our societies and for us as individuals? How can we assess the economic risks of climate change in the context of deep uncertainties in the physical and societal impacts? How can we cope with the limited relevance of look-and-see approaches for understanding the social consequences of future climate change? What conclusions can be drawn from how-it-functions approaches and things 'known beyond reasonable doubt' in climate economics? How do different people value the environment and the future in the context of the scale of anticipated changes in climate? How can we empower diverse communities to contribute to debates over the relative costs of climate change policy and climate change outcomes?

24
Making it personal

24.1 What's a big deal?

Ultimately climate change is a social issue and addressing it requires political and social engagement. It's important, therefore, to consider it in terms of human welfare and the plethora of things that humans value. Indeed, for us to even debate how climate change fits into the diverse collection of demands on our societies we need to be able to paint pictures of what it means for individuals. This involves not just pictures of changes in physical climate but descriptions of what the wider consequences of climate change may be for our communities and for the facilities and opportunities we access. These pictures are essential for enabling individuals to make informed personal judgements about how society should act: judgements constructed in the context of what they individually care about and their personal priorities.

I wish this chapter were about how we paint such pictures and make such connections, but the state of climate change research is still a very long way from being able to these links with the fallout for individuals. We can, however, take some steps in that direction using the tools of economics. Doing so reveals what we need to know to assess the scale of the issue from a slightly more human perspective than global average temperature. It requires us to connect physical climate science with social climate science, and to grasp the many nettles of physical and social science uncertainties. The starting point is to look at the economic consequences of climate change at a global level. This might seem distant from individuals' concerns but it turns out to be helpful in illuminating why we might come to different conclusions regarding the scale of the issue.

Examining the physics-economics link, even at the global level, requires us, as ever, to think carefully about what we know. If we can't differentiate between the robust and the debatable in physical climate predictions then any discussions we have about social consequences and climate policy are likely to be fraught with confusion because each of us will have made our assessments based on different foundations. In such a situation disagreements over what actions we should take will arise simply out of confusion over what we know from physical science. If you believe that climate sensitivity can't be over 5°C and I believe there's a 10% likelihood that it could, then I will assess the risks to society to be greater than you do, and therefore that stronger action is more justified. This is an example of why being clear about the foundations of climate science knowledge really matters: it is necessary to support coherent discussions about climate policy.

The message of the last paragraph has strange and wide-ranging consequences. It implies, for instance, that differing beliefs about the ability of computer models to represent the complexities of physical reality—a conceptual, even philosophical, issue—could fundamentally affect our judgement about how mankind should respond to the threats from climate change. That's a deeply worrying thought because it means that academics' ability or inability to reflect on—and then communicate—the limitations of computational technology could profoundly affect the future of human society. This concern was one of my motivations for writing this book.

The context here is that computer simulations have opened up a completely new set of methods in scientific and economic research. Our current very limited consideration of when those methods are reliable and when they can mislead, could itself fundamentally affect how we design our futures because it skews the information fed in to the policy debate. This is why the conceptual foundations of climate predictions matter and have dominated this book, while the challenges in the social sciences are only now rearing their head in Chapter 24. The economic consequences are of critical importance in understanding and responding to climate change but they have to be built on the foundations of physical science. Get the physical science wrong—or simply fail to question the limits of physical science methods—and the economic assessments may well be misleading and potentially substantially underestimate the scale of the risks we face.

This is important as context but I'm going to set it aside for now and focus on the uncertainties in the social sciences. Even if there were no difficulties associated with physical science knowledge—even if we could make perfect predictions of physical climate which told us exactly what the future climate distributions are going to be[1]—there would still be uncertainty in what the consequences would be for our societies due to the complexities, sensitivities, and nonlinearities in the global socioeconomic system. I'm going to call these consequences the 'damages' which result from physical climate change because all credible assessments show that in aggregate these consequences are negative—strongly negative. But let's set these uncertainties aside too. Consider a situation where for a given climate policy we know well what the changes in physical climate would be and we also know well what the damages—the consequences for our societies—would be. Even in this situation we might well disagree about the value we are willing to put on those social consequences and hence about the best climate policies, that's to say how much effort is worth putting into reducing emissions. You and I almost certainly have different cares and different priorities, not just with regard to the impacts of climate change on us personally but also in relation to how we value human welfare, human societies, and the natural environment. These differences will lead us to different conclusions about how much action and investment is worthwhile to avoid a certain proportion of those future damages.[2]

---

[1] Conditioned, of course, on assumptions about future greenhouse gas emissions.
[2] We can't avoid all future damages because climate change is already well under way.

Having said that, it may be that the scale of the damages is so large that we don't differ very much—this might be the case if we're talking about the potential collapse of global society. Or it might be that we differ a lot because the situation is more subtle and nuanced than that. An important challenge for the social science and economics side of climate change research is to help us make the foundations for our conclusions clear so that we can tell whether we disagree simply because we have different values or because one or each of us is missing an important piece of understanding—a piece of understanding that might be a 'basic expectation' or even a 'known beyond reasonable doubt'.

We need to better understand and to be more transparent about how physical changes lead to damages to society and also about how we—individuals, communities, nations—value those things we are at risk of losing. Both these elements—damages and the way we value the damages—have a fair degree of subjective uncertainty; experts disagree about the scale of the damages and we all value our societies in different ways. Both also suffer from the limited ability of look-and-see approaches to constrain the answers because of the extrapolatory nature of climate prediction:[3] we can't just look at how we've valued other things in the past.

Addressing these issues at local or national scales, which is what many of us are most interested in, can only be done in the context of global changes. Just as future changes in physical climate in Australia or Argentina can't be assessed without a good knowledge of the potential change in global average temperature, so the damages to the Australian or Argentinian economy and social structures can't be assessed without out a sense of the scale of the impact on the global economy. That's why we need to consider the global scale. We need to translate predictions of global average temperature into predictions of global social impact. We want to know what change in global average temperature is a big deal for humans and human societies. We also want to know the implications of our physical science uncertainties for our conclusions about society. In short, we want to make climate change real and focused on what matters to us. To the extent we care about global society, this chapter is about making it personal.

The starting point is to reflect on what we value.

24.2 The things that matter

Any debate about climate change is intrinsically linked to a consideration of what we care about, what we value. If you place little value on the state of the world in the future then action to improve it will have little worth to you. At the other extreme if you value the future as much as, or perhaps more than today then actions today to make the future better will have very significant worth. An important element of climate change research and a key component of climate economics, is therefore

---

[3] This is an even greater problem in the social sciences than in the physical sciences because they tend to be even more strongly founded on empirical, look-and-see approaches.

the study of how we value the future in the light of changes predicted by science. What though are we trying to value? Loss of ecosystems and biodiversity? Changing water and food availability? Personal vulnerability to floods, droughts, heatwaves, wildfires, coastal inundation, and other extreme events? Remote peoples' vulnerability to the same events? Local conflict? Remote conflict? Job security? Health consequences such as vector-borne diseases like malaria, or the changing health impact of heatwaves and cold spells? The ability or inability to hop on a plane and visit friends or relatives in faraway locations? The pleasures of exploring the cultures and geographies of the world, or maybe just travelling somewhere to lie on a beach in the sun or go skiing? Personal income? The value of assets, property, and pensions? National wealth and the state of the global or your national economy? Having national or local resources to spend on health provision, education, infrastructure?

Climate change will affect most of these things. Action to combat climate change will also affect many of these things. We should not, however, assume that the impacts of climate change and our actions to limit those impacts are automatically in conflict. Action to combat climate change is not necessarily an onerous cost that we have to bear in order to protect things that we value. In some—perhaps many—situations, the actions might be desirable regardless of their value in mitigating climate change. For instance, from a national or international perspective a shift from fossil fuel industries to renewable energy industries could be one of replacing one type of job with another. The new jobs could potentially be better, safer, and more plentiful than those they replace. They may also bring benefits in terms of energy security and therefore perhaps global stability. Overall the state of society and employment opportunities could improve as a consequence of combating climate change, over and above the climate benefits of doing so.

The crucial word here though is 'overall'. The societal challenge, in this example, is to ensure that those communities that might lose out from a shrinking fossil fuel industry benefit from the opportunities of an expanding renewable energy industry. More broadly the societal challenge is to ensure that those who might lose out from society deciding to combat climate change, also see the benefits of doing so. In the jargon of climate policy, the challenge is to ensure a 'just transition'.

It's not just about how things will turn out in the long term though—it's also about how it feels in the short term. If it doesn't feel good or if the benefits aren't clear then there will be resistance. Those in a comfortable position now but faced with personal, business, or national disruption imposed by government targets on climate change, might not anticipate a win-win situation. Indeed it may not be a win-win situation if our governments don't manage well any transition to a low-carbon society. If it's managed badly, it could be a story of winners and losers instead of the equally possible story of winners and mostly winners.

In any case change can be disconcerting. Even if the transition is managed well, the changes may still be unwelcome for many, at least in anticipation if not in practice. Furthermore, some changes will be seen as positive by some but negative by others.

The point is simply that we shouldn't expect everyone to be happy about the actions proposed to combat climate change.

If we have leaders with vision and ability, the response to the threats of climate change could be an opportunity to redesign and improve many aspects of our societies. There will, though, inevitably be costs along the way so getting widespread agreement for the proposed responses is unlikely to be a trivial matter. At the moment,[a] there is substantial popular global support for action to combat climate change but that support is founded on the goal of avoiding a problem. When it comes to specific activities and investments, the support may not always be as enthusiastic. In the United Kingdom, for instance, despite significant awareness of climate change and significant support for the idea of policies to tackle it, attempts to build wind and solar farms often face opposition.

This is the context for climate policy: we may agree that we want to avoid the problem but when it comes to taking action that impacts ourselves we are often less enthusiastic. It's important therefore to discuss climate policies within a wider picture of the consequences of action versus the consequences of inaction, or the consequences of greater action versus lesser action. It's not just about building a windfarm or imposing a carbon tax or regulating for a shift to electric cars or putting investment into public transport—it's about putting that change in the context of the benefits it leads to.

It's also worth remembering that governmental action on climate change will usually involve costs and this inevitably puts it in competition with other demands on government budgets. For climate change policy to take its appropriate seat at the decision-making table, therefore, the consequences of tackling climate change need to be put in the context of the consequences of not doing so. We need to be clear what we get from investing in climate policies, particularly if that takes money away from other societal priorities such as education, health, development, and global equality.

To put climate change in context we need to be able to paint pictures of both the consequences of climate change and the consequences of action to combat climate change. I've already talked about assessing the damages but in order to choose climate policies that reflect our wishes these need to be contrasted with the consequences of mitigating action. A national climate policy might lead to substantial changes in society, perhaps more than some would like, but it can only be judged in comparison with the scale of changes anticipated from climate change. Even large changes to society from climate policy might be small by comparison with the avoided damages. There is a balance to be struck between the value to us of the damages from climate change and the value to us of the costs of action. The question is whether we're getting the balance anywhere near right.

Answering this question is usually left to economists but picking apart the components of this balance should be at the forefront of the public debate on climate change. It involves us all and it is about what **we** value. A major role for economists and other academics should be to provide the tools necessary to help us think about the problem and structure our debates, rather than simply producing academic papers that give their version of the answer.

24.3 Getting the level of action right

Addressing climate change is not about achieving the optimal solution. Rather it's about getting an idea of the scale of action that is justified given the scale of the anticipated damages. The essence of the issue is one of balance.

If we shut down all use of fossil fuels overnight and stop using fertilizers to grow food crops and stop using cement (which would be easy because we would have removed most means of transporting it) then we would have done a great job of tackling climate change but starvation would be rife and the whole global economy would shut down along with our ability to provide social care for the vulnerable, education, national and international sports competitions, and many other things we value.

On the other hand if we do nothing, we might reasonably expect global average temperature to reach somewhere in the range of 2.8 to 5.7°C[b] above pre-industrial levels by the last couple of decades of the twenty-first century, and substantially more thereafter. This would likely lead to huge damages to our societies; species and ecosystem loss; widespread environmental degradation; very substantial increases in risks from floods and heatwaves; and arguably starvation would again be rife. And of course this would also impact our ability to provide other things that we value in society: health services, education, sports competitions, and so on.

What about in between? If for instance by 2100 global society achieves 100% renewable energy generation and addresses the emissions from agriculture either directly or indirectly,[4] would that be sufficient? Is this scenario of 'global NetZero' by 2100 enough? After all there would still be climate change and consequential impacts. What would be the damages due to the remaining climate change? Would they be acceptable or do they imply that we should act more quickly and more strongly? What about NetZero by 2050 for the whole world—that's a tall order, but would it be sufficient? What would the remaining damages be and what would the implications be for society of taking such rapid action?

What is roughly the right level of aspiration? What is roughly the right level of action to tackle climate change given the inevitable residual damages from climate change associated with any climate policy?

An optimal policy would be one where the total combined costs of climate change and of the actions to tackle it are as small as possible: we want to stop making our climate policy more stringent at the point where spending a bit more on climate mitigation reduces the damages by less than the value of that extra bit of spending. This is the idea behind many economic assessments of climate change.[5]

I'm using the terms 'costs' and 'damages' but in practice economists make allowance for how much those costs mean to us—what value they have to us—it's not just about adding up the dollars, pounds, or euros spent. I'll come back to this later in the chapter.

---

[4] By carbon capture and storage for instance.
[5] It's often expressed in terms of what's called 'the social cost of carbon'.

Finding the optimal balance requires not only accurately predicting the changes in physical climate resulting from a particular emissions scenario, but also accurately evaluating the damages to society as a result of those changes. It also requires an ability to accurately evaluate the costs of achieving that scenario. The diverse challenges of this book so far all suggest that we aren't in a position to do this—not on the physical climate side but also, and for similar reasons, not on the social science side either. The optimal solution is not something we can know and assessments framed in such terms are likely to misrepresent the consequences of deep uncertainties and partially-known information. The question is not therefore whether we are on track for achieving the optimal solution but whether we are even roughly achieving a suitable level of aspiration. This is a task that can much more easily embrace uncertainties, deep uncertainties and unknowns but it is far less academically satisfying and doesn't fit well with the structures that incentivize researchers.

It is, also, a matter of perspective whether looking for the right level of action—either optimally or just approximately—is the right framing for the issue at all. Finding roughly the right level might be described as a 'normative' view. It approaches the problem from the perspective of trying to find the most desirable outcome. It may, however, be that such an outcome is simply unachievable given the state of human society in the early twenty-first century. An alternative view is to ask what changes our societies might credibly achieve. The question from this perspective is not what **should** be achieved but what **can** be achieved. In the jargon of social science this is more akin to a 'positive' take on the problem. For both types of study, however, it is important to have a grasp on the scale of, and the uncertainties in, the costs of actions and damages.

Consider a 2°C global warming target. Without an idea of what that 2°C means for what we care about we can't say how much effort we should expend in achieving it. However, even if we were able to make the judgement that 2°C of warming were indeed a sensible target in itself—an acceptable level of climate change—the uncertainties in the physical system mean that taking action to make it the most likely outcome could nevertheless result in substantially greater warming. Remember, there are large uncertainties in the outcome of any given scenario of future greenhouse gas emissions. This uncertainty should itself influence our choice of target. If the risks associated with higher outcomes are high (say 10% of more than 3°C, even though 2°C is the central estimate), then we might consider a 2°C target to be too high even if we think that a 2°C warmer world would in itself be acceptable. This is the type of complexity that needs to go into making a judgement of how much action is justified. Climate change is a question of how we, as a global society, respond to risk, and that has to be factored in to the normative perspective.

On the other hand, the positive perspective is about evaluating what is achievable given the current state of national and international politics. In this case it might be that achieving the maximum we consider possible to limit climate change is still hopelessly inadequate to maintain a stable global society. Again understanding the scale of the implications is essential. In this case the implication would be that we need to consider more radical options—options that might have initially been considered infeasible and therefore off the table. We would need to face up to the fact that there is a gap between the maximum we think is possible and the minimum we think is necessary.

Imagine I'm driving a car and have to swerve to avoid an obstacle and I find myself heading towards a cliff edge. In that situation I would slam on the brakes. But if somehow I knew that slamming on the brakes wouldn't be enough to stop me from going over the cliff, then I might try to jump out of the moving vehicle despite the risk of severe injury. Slamming on the brakes represents what we know we can do; jumping out of the car is the more radical option. For climate change, having a good handle on the scale and scope of the damages is central to judging whether our current targets and actions are potentially sufficient or whether something more drastic needs to be considered.

An additional factor in such an analysis is how rapidly the undesirable consequences might escalate if we miss our targets. In the car it's about how confident I am about the brakes whereas for climate it's about how sensitive the combined social/physical climate system is. It might be that our targets are in some sense sufficient—maybe 2°C warming is acceptable and the risk of 3°C is also acceptable—but if it becomes clear that we are failing to take enough action then we want to know how bad things could get. We want to know not just what are roughly the best actions, but also what the implications of failing to meet our aspirations would be. How much leeway do we have? These are all urgent questions to address because climate prediction is not just an academic problem.

Understanding the costs associated with anticipated damages is therefore of critical importance for setting our aspirations appropriately. However, since it is such a difficult and urgent problem, we are unlikely to have a robust answer so it's also important to understand why people, including experts in the various relevant fields, might disagree. Expert A and expert B might disagree on the scale of climate change policy aspirations but what is important is not that they disagree but **why** they disagree. Is it because they place different values on the future? Is it because they disagree on the likely scale of impacts on society for any given level of warming? Is it because they have different beliefs about the scale of physical climate feedbacks, the risk of a highly sensitive climate or a low effective heat capacity of the system? Is it because they have different views on the prospects for technological innovation to solve the problems at a reasonable cost in the future? Some of the differences might be due to them having different awareness of the state of research while others might reflect differences in judgement, priorities, or values. Identifying these differences is crucial for achieving constructive dialogue on the issue and for enabling us to form our own personal coherent view of what climate policies we'd like to see.

Good climate policies cannot be deduced simply from physical climate predictions but neither can they be achieved without a good understanding of the foundations and limitations of those predictions. Climate economics is fundamental to understanding the societal challenges of climate change but the conceptual challenges in nonlinear systems, philosophy, and computer modelling are themselves central to developing robust approaches in climate economics. It is essentially, deeply and completely, a multidisciplinary problem.

To see how much physical science uncertainties matter for the economics of climate change, we first need to look at how climate economics addresses the costs of damages resulting from climate change. There are two really big areas of uncertainty in this process. First, how much impact climate change will have on what we value—the damages—and second, how we compare the value of things at different points in time. We need to delve into these details.

24.4 What's the damage?

What are the damages to society associated with a particular level of climate change? This question is huge, complex, and difficult to get a handle on. Describing physical climate change involves changing local distributions of temperature, rainfall, humidity, etc. at all places around the globe, as well as how these variables and distributions relate to each other. In the same way, the aggregate effects of physical climate change for human society involves the varying consequences around the globe for buildings, infrastructure, health, livelihoods, agriculture, ecosystems, investments, trade, and so on, as well as how they too are all linked. Furthermore, just as observations are of limited use in tying down how physical climate will respond to changing levels of greenhouse gases, so they are of limited use in tying down how social systems will respond to changes in physical climate: the extrapolatory nature of the climate change problem undermines the reliability of many current research methods.

To cut through the complexity, economic assessments of climate change often take the globally aggregated damages as being simply related to changes in global average temperature. That's to say, the size of the global economy in the future (measured, for instance, by the output of all human economic activity) is assumed to be reduced from what it would otherwise be, by an amount that depends on the change in global average temperature. The relationship between the reduction in global economic activity and global average temperature change is known as the damage function. What the damage function looks like is central to assessing the economic consequences of climate change and is one of the key uncertainties in evaluating the issue's importance.

Figure 24.1 shows two damage functions found in economic studies. The big thing to note about them is simply that the two curves are very very different. One is saying that with 10°C of global warming, over 95% of the global economy will be wiped out, while the other is saying it will be less than 19%. This difference illustrates just how little we know about the consequences of climate change on human society: the assumptions made about economic impacts can be hugely different.

There are a growing number of studies trying to understand these damages better using computer models of impacts[d] and historic observations of social systems.[e] The computer impact models need to account for the same concerns regarding sensitivity, nonlinearity, and limited scope, as the physical climate models. The observational

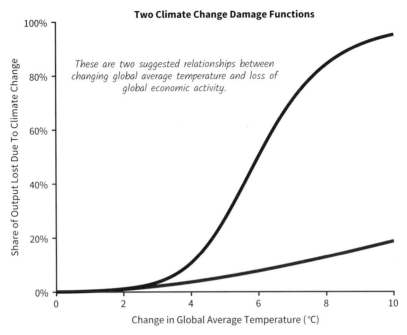

Figure 24.1 The Nordhaus (lower) and Weitzman (upper) damage functions.[c]

ones are look-and-see approaches and face all the same limitations as the constraints on physical climate derived from physical observations. These sources of
information can help us think about the issues but they can't provide the answer:
there is every reason to expect that the information we are looking for is simply
not in the data sources we have. We don't have observations of how global economic behaviour responds to changing global average temperature of 10°C, or 8°C,
or even of 2°C or 1.5°C, and small historic changes are unlikely to be a good guide
to much larger changes in the future because the relationship is likely to be highly
nonlinear. This is not to dismiss such studies as irrelevant—far from it—but it does
imply that they can't tell us everything we want to know. We need to be more
imaginative.

Of course this need to go beyond current approaches is not new or even limited to climate change. Back in Chapter 7 I mentioned the comments of the British
Academy's 'Global Financial Crisis Forum': they put down the failure to predict the
2008 financial crisis to a 'failure of the collective imagination of many bright people
[. . .] to understand the risks to the system as a whole'. In assessing the economic
impacts of climate change, we urgently need to stimulate the imagination of many
bright people to understand the risks to the system as a whole. It's a matter of stepping back from the urge to predict the economy under climate change and instead
reflect more carefully on the processes and connections by which climate change
could impact the economy. It would be useful to much better understand what **could**

happen before we begin to approach predictions of what **will** happen. Computer models and historic observations are valuable inputs to this task but they are only starting points for the imagination—not end results ready to be fed into policy.

While thinking about damages and the impacts on society, it's also worth commenting on the gulf between physical scientists and economists in this respect. In my experience most physical climate scientists give little thought to the economic consequences of climate change. It's nevertheless interesting to see how they respond to the curves of Figure 24.1. On several occasions I've shown groups of physical scientists the damage function which indicates an 18% decrease in the size of the global economy as a consequence of 10°C increase in global average temperature. Their initial reaction tends to be nervous laughter and an assumption that I'm joking. On realizing that this is an assumption which is widely used in the economics literature, they respond with incredulity. The idea that a 10°C rise in global average temperature would not completely decimate the global economy is considered by many, including me, to be beyond credibility.[6] Perhaps this physical-scientist's perspective is simply wrong but there is certainly a need for more interactions, even arguments, between economists and physical scientists to reflect on what are credible and what are incredible assumptions about damages.

I should acknowledge at this point that William Nordhaus, who came up with the lower curve in Figure 24.1, didn't consider it to be reliable for high levels of change. He knew about the lack of evidence. Nevertheless it has been widely used.

Perhaps, though, you might think that the differences between the damage functions in Figure 24.1 don't actually matter. After all both curves show similar, rather small, levels of damage at 2°C warming and 2°C is the Paris Agreement target, so does it matter at all what we assume for societal impacts at higher levels of warming? Unfortunately the answer is a big yes. It matters because of the risk of overshooting the target. This could happen because nations around the world fail to meet their targets. It could also happen as a result of uncertainty in the behaviour of the physical system. We don't know the value of climate sensitivity or the heat capacity of the climate system—we don't even know the probability distributions for their uncertainty. Furthermore we don't know how climate feedbacks and the take-up of heat by the oceans could change as the planet warms. All these things mean there is the possibility of substantial levels of warming even if there were to be significant action to tackle greenhouse gas emissions. The consequence of these uncertainties is that assessing the economics of climate change has to be done in terms of risk.

Our primary interest is not in the consequences of 2°C of warming but in the consequences of reducing greenhouse gas emissions such that the probability distribution of future global average temperature is shifted towards lower values, perhaps with 2°C being the central estimate for the outcome. As with physical climate predictions this is all about probability distributions and our limited ability to specify

---

[6] Of course there is a time element here. A large change over thousands of years might have little impact but here we are considering changes over the next one, two, or three centuries.

them. Since the probability distributions for physical change can often include a risk of large changes in global average temperature, the shape of the damage function for these high changes can very much affect our conclusions about what the best actions are: if 10°C reduces the size of the global economy by 95% then we may be willing to accept a far smaller probability of its occurrence than if it 'only' shrinks it by 19%.

But of course it's not **just** the damages at high values that matter. Another notable point about the two damage functions in Figure 24.1 is that they are very similar for low levels of warming. The question then is whether they are both right or both wrong. Remember, 2°C of warming is still very large in terms of historic variability of the physical climate system. Is it reasonable to expect that its impact on the global economy is a reduction of only about 1%—as both damage functions imply? Is this too low? Is it way too low? Could it be too high? Understanding the uncertainties and their implications should be a high priority.

To grasp the economic reality of big picture assessments, we need to be able to picture what they mean. This is the other side of the imagination coin. On the one hand imagination and speculation are critical to exploring what could happen and therefore informing us about the possible shape of the damage function. On the other hand any assessments at the global scale needs to be accompanied by descriptions of what they imply for our societies: they need to be connected to people's lives because climate change isn't just an academic subject. How do the economic damages associated with a particular global average temperature change come about? What do they mean for individuals and individual nations? To answer these questions requires approaches akin to the tales of Chapter 21; the extrapolation characteristic limits what we can conclude directly from the many economic studies examining the past and present behaviour of societies, markets, and people. Connecting the global scale assessments to pictures of what they mean for our societies is one of the little-addressed but core challenges in climate economics; it is about making the consequences of climate change real without falling in to the trap of simply looking at assessments of the present-day—or the past—or the outputs of computer models.

One thing, though, is for sure: the assumptions we make about the shape of these damage functions profoundly affect our conclusions regarding the value of actions to tackle climate change.

24.5 Valuing the future

The damage functions give us the reduction in global economic output. Whether global economic output is a good starting point for measuring what we value is something for other books to discuss but to some degree the size of the global economy is of value to us in the same way that personal income is of value to us. When we're out of a job and have no income, we are likely to be less happy, while to some degree the more income we have the happier we tend to be. Up to a point. If the world as a whole has more income then many individuals are also likely to have more income, so to some degree global economic activity is part of what we value.[7]

If physical science can give us a time series for future global average temperature and economic assessments can give us a time series for the future size of the global economy without climate change, then these can be combined, using the damage function, to provide a timeseries for the economic losses arising from climate change in each future year.

Assessing climate policies is about balancing the costs of action to tackle climate change against the future losses that are avoided by taking such action. This is how we assess whether our climate policies are getting the roughly right level of aspiration. However, the costs of action are mostly incurred now and in the near future, while the losses they avoid would usually be experienced years or decades later. So this raises the question of how we compare damages in the future with costs today.

Humans tend to value 'the future' less than we value 'today'. This simply reflects our preferences. If I were to offer you something that you wanted, maybe a guitar or piano and I said you could have a top-of-the-range model today or a top-of-the-range model next year, most people would prefer to have one today. That preference implies a greater value on commodities today than in the future. There is a corollary to this. While we prefer to get things we like sooner, we prefer to put off things we don't like, such as expenditure, till later. If you had to pay $1000 for the guitar/piano but you could choose to pay for it this year or next year, most people would choose next year.

A central question in the economics of climate change is how much less we value the future than today. If giving you $1000 today can give you a certain amount of happiness, perhaps by providing the ability to buy that guitar, how much less happiness would you get from knowing you were going to get $1000 in a year's time? We are interested in the value of these two options to you—something that might be discussed in economics as their 'utility'. Here I am associating it with money in order to talk about it in numbers, but I don't mean to imply that everything can be measured in monetary terms. The concept of how we value the future is real whether we express it in monetary terms or some other measure of what we value.

I would prefer to receive $1000 now to $1000 next year, but $1100 in a year's time might for me be equivalent to $1000 today: given the choice of a $1000 today or $1100 in a years' time, I might find it difficult to choose. On the other hand $1500 in a year's time might beat hands down $1000 today: the future is valued less but given enough incentive we would still choose a delayed reward.

With climate change, of course, it is not nice things but nasty things that we are considering; negatives not positives. The greatest damages are expected in the future but what are they worth in today's terms? Knowing this would tell us how much we should be willing to spend today to avoid those damages in the future.

---

[7] Yes—there are many many complicating factors including population growth and the consequences of inequality, but here I am still trying to get a handle on the scale of the issue, so I'm going to set these factors to one side.

We express how much less we value the future in terms of a 'discount rate'. The discount rate is like the interest rate you get on a savings account. Say you can get 10% interest, then $1000 this year becomes $1100 next year and $1210 the year after. In terms of valuing the future, if we use a discount rate of 10% then $1210 of damages in two years' time would be equivalent to $1000 of damages today. It would say we should be willing to pay up to $1000 today to avoid $1210 of damages in two years' time. That's the principle, but what is the right discount rate to use? This is one of the biggest questions in climate change economics. Indeed it is one of the biggest questions in the subject of climate change.

I don't have an answer, of course. One of the difficulties is that our perspective on valuing the environment is likely to be different to valuing a guitar. Buying a guitar is a small change to an individual's level of happiness, whereas our environment is the background and context for our societies. A guitar is something one might save up for and purchase, but our environment is a huge resource which was handed to us on a plate when we were born. We probably don't buy a guitar in anticipation of handing it down to our children, but a stable environment has been something that has been handed down from generation to generation, so damaging it is a bigger deal. In any case, a musical instrument is a small, marginal change to our happiness, whereas environmental degradation on a global scale, and potential instability of our societies, is a wholesale change to our state of well-being.

The principles of valuing the present differently to the future are still important when considering the environment or the stability of future society, but how much is a difficult question to answer. It's not something that can be easily assessed by looking at how we value other commodities. Again the look-and-see approach is likely to let us down because we haven't experienced anything similar to the changes we foresee.

We can, however, break down the question of the discount rate a little further. Why do we value the present more than the future? One answer is that we just do. We live in the present and get benefits from the present whereas quite what will happen in the future is intrinsically uncertain and unreliable. The component of the discount rate related to this intrinsic preference for having good things sooner and delaying bad things, has a name. It's called the 'pure rate of time preference'.

There is another component. The value of $1000, whether you receive it or are required to pay it, depends on how rich you are. For someone who owns no substantial assets and earns, say $10,000 per year, it has a very high value. For one of the world's billionaires, it has almost no value at all; it might not be worth the hassle of picking it up and putting it in their wallet. The point is that the value of costs and damages depends on your current level of wealth and it is generally assumed that the global economy will grow in the future so that on average we will become richer. This is another separate reason to consider costs or damages in the future to be worth less in today's money: costs in the future are worth less than the same costs today because we expect to be richer in the future.

In terms of climate change, this second component can be interpreted as saying that, up to a certain point, the costs of climate change will be more easily borne by future governments and generations because they will be richer. The discount rate represents a way of defining what we mean by 'up to a certain point'.

The first component of the discount rate, the pure rate of time preference, is sometimes taken to be 1.5% per year, although many argue that when applied to climate change, a value of almost zero—say 0.1% per year—would be much more appropriate. The second component, relating to the expectation that we will be richer in the future, is not something that is just assumed in itself but rather arises out of assumptions regarding the expected growth rate of the global economy, the expected increase in global population and how the value we place on gains or losses decreases with increasing income; this last component is called the utility function.

Of course while the value of money may decrease with increasing income the same is unlikely to be the case for the value we put on our environment, our future environment, and the future stability of society. One might argue that both a billionaire and a person on average income might both value these things to a similar degree even though they may value $1000 quite differently. They might both be willing to give up, say, 20% of their income to achieve some particular future benefits to the environment and to societal stability. In that case, we would not want to value many aspects of the damages from climate change in the future less, even if we will be richer then. Of course for someone on minimum wage or on the poverty line, the value they can put on the future environment in terms of the resources they have today may be minimal. This suggests we need to handle this effect in a more complex way than is currently the case—a way tailored to the climate change problem rather than adopting methods that are already commonplace in economics. It requires imagination.

Differing assumptions lead to different conclusions about the appropriate discount rate to use in economic evaluations and hence different assessments regarding how worthwhile actions to limit climate change are. Yet assumptions about the discount rate are a matter of judgement, opinion, and moral perspective. They are not things which can necessarily be deduced from observing people and social systems, thanks again to the extrapolatory nature of the problem. We shouldn't therefore expect academic experts to resolve these issues. Rather these issues are subjects which should be the subject of mainstream debate. To be able to formulate our own personal coherent perspectives on climate change, we need information to help us judge how we want to value the future in relation to today. What would we be willing to give up, or perhaps what tax rises would we be willing to pay, to avoid some degree of future instability in our societies or loss of ecosystems?

The damage function and the discount rate are to some extent a matter of judgement but the risks associated with physical uncertainties will influence your conclusions whatever your ethical perspective regarding future generations. There has been relatively little study of the interactions between uncertainties in climate physics and climate economics but there is enough to show that they do matter. They matter a lot. The next chapter demonstrates just how much this is the case.

25
Where physics and economics meet

25.1 The costs of what we don't know

A number of computer models, known as integrated assessment models (IAMs), have been built to evaluate how much action is justified in tackling climate change. Often they look for the optimal balance[1] but they can also be used to give an idea of the benefits of one scenario of greenhouse gas emissions over another. To illustrate the economic implications of physical science uncertainties I'd like to give you some results from such a study but first let me explain a bit more about what's in these models.

These models represent various components of the global economy including economic growth, population growth, the damages due to climate change, the costs associated with tackling climate change, the discount rate associated with how we value the future intrinsically (the pure rate of time preference), and a description of how the value we associate with costs and damages changes as we get wealthier or poorer. These are the key economic aspects of any assessment. But of course to calculate future damages these models also need to know how global average temperature will change in the future. So they also have components related to physical climate.

They simulate greenhouse gas emissions based on the size of the global economy and various assumptions about how economic activity is related to emissions. These emissions are then used to simulate changes in global average temperature in a two-stage process which aims to capture our understanding of the physical climate system. First there is a representation of how emissions translate into changing atmospheric carbon dioxide concentrations; this is a simple representation of the global carbon cycle.[2] Second there is a representation of how changing atmospheric greenhouse gas concentrations lead to changing global average temperature. This second stage uses the one equation of this book from Chapter 23.[3]

This setup gives us a route to investigating the implications of uncertainties in the physical system for economic conclusions. We can take the parameters of the one equation and vary them within one of these IAMs.[4] One study to do just that was

---

[1] Typically expressed in terms the social cost of carbon.

[2] It gives us F in the one equation.

[3] Actually some IAMs—including the one whose results are presented in the coming pages—use a slightly more complicated version of that equation which accounts for the relatively fast mixing of heat within the upper oceans but the relatively slow transfer of heat to the deep oceans. In terms of understanding what's going on and the implications of physical science uncertainties, we don't need to get into such details here.

[4] We ought to also look at uncertainty in the simple equations representing the carbon cycle but neither the science nor the economics is yet sufficiently developed to explore that aspect very well. It is one of

carried out by Prof Raphael Calel of Georgetown University in Washington DC, and colleagues.[5] They took one of these IAMs and studied two scenarios. In the first, emissions continue without efforts to reduce them for the next 250 years. This is called the business-as-usual scenario. In the other it is assumed that policies are put in place to reduce emissions in such a way that concentrations of carbon dioxide in the atmosphere never exceed 500 parts per million. That's a pretty strong and difficult-to-achieve target; remember that as of 2019 we are well over the equivalent of 460 ppm.[a] The 500 ppm level would lead ultimately to about 2.3°C of warming above pre-industrial levels based on what is today considered the 'best estimate' for climate sensitivity.[b]

The question that Prof Calel and colleagues addressed was how much better is a world limited to 500 ppm than one left to its own devices? The idea was to first look at a world where there's no effort to limit climate change and to ask what the total value of the future global economy would be for us today, allowing for how we value the future increasingly less as it gets increasingly distant. Doing this required an assessment of the damages from climate change associated with that scenario. Next was to do the same thing for a world where atmospheric greenhouse gas concentrations were limited to no more than 500 ppm. In this case the damages were less because the warming was less, but there were additional costs associated with adapting society to achieve this scenario. Finally the two scenarios were compared by looking at how much larger or smaller the total value of the future global economy would be if we act to tackle climate change in this way. This was expressed as a representative change in the size of the future global economy, for all years going forward.[c]

The point of this exercise was simply to give a sense of the scale of the issue. If a 500 ppm scenario leads to a future global economy which is 1% larger than the business-as-usual scenario then this indicates that even after allowing for the costs of reducing emissions, the future is improved in a way that is equivalent to increasing the size of the global economy by 1% forever. Setting aside issues of equality, think of it as us being 1% wealthier than we would otherwise be; or perhaps avoiding being 1% poorer. The aim is to evaluate this change. Is it 1%? Or minus 1%? Or 20%? Or what?

The answer, of course, depends on a number of assumptions. On the economics side the really important assumptions are about the damage function, the pure rate of time preference, and how value is related to the size of the economy. The results coming up use a damage function more similar to the higher one of the two in Figure 24.1, a pure rate of time preference of 1.5%/year and assumptions about economic growth and how we value our incomes that in the absence of climate change would add more than 2.5%/year to this, giving a total discount rate of more than 4%. These are not unusual numbers to use for economic assessments of climate change, but 4%

the many areas which needs much further work: a gap in the link between climate physics and climate economics.

[5] Just so you don't think I'm hiding anything, the colleagues were Prof Simon Dietz and me.

is quite high and higher discount rates reflect the idea that we value the future less, so these numbers are highly contested.

To see why they are contested and why they matter, take a look at Figure 25.1. It shows how each year in the future is valued by us today based on different assumptions for the discount rate. Many would say that in the context of climate change and environmental degradation, we actually value the future much more highly than implied by a 4% or higher overall discount rate. I would agree but the issues I want to describe here are not about the economic assumptions but about the consequences of physical science uncertainties for economic assessments. These economic details are just letting you know the context for this assessment.

The starting point for representing physical science uncertainty in the Calel study was to do many many assessments of the type outlined in the last few paragraphs, but with each one using a different value of the feedback parameter, λ. Actually, economists talk about it in terms of climate sensitivity, as indeed do many climate scientists, so that's what was used here. This is not ideal because, as mentioned in Chapter 23, the climate sensitivity and the feedback parameter are only related when the climate comes to equilibrium at a new temperature, not while it's getting there. Let's ignore this limitation though and assume we can use uncertainty in climate sensitivity as a good representation of uncertainty in the feedback parameter.

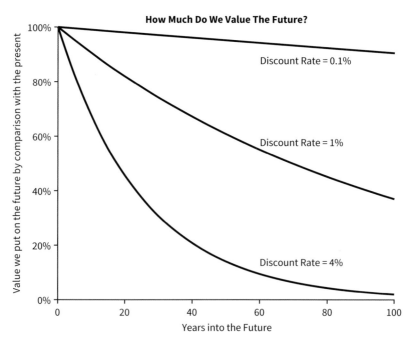

Figure 25.1 How we value the future according to different discount rates. The figure shows the value we give things in the future as a per cent of the value we place on them today.

In this study, many assessments were done with different values of climate sensitivity sampled from a probability distribution somewhat like that in Figure 19.2. When I say 'sampled from' I mean that if the probability distribution had a 67% probability of climate sensitivity being between 2°C and 4.5°C, then 67% of the climate sensitivity values chosen would have been in that range. If it gave 15% probability of climate sensitivity being between 4.5°C and 6°C, then 15% of the values chosen would have been in that range.

This gave many assessments for each of the two scenarios, but in each assessment a different value of climate sensitivity was assumed and the values of climate sensitivity selected reflected the probability distribution. For each scenario (500 ppm and business-as-usual) the representative sizes of the future economy, from all the different assessments, were averaged. This gave a representative value for the size of the future global economy under each scenario, expressed in today's money and allowing for uncertainty in climate sensitivity. Finally they calculated the difference between the two scenarios. The result was an evaluation of how much better a 500 ppm world would be than a business-as-usual world, allowing for some uncertainty in physical understanding.

And the answer was . . . just under half a per cent. The 500 parts per million scenario is just under half a per cent better than a business-as-usual scenario. That doesn't sound much but even a half a per cent loss of the global economy, now and forever, is not negligible and in any case it is saying that this can be avoided by taking action to limit atmospheric carbon dioxide concentrations to 500 parts per million. It is saying that action to tackle climate change is worthwhile and better than not doing anything.

But that's not the end of the story. Far from it. The study so far had included uncertainty in the physical science but not deep uncertainty. It had assumed we knew the probability distribution for climate sensitivity. For that particular distribution the answer was half a per cent but when they took another distribution they got three-quarters of a per cent, while a third distribution gave an answer of 77%. Whooa, hold on a minute, 77%?! That's quite a different message.

This is saying that the deep uncertainty in our understanding of the physical system—our lack of knowledge over which is the right probability distribution for climate sensitivity—can change our economic assessment of the benefits of reducing emissions from a half a per cent increase in the size of the global economy—already pretty significant—to a 77% increase. The former says do all we can to limit climate change because it is worthwhile but don't necessarily jump out of the car if things are going worse than expected, while the second says absolutely jump out of the car, do everything possible and a few things that we didn't think were possible as well. It's a completely different economic message and it arises simply from different assumptions about uncertainties in the feedbacks of the physical climate system. This really demonstrates just how deeply this is a multi-disciplinary problem.

An aspect of particular interest in these results is that the three distributions for climate sensitivity were actually all very similar. They all gave 67% probability that climate sensitivity was between 2°C and 4.5°C and even the shape of the probability distribution below 4.5°C was exactly the same. It was only the probabilities above

4.5°C that were different. For these higher climate sensitivities all the distributions agreed, for instance, that the probability of being above 8°C was very small, but it was nonetheless a lot smaller in the first two than in the third; the probabilities of very high sensitivities were very small in all three but nevertheless larger in the third distribution.[6]

This is interesting because it tells us what from the physical side is important for us to understand if we are to grasp the scale of the economic threats from climate change. In terms of climate sensitivity, it is the tail of the probability distribution. The best estimate for climate sensitivity is rather unimportant but what is tremendously important is to know how much probability there is of it being above 6°C or above 7°C or 8°C. Unfortunately, this doesn't tend to be the focus of climate science assessments of climate sensitivity. This is one of many areas where climate science is failing to focus on what is important for climate economics and climate policy and is thereby failing society.

It's worth pausing for a moment to think about what the 77% impact means. What this is saying is not that acting on climate change releases a massive increase in growth and prosperity. What it is saying is that acting on climate change protects the growth and prosperity that we have already gained and aspire to build on, while failing to act has a risk of leading to a collapse of global society and the global economy. It is all about risk, again. This assessment used a damage function more similar to the higher one in Figure 24.1 than the lower one, so when combined with high values for climate sensitivity and a business-as-usual scenario which gave high greenhouse gas concentrations, the damages became very big indeed and society collapsed. Hence the value of a world with a target of 500 ppm was much larger than a business-as-usual world because the chance of this collapse was much less.

There's an interesting footnote to this story. This assessment, in line with most studies using these sorts of IAMs, used a rather high value of the effective heat capacity. The consequence of this was that the planet was assumed to warm more slowly than observations would suggest is most likely. As a result the highest changes in global average temperature, and therefore the worst damages, were felt later than would otherwise be the case. This reduced the value placed on them because the further into the future we go, the less we value the damages. When the effective heat capacity was reduced to a value more consistent with observations of the past, the evaluations based on the different climate sensitivity distributions changed. They all increased. A lot. The benefits of achieving 500 ppm were even more starkly clear. The point is that the discount rate interacts with the uncertainty in the effective heat capacity and the feedback parameter to substantially affect our economic conclusions.

For me the main message of this research is that physical science uncertainties—and deep uncertainties—really matter in assessing the impacts of climate change on our societies and on our economies. Until the physical sciences and the economic sciences work together much more closely, we will be missing some very substantial risks and misrepresenting the integrated aspects of the issue. Grasping the scale of the climate change issue is fundamentally a multidisciplinary problem. Assessments

---

[6] In the jargon of mathematics the third one has what is called a fat tail.

are sensitive to our assumptions and uncertainties about climate sensitivity/climate feedbacks, the effective heat capacity, the damage function, the pure rate of time preference, and also our utility function—which further influences how we value the future. None of these are well known. For many of them it is not even clear whether the way we think of them is appropriate to the climate change issue. They are, for instance, usually treated as constant over time which in at least some cases (climate feedbacks) we know is incorrect. Furthermore the damages are related simply to global average temperature whereas a link to the rate of change rather than just absolute change is also known to be important. There is much to do. It is important to be aware of the limitations of current approaches but we do have many useful tools in our tool boxes—tools with which we can get an idea of the scale of the issues and how we might tackle the questions better.

Before leaving the economics of climate change, though, there is one more aspect to consider. So far the uncertainties I've focused on have all been about our understanding of the system,[7] but back in Chapters 14, 15, and 16, when thinking about definitions of climate, it was all about acknowledging the intrinsic, unavoidable uncertainty in what will happen.[8] Chapter 23 also mentioned the importance of intrinsic variability, in that case in relation to global average temperature specifically. Does this uncertainty also affect economic assessments? It turns out the answer is yes; it further magnifies the risks associated with climate change.

25.2 The costs of what we can't know

Chapter 16 showed how the climate of the future is a distribution of possibilities, but the one equation of Chapter 23 only represents the average of that distribution. Heat flows around the system—sometimes more is held in the sub-surface oceans and sometimes less—so even without climate change we expect year-to-year and decade-to-decade variability in the global average temperature. As a result the time series is pretty wiggly. This is seen in observations of the past and also in the results from global climate models but the one simple equation of Chapter 23 doesn't capture it (Figure 25.2). That means that when we try to evaluate the economic consequences of the future, we aren't looking at realistic time series but rather highly idealized ones that only represent the central estimate of future changes without the inherent, natural variability.

A little thought suggests that year-to-year and decade-to-decade variability is likely to affect our economic assessments because the damages due to climate change are related to global average temperature. The question is how much. To answer this we need an equivalent of the one equation of Chapter 23 that generates more realistic time series for global average temperature. It turns out that this is not too difficult to achieve but it does mean that I have to shame-facedly admit that I fibbed about there

---

[7] This is called epistemic uncertainty.
[8] This is called aleatory uncertainty.

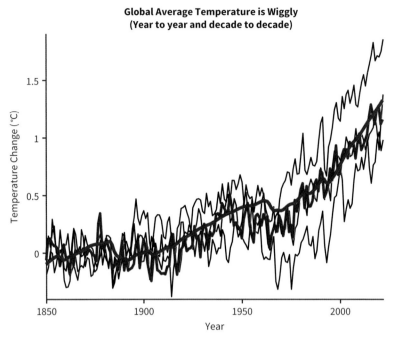

Figure 25.2 Even global average temperature has a lot of variability—shown here in observations[d] (blue) and results from three global climate models.[e] The one equation of this book doesn't capture this variability (red).

being only one equation in this book. There are actually two—although they're very similar and closely related. To convert the first equation into one that generates more realistic-looking time series with natural variability, we simply need to add a random term onto the end (Table 25.1). It's called a 'stochastic term'.

This new equation allows us to step forward in time just as the original one does. The only difference is that at each timestep the change in global average temperature has a random number added to it. The random number can be positive or negative. Quite how it is chosen opens up a whole new series of interesting challenges which have yet to be addressed, but to see whether variability matters it is sufficient to simply select it from the archetypal probability distribution: a bell curve (see Figure 13.2). There is nevertheless a subtle but important detail to be aware of here. Although the random number added at each timestep is unrelated to any previous ones—it's like a new roll of a dice—the variations it generates in the time series of global average temperature <u>are</u> related to the previous ones.[9] As a consequence there are often multiple years, even multiple decades, sometimes even periods of more than 50 years, where the global average temperature is above or below what we expect on average

---

[9] To put this another way we are not adding random numbers to the smooth curve, we are adding random numbers to the previous timestep's change in temperature. In doing this the randomness interacts with the physics creating a connection to the variations from previous timesteps.

Table 25.1 This is the second, and honestly the last equation of this book. It is due to a Noble prize winner, Klaus Hasslemann,[f] and is known as the Hasselmann model

The Only Second Equation In This Book			
Ocean heat uptake	Global society and the carbon cycle	Climate feedbacks	Random variability
Effective heat capacity (C_{eff}) × Change in global average temperature (per second)	= Extra energy trapped by human-emitted greenhouse gases (per second)	− The feedback parameter × change in global average temperature from pre-industrial levels	+ A random inflow or outflow of heat from the surface—mostly into or out of the oceans

$$C_{eff} \times \frac{d(\Delta GMT)}{dt} = F - \lambda \times \Delta GMT + \sigma$$
where:
σ = A random addition or extraction of heat from the surface and lower atmosphere.
All other terms are as in the only equation of this book—Table 23.1

(the result of the original equation) (Figure 25.3). This aspect of the variability is seen in historic observations—it is realistic.

The future is a one-shot-bet but with this update to the equation we can produce thousands or millions of possible future time series, all based on the same values for the feedback parameter, and the effective heat capacity, and the future concentrations of greenhouse gases. Each one represents a different outcome of natural variability, a different possible outcome of the one-shot-bet.

How does this affect the damages resulting from climate change? We can answer that by examining these thousands of time series. First convert each one to a time series of economic damages using a damage function. Next apply a discount factor to each year to account for us valuing the future less than today. Now add up all the values in each time series to give a single number for the value of all future damages in each time series. The result is a probability distribution for future damages resulting from climate change expressed in terms of today's money. This analysis has been done[h] and the results, for three very different scenarios of future greenhouse gas emissions, are shown in Figure 25.4.

The message I'd like you to take from Figure 25.4 is that the distributions are wide. Unavoidable, intrinsic uncertainty has a big impact on the economic damages resulting from climate change. With high future emissions—a scenario which assumes we do little to respond to climate change—the total value of all future damages without variability is $486 trillion but when we allow for variability we get a distribution which gives 90% probability of it being in the range of $421Tr to $563Tr. To put this in context, the size of the global economy in 2020 was roughly $80Tr dollars so this is saying that the consequences of intrinsic, unavoidable uncertainty in the physical climate system means that the value today of future damages from climate change could be higher than expected by roughly the size of the whole global economy today.

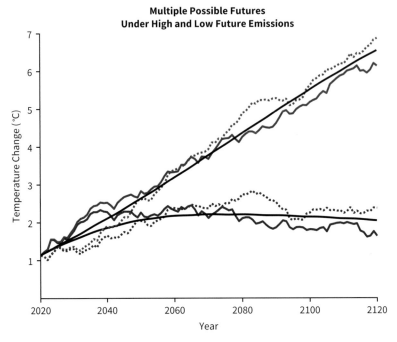

Figure 25.3 Under a high (red) and a low (blue) emissions scenario and assuming we know the feedback parameter and the effective heat capacity, the future physical climate is still uncertain. The solid black lines come from the one equation. The coloured lines are examples of the wide range of possibilities from the updated equation of Table 25.1—the Hasselmann model.[g]

Or the damages could be smaller than expected by roughly four-fifths of the whole global economy today.

In a scenario where we do a lot to combat climate change—a scenario in which it's as likely as not that global average temperature is kept below 2°C above pre-industrial levels—the value of all future damages in today's money is 'only' $30 trillion dollars. The uncertainty due to physical climate variability in this case is, with 90% probability, $21Tr to $45Tr. The benefit of reducing emissions—in terms of reduced climate change damages—is clear but the remaining damages are still substantial and the uncertainty as a fraction of what is 'expected' is even more than in the high scenario: the 90% probability range goes from 30% below the central value to 50% above.

The conclusion is that intrinsic uncertainty matters and should be built into how we respond to the issue. It means that responding to climate change is partly a matter of responding to a known threat but partly a matter of responding to risk. Furthermore, the risk comes partly from limited understanding of how the climate system works—for instance, climate feedbacks, the rate of ocean heat uptake and how the carbon cycle might change—but partly from unavoidable variability in the climate system. This latter part is never going away, however much our scientific understanding improves.

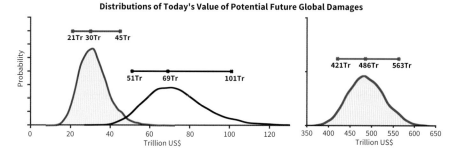

Figure 25.4 Distributions of potential future damages expressed in terms of trillions of dollars today. Horizontal lines give the average (median) and the values for which 5% of the distribution is lower and 5% of the distribution is higher - the 5–95% range.[i]

Valuing risk is, however, something economists know quite a lot about and we can ask how much the variability in damages is worth: how much should we be willing to pay to avoid just the uncertainty in the outcome?[10] The study on which this section is based[j] calculated this number; it came up with $9Tr for the low emissions scenario and $46Tr for the high one. But of course we can't simply pay to avoid this intrinsic uncertainty so what does this mean? One interpretation is that it gives us a scale for how much we should be willing to spend to make our societies more resilient to climate within a context of climate change. The thing is that while reducing emissions reduces the expected damages—in this case by 94% from $486 to $30Tr—the costs associated with variability within a warming world decrease less rapidly—in this case by 80% from $46Tr to $9Tr. The message is that reducing emissions is not enough to tackle variability: climate variability will be an important risk factor within a warming world, for any credible level of action we take to limit climate change. The only way for us to address it is by adapting our societies to such variability—well, adapting where we can and insuring against the consequences where we can't. These numbers give us an idea of the scale to which this is worthwhile. It's saying that it is worthwhile spending trillions of dollars making our systems more resilient, on top of efforts to reduce emissions.

It will probably not have escaped your notice, though, that this assessment has to be based on all sorts of assumptions. It is. The numbers are based on a particular damage function (the higher one in Figure 24.1), a particular pure rate of time preference (1.5%/year) and assumptions about growth and how we value income that increase the discount rate further (from 1.5% to 4.25%/year). It also uses particular values for climate sensitivity, the probability distribution for climate sensitivity, the effective heat capacity, and the size and character of the random fluctuations. Changing any of these assumptions can change the numbers, sometimes by a lot.

---

[10] Not that that's something we can do but if we could how much would it be worth? Or how much should we be willing to pay to insure against it?

In this study the message of concern remains when the assumptions change, but the numbers' sensitivity to the assumptions highlights again the importance of understanding the connections between climate physics and climate economics. It would be useful for societal planning if we could refine what numbers are credible in this kind of study and what aren't. In some sense this requires an effort to tie down what might be considered 'fundamental expectations' in climate economics: 'fundamental expectations' not regarding the final conclusion but regarding what underlying assumptions are credible.

This is not, however, just about understanding the economics better—it is also about using its approaches to improve the quality of climate change debates. How can you or I use these concepts to help grasp how action to tackle climate change fits within our personal lives and values? What policies would achieve the right balance for us? What are the right numbers to represent our perspective? This requires a shift in climate economics and climate science towards their use as a tool for social engagement and away from trying to give answers.

The importance of variability also raises a series of interesting challenges in the boundary lands between maths, physics, and economics. The addition of a random element to the equation (Table 25.1) provides a connection to an existing field of study on what are known as stochastic processes. There are many fascinating questions regarding what the most realistic size and form of the randomness should be, and how we can use observations of the past to tell us about it. Those questions, though, are for another time or another book.

25.3 What's the target?

It is tempting, so tempting, to respond to the issues of this chapter, and indeed this book, by trying to quantify everything. For many people, particularly the type of researcher or academic who has spent their career processing and analysing data, there is a huge desire to find 'the answer'. The response to the issues raised is to somehow measure or model or otherwise identify suitable values for all the uncertain quantities, and to use them to find the best possible solution. If perhaps that is acknowledged as impossible then it becomes a matter of finding the 'right' probability distributions for those values; of accurately representing our uncertainty. In either case, such an approach encourages a normative view, an approach that says 'we understand what's going on and this is the right thing to do'. Many people studying climate change, particularly in the physical and economic sciences, have been trained with problems of this nature so this approach comes naturally to them.

Unfortunately, the characteristics of climate change undermine our ability to achieve this in both the physical sciences and the economic and social sciences. Deep uncertainties abound. We don't have a sound basis for giving reliable probabilities for many of the uncertain quantities. We don't even have a sound basis for capturing the broad and diverse range of human knowledge on the subject: people with different but still highly relevant expertise often come up with very different answers. And yet at the same time, it is certainly not a free for all. Anything, absolutely does not go.

The target of integrated climate research, at least in terms of how it interacts with society, should arguably be the provision of tools for debate and discussion. It's about painting a picture or telling a story about how the variety of different physical and economic systems interact and what sort of outcomes are possible. What could future society look like if we take certain actions? What could it look like if we don't? To expect no warming or no serious impacts on society is simply not credible, but we do want to be clear about what the domain of credible outcomes is.

Some of the uncertainties we have relate to personal perspectives and moral judgements. The pure rate of time preference and how we value money, social goods, and the natural environment, are not universal truths. They depend on who we are, the societies we live in and our position in those societies. The same can be said about the way we respond to and put a value on risk—something I've glossed over in this chapter but which also raises serious challenges of understanding and communication. These elements of the puzzle reflect individual or societal preferences, and economists look to societal behaviour to reveal clues about these preferences, but the unfamiliarity of the problem along with its one-shot-bet and extrapolatory nature severely limits the reliability of such methods. Ultimately, like the one equation of Chapter 23, these concepts provide tools for discussion; they provide a way of structuring and improving debate and avoiding contradictions in our personal or political assessments.

I may, for instance, say that I value significantly the avoidance of the risk of the die-back of the Amazon rain forest within the next 200 years but that implies a high value on the future and arguably therefore a low discount rate.[11] Given this I should support policies that involve substantial action to tackle climate change and this might involve me supporting policies which involve substantial investments in new technology and perhaps increasing taxes or decreasing spending on other public goods. This chain of connections means it might be inconsistent for me to not support such policies while still claiming that I passionately want to avoid the die-back of the Amazon rainforest. The role of studies which integrate climate physics and economics is not necessarily to find the best answers but to illustrate how the different elements of the problem are connected and to demonstrate which personal views and/or policy statements are self-consistent and which are contradictory.

Part of this process is about ensuring that policy positions are based on a justifiable interpretation of the science. We want to know that economic assessments are representing well the best scientific understanding from across multiple physical science perspectives. We also want scientific research to address the aspects that are most influential for economic, social, and policy assessments.

Consider IAMs, for instance. To date most of these have used a value for the effective heat capacity of the climate system which is very much at the high end of what is consistent with historic observations.[k] Of course the actual value may change in the

---

[11] Which implies a low pure rate of time preference and certain features of my utility function.

future and in any case the value they use can be justified—it's only that it is very much on the high side. This assumption means they all warm up more slowly than would be the case if a different (perhaps more justifiable) value were used. Hence they assess the damages to be smaller than they might otherwise be because they occur later and are subject to greater discounting of the future. In this case one could argue that economic assessments are not using the best scientific understanding. But this is almost inevitable, or at least understandable, because scientific assessments rarely provide information about this parameter. The problem is the multidisciplinary disconnect, not the failings of any individual field.

Considering what is credible is crucial for constructive debate about climate change. For instance, if someone were to argue that climate change is not a serious issue even though they value the future highly, it may be because they believe that the future changes will be small. This could be associated with an assumption that climate sensitivity is very small, say 1.0°C. Such an argument should certainly be part of the debate, but it should be put in the context of the plethora of evidence indicating that much higher values for climate sensitivity are possible and indeed much more likely. The argument based on very low climate sensitivity is highly selective. That's not a reason to exclude it entirely from the debate, but climate change is about risk so it's essential we consider the whole range of possibilities. A narrative describing a future based on a very low climate sensitivity is part of the story but it has to be put in a context of its extremely unlikely status,[12] and in the context of many other narratives with more dramatic or even catastrophic impacts.

And then there's the damage function. In economic calculations this is central to defining the scale of the issue. How it accounts for the diverse aspects of our societies and our environment that we value could helpfully be a much more open part of climate change discussions. Climate change is not just a problem for academics to solve, but one for individuals, society, and politicians to have informed opinions about.

One of the biggest challenges in climate prediction is therefore about relating physical climate predictions—along with the large uncertainties in them—to what they mean for the societies in which we live. This includes an assessment of the value that we might put on ethereal things such as knowing that there are healthy coral reefs in the oceans. I might never see or experience them but I might still value them. In practice, they may have direct value to me through keeping the oceans healthy and supporting aquatic life, which might directly support my own life—but even if they didn't I might value them because I like the idea of such beauty existing on our planet.

When we consider the impacts of climate change on our societies, we need to include such things in addition to increased physical risks (e.g. floods and storms) and their consequences for trade, supply chains, and our ability to get the products we enjoy. Furthermore, we might also want to consider the consequences of climate change for the availability of funds to support other desirable things such as health,

---

[12] We can know that some things are unlikely, or much less likely, without necessarily being able to assess the probabilities of how much this is the case.

education and community facilities: if the damages due to climate change mean that in the future we need to spend more keeping our basic infrastructure running, then we will have less to spend on these other things. This is not generally part of the picture that's painted of what climate change means for us and for future generations.

We certainly need to understand the possible range of damage functions better, but even more important is to relate them to a picture of what they mean for our societies and livelihoods. It's about making the physical climate/societal impact link real. It's about really grasping the question of how big a deal climate change is for us and what we value.

Challenge 9: How can we build physical and social science that is up to the task of informing society about what matters for society?

How can we push climate research in a direction that is useful for society? How can we structure academia to enable diverse disciplines to be interactive and responsive to societal needs? How can we ensure that those directing and funding research understand both what is useful to society AND whether what they are asking for is credibly achievable? How can we make a market for diverse approaches to climate predictions and avoid putting most of our eggs in one basket? How can we stimulate innovation in climate science? How can we create drivers for individual researchers to produce societally relevant information? How can we ensure that what is considered societally relevant is also scientifically credible, fit for purpose and actually what policymakers, businesses, and the public can use, rather than what specialist researchers and funders think they can use? This challenge is about achieving the right level of humility without underplaying what is known robustly within any individual discipline.

26
Controlling factors

26.1 Designing better research

Predicting climate entails challenges in mathematics, physics, economics, modelling, philosophy of science, statistics, and a vast range of somewhat related fields. There are also, however, challenges related not directly to the problems of climate prediction but to how we create an environment which is capable of addressing them: challenges in how we get the smartest people in our societies working on these problems in an effective and useful way. These challenges relate to the structures of how research is organized, judged, and supported.

For centuries—millennia even—scientific research has trundled along at its own pace. When breakthroughs have been made that are seen as beneficial to society they have been adopted, but for the most part scientific progress hasn't been planned, prioritized, or designed. In the context of climate change, that ad hoc approach isn't good enough. Academic research in climate science—both physical and social—is a source of information on which we are founding the future of human society. It is no longer about providing an opportunity to do things better than before—take it or leave it—it's about providing guidance on how to avoid unwanted, totally unfamiliar, and possibly catastrophic outcomes. It's not about providing new opportunities such as mobile phones, refrigerators, satellite-based internet, or ZX Spectrums, but about pointing out that we may be heading for a cliff, laying out the consequences of societal decisions, and guiding the choices humans face. Academic understanding is of itself—and to a significant extent—a driver of social change. What leaders and populations understand—or what they think they understand—about climate change is affecting all our futures. The urgency characteristic of climate prediction, together with the multidisciplinary and curse-of-computing characteristics, mean we need to question whether the structures we have for carrying out research are fit for the purpose of carrying out research which supports society in this way. There are many reasons to think they are not.

We all respond to contextual drivers: aspects of our society and environment that encourage us to behave in one way or another. These include everything from placing snacks at eye-level in the supermarket to encourage us to buy a particular brand, to decreasing national bank interest rates to encourage businesses to borrow and invest more. With climate change the contextual drivers are central to the outcomes we will experience. Get them wrong and we will fail to achieve our goals, because they matter for reducing greenhouse gas emissions. We need to think carefully about what contextual drivers encourage people to change behaviour in ways that reduce greenhouse gas emissions; what contextual drivers on industries encourage investment decisions that help address climate change; and what contextual drivers on politicians enable them to consider how this issue connects to the many other concerns they may be more interested in.

But contextual drivers are not just important for reducing emission: drivers in academia influence what research is done, how influential it is and how it is presented; drivers on the media and journalists influence what research gains public and political attention; drivers on modelling centres influence how models are built and what types of experiments are run.

The contextual drivers of research are often not well suited to the problems of climate prediction and climate change: the structures of academia work against multidisciplinary collaboration and research; funders often ask for answers rather than first asking whether questions are answerable; individual researchers gain most from addressing a complicated and acknowledged problem in their field, rather than a less complicated one that is more relevant to society, or one in which the complexity derives from multidisciplinary interactions. The drivers of climate change research and researchers are simply not well focused on the most important problems.

How we address these structural research challenges requires us to bring in yet more disciplines. Psychology and Operations Research stand out for me as having potentially valuable contributions to make regarding how we could better organize climate research and achieve better information for society. There would be significant value in examining how research funders, universities, and academic journal publishers each contribute to influencing research priorities and limiting progress, through their essential structures. The problem is an organizational one. The system behaves in a certain way which limits its ability to achieve what many see as its goals; the challenge is to build a better system. Or maybe just to acknowledge and reflect on its failings.

The challenges related to misdirected contextual drivers and unhelpful organizational structures are the subject of this chapter.

26.2 Drivers of research

'There are known knowns; there are things we know we know. We also know there are known unknowns; that is to say we know there are some things we do not know. But there are also unknown unknowns—the ones we don't know we don't know.'

Donald Rumsfeld, US Secretary of Defence, 2002

Discussions of risk and uncertainty often reference Donald Rumsfeld thanks to this famous sound bite on the subject. In 2002 the then US Secretary of Defence made this statement in a press conference relating to military intelligence and weapons of mass destruction in Iraq. It was widely quoted and often ridiculed at the time but has since provided a hook on which to hang many diverse debates and discussions about uncertainty. It provides a particularly useful hook on which to hang a consideration of climate prediction uncertainties.

The meanings of known knowns and unknown unknowns are both fairly obvious. When applied to climate predictions, the third category—known unknowns—covers many of the issues discussed in earlier chapters. It describes uncertainties that we might try to quantify with probability distributions. It covers situations where we can't know what the outcome will be because it is intrinsically unpredictable (for instance, the consequences of initial condition uncertainty in Chapters 14, 15, and 16, and inherent variability in Chapter 25), as well as situations where we don't know the outcome because we don't fully understand how reality—the climate system— works (for instance the consequences of model uncertainties in Chapters 18, 19, and 20, and the uncertainty in the value of climate sensitivity in Chapter 23). In this second set, even if we don't know what the probability distributions look like, that is to say there is deep uncertainty, we usually still know what it is that is unknown. We might even know something about why and how uncertain it is and possibly even have some idea of the relative likelihood of different possibilities. These are known unknowns. They have been the subject of most of this book.

But what about the fourth, unmentioned category, the unknown knowns? This category is particularly relevant for climate predictions. It is much less discussed but has been associated with the things that we don't know we know, or that we refuse to accept,[a] or alternatively the things that other people know about us but we don't know about ourselves.[b] For climate predictions though I want to associate it with the fundamental challenges of multidisciplinarity: there are things that are reliably known but not to us. We may not even know that knowledge on a certain subject exists, or that if it did that it might be relevant to our own endeavours.

There is often understanding within the domain of human knowledge that would be useful to some researcher or politician or decision maker, but is not known by them. This happens all the time, of course. None of us knows everything. In practice all of us know very little. But that's not profound and it's not the point here. The point here is that the contextual drivers of researchers on climate change encourage them not to know things that would be tremendously helpful in better understanding the issue.

Specialists gain increasing amounts of knowledge **within their own disciplines** as they spend years working on their subject. As a consequence 'experts' often know a huge amount about both the foundational principles in their wider field and their own very particular domain of specialization. Most of them also know when they need to seek further information, and who or where to go to find it, but this familiarity with the limits of their knowledge is mostly limited to within their own discipline— it perhaps extends to nearby, closely related disciplines but rarely beyond. A climate economist might know where to go or who to talk to, to get the latest thinking on how people respond to risk. A cloud physicist will know where to go or who to talk to about the latest thinking on how the Atlantic thermohaline circulation will respond to climate change. The information they are seeking is not exactly in their own discipline but it is in a closely related discipline. Because of this when they find the right person to talk to they will speak the same language, or at least a dialect that with a little effort can be easily translated. It's like a New Yorker travelling to Texas or Johannesburg, a Parisian to Marseille, or a person from Brisbane to Glasgow. They'll certainly notice the different dialect but they'll be able to communicate even if it takes a little effort. Most important of all, though, is that they have a broad enough understanding of their own subject to have had an inkling that what they didn't know in a related area might matter for what they are trying to achieve. They know that what they don't know in their own or similar fields could matter. They therefore make the effort to go and find out what they don't know.

The problem for climate prediction is that this happens mostly within disciplines or between very closely related disciplines. Across very different disciplines it happens only rarely. A physicist is unlikely to ask what aspects of the climate sensitivity distribution may be of importance in economic assessments of climate change. They may well simply be unaware that it is a relevant question to ask. And if they did ask the question it might take weeks or months of discussion to understand the answer, or to understand the context in which the answer sits, which can be just as important. Similarly an economist may well be unaware that a physicist or modeller could provide constraints on the effective heat capacity that would influence their calculations; they may not even be aware that one of the parameters in their model represents a physically observable quantity such as effective heat capacity. A mathematician may focus on the behaviour of a certain type of idealized system without realizing that a small change in their research approach could enable it to have a significant contribution to debates in the philosophy of climate science[1] and to the design of climate prediction systems.

This locking of knowledge within disciplines leads to unknown knowns: information that is known but not to many of those who could make use of it or for whom it would change their research direction or policy conclusions. They are important. Really important.

Of course it is not that no information at all passes between disciplines. Far from it, but the information that flows, flows through a limited set of routes and is often mediated by data rather than discussions and concepts.

So where does that leave us with the Rumsfeldian uncertainties (Table 26.1)? The known knowns are the robust conclusions of Chapters 8 and 22. The known unknowns are all those things that are currently—to some, often limited, extent—researched within mainstream climate science and climate social science. If there is behaviour that we have no reason to expect but it nevertheless turns out to be the case, well these are the unknown unknowns, but hey, c'est la vie. Investing in broader and better knowledge may illuminate some of these unknown unknowns and may well be worthwhile but if, at the moment, human knowledge is not at that level then I'm not inclined to spend too much time worrying about these unknown unknowns. However, if 'we', the combined collective of human expertise, know

---

[1] For instance, to what extent do climate model simulations represent evidence about the behaviour of reality?

something which is somehow critical for how human society should prepare for the future then that information is not an unknown unknown. And yet if most or all of those who could use the information do not know it, then in some sense it is still an unknown. It might even be that no one has put all the pieces together even though to do so may be trivially easy. It is therefore not an unknown unknown in the deep and fundamentally problematic way that that phrase seems to imply; it's just that nobody has bothered to ask the right question. This is the world of the unknown known.

This lack of knowledge is not due to our limited ability to quantify uncertainty (known unknowns), nor due to the limited overall knowledge of humanity (unknown unknowns). Rather it is unknown to those who might be able to use it because we have created systems that are so strongly formalized in the way we hold our knowledge that we simply don't or can't access the information that is sitting there ready to be used.

This is the curse of multidisciplinarity. It is widespread in academic research, and beyond, and it represents some of the key constraints on the usefulness of climate predictions.

Why, though, are there unknown knowns? Surely it is the job of researchers to tackle these issues and bring all the relevant knowledge together? Are we being failed by our researchers? The truth is that many researchers desire to do just this, to bring multiple lines of knowledge together spanning diverse disciplines, but the contextual drivers on researchers, the framework of the academic system, makes it almost impossible. Why? The answers relate to the peculiarities of the academic system. Climate prediction is mostly not about producing a saleable product, so the vast majority of research takes place within the academic system and the academic system's organizational structure doesn't lend itself to addressing unknown knowns.

The problems start with universities and research institutes. A successful academic career is almost always founded in a university department or a research institute with a narrow and focused specialization: for example, a geography department, a maths department, an international development department, or an oceanography or economics institute. Job security and promotion is founded on teaching, obtaining research funding, and the production of academic publications **in the subject of the department or institute**. Often there will be a particular list of discipline-specific journals that are considered worthy enough for promotion or to secure tenure and a stable job. If a researcher produces a set of publications in different journals or that spans different disciplines, it can be difficult for colleagues to assess them and in any case only a part of each one is considered relevant and valuable through the eyes of the organization.

Multidisciplinary research is about seeing more by combining approaches from multiple disciplines. Often though the approaches are not at the cutting edge of those disciplines; indeed it may be better if they are not, because then they are more likely to have been thoroughly tried and tested and therefore more demonstrably reliable. This makes them a sound starting point for creating robust foundations for new work at the interface between disciplines. Unfortunately, it also means that the contribution

Known knowns	Known unknowns
Things we understand well such as the robust knowledge of chapters 8 and 22.	Things we have begun to think about but which are either quantifiably uncertain or are unquantifiably uncertain (known as 'deep' or 'Knightian' uncertainty). Even if they are potentially quantifiable they may not have been quantified yet. Most research sits in this box.
Unknown knowns	**Unknown unknowns**
Things that we don't know, or that we don't understand, but that others do. We may not even know that knowledge on the subject exists and we certainly don't know that it could affect our own analysis and conclusions. The information therefore exists (it is known) but people whose conclusions and actions would be changed by it do not know about it (it is unknown). This affects researchers and academics just as much as policy makers and the general public. This is where the limited ability to pursue multidisciplinary research holds back our understanding of complex problems like climate change.	Things we haven't thought about yet or behaviour which we currently have no reason to expect but which may turn out to be real possibilities.

Figure 26.1 Rumsfeldian uncertainties as they apply to our understanding of climate change.

to each discipline individually may be minor. Viewed through the blinkered lens of a single discipline, the research can seem pedestrian. The upshot is that any researcher who cares about their career is likely to be wasting their time on multidisciplinary collaborations because few organizations have structures which are able to assess and value them.

This is the case despite the fact that most research organizations would claim that they are keen to support multidisciplinary work. Certainly it is true that many consider it a very good thing—just not a priority. It is good and encouraged but only so long as it comes after achieving all the other discipline-specific goals that come with your job. It's fine to do multidisciplinary research if you've got some free time, which means it's often pursued by young, early-career researchers, who haven't yet realized what's the best thing for their career, but for success and job security in an academic position the unacknowledged drivers are to steer well clear.

In this way the drivers from research employers are firmly against multidisciplinarity, but the problems don't stop there. It is also harder to get funding and harder to publish multidisciplinary research. Part of this comes from the limitations of the peer-review process.

In the UK House of Commons in 1947 Sir Winston Churchill said: '[I]t has been said that democracy is the worst form of Government except all those other forms that have been tried from time to time'.[c] In academia, the role of 'peer review' in academic publishing and in deciding what projects get the go-ahead is viewed in much the same way. Peer review is the process by which academic papers or funding proposals are sent out for assessment by other researchers working in the same or similar fields. It is commonly acknowledged to be flawed and to sometimes lead to unhelpful or perverse outcomes, but it's often taken as a matter of faith that no better alternative exists.[d] This may be the case, but even if no better alternative exists, the way it fails should really be the subject of constant and heated discussion rather than a shrug of the shoulders and meek acceptance. The challenge is to design a system that intrinsically reflects on its own limitations. Even if we believe that peer review is the foundation of the best possible system of scientific management, we should still think carefully about just how it's implemented to avoid some of its most unhelpful characteristics.

Peer review is particularly problematic for multidisciplinary research because almost by definition there are few if any peers who are familiar with all the different aspects of the work. Research funding is tight and increasingly competitive; only projects which are assessed to be the best of the best are successful. A multidisciplinary project will often be assessed by multiple individuals from different specific disciplines, so for them each to give it top marks it has to be groundbreaking in each discipline individually as well as being a valuable integration of knowledge. A project spanning maths, hydrology, and finance, or spanning physics, computer science, and policy has, therefore, to be exemplary in all three. Inevitably though, it will involve less research into the details of each one than an equivalent single-discipline proposal. Reviewers inevitably pick up on this and highlight how the project could do more in their own subject area. They often also comment on how they don't fully understand the aspects of the research related to other disciplines. This is often because the project proposal has to cover the background to multiple lines of research across multiple disciplines within the same limited space allowed for a single-discipline proposal. It therefore inevitably has much less information about each particular strand and consequently comes across as lighter weight. The upshot is that a multidisciplinary proposal is much more likely to receive reviews along the lines of: 'this is great but it fails to also explore this or that' or 'this is great but I can see lots of ways that it could do more in my subject area', or 'this is great but it isn't totally clear' which can often mean 'it hasn't taught me the foundations of a whole other discipline'. Since only the best of the best get funding, the presence of many 'buts' and anything less than 'this is brilliant full stop', is often enough to kill the prospect of getting funding.

Multidisciplinary collaborations face similar peer-review barriers to publishing results in the top research journals and thereby gaining the widest academic impact. Together the difficulties in finding funding and publishing results create an additional nudge away from pursuing multidisciplinary collaborations.

To get around these issues, research funders sometimes create specific funding opportunities which require multidisciplinary collaborations. They are often focused on a particular topic. This somewhat addresses the peer-review problem because all the proposals then face the same issues so it is a more level playing field. Unfortunately it raises another one: the problem of answer driven research. This problem arises because funders regularly misinterpret multidisciplinary research as synonymous with applied research. It is seen as a matter of simply bringing social science questions together with natural science expertise. The tendency is therefore to ask for answers to societal or industrial questions, rather than asking broader questions of what is answerable or how a problem should be addressed. This focus on providing an answer is dangerous. It encourages a limited perspective which is centred on generating data that appears to answer the question but may in fact fail to consider deeper questions of reliability. When funders offer such funding it often leads to studies founded on computer models, because these studies easily generate data that look like an answer without necessarily reflecting on whether that answer is robust—or indeed whether it addresses the underlying question.

These funding structures create an environment which discourages large, conceptual, multidisciplinary projects aimed at understanding a problem and all the assumptions required to address it. Rather they encourage consideration of small elements of a problem or, worse, large computer-based projects which give apparent solutions but actually contain a lot of potentially spurious detail (the curse of computing). They may be multidisciplinary but only to a limited extent. For instance, they might cover maths and water planning, or climate models and the finance industry, but not maths and climate models and philosophy and how they can be combined to guide an understanding of impacts on water planning and knock-on consequences for the finance industry. They also tend to support small, short projects. This is partly because multidisciplinary projects are undervalued and therefore small, and partly because they are conceived from the perspective of finding an answer to a defined and limited problem rather than pushing back the limits of scientific understanding. Funders too often consider multidisciplinary problems as being ones which require turning a research handle, rather than developing an understanding of the complexity of an issue—possibly without even generating an answer at all in the short term. These funding environments create contextual drivers on researchers which work against attempts to understand the complexities of climate prediction and reduce the unknown knowns.

Researchers are also driven, by both their employers and a desire for influence, to publish their results in 'top journals'—the sort of journals you hear quoted on the evening news: *Science*, *Nature*, and the like. But like the funders, these journals are also more interested in answers than conceptual understanding. An analysis of the latest set of climate models leading to a new model-based assessment of the likelihood of the thermohaline circulation in the Atlantic collapsing in response to climate change, may well generate a high-profile, career-making paper in a top journal. An analysis of the inherent assumptions and dependencies in such an approach, or an experimental design which over the next ten years could generate a reliable assessment of it, would be very unlikely to get the same profile. As a result, jumping on the data or the models to get a conclusion about some aspect of climate prediction is a good career move; spending time gathering information about unknown knowns from multiple disciplines is not.

One might argue that what climate science needs most at present is the back and forth of academic debate and disagreement over proposals for new and different designs for climate prediction experiments. Yet the top journals don't tend to publish such 'design' papers, and the funders rarely fund such design projects. Trying to build a research career around laying solid foundations for better climate information is therefore, to say the least, a risky strategy on a personal level.

The expertise which is so important for designing the future of our societies under climate change is consequently located within a research environment which encourages quick answers rather than deep understanding. In scientific research historically, it has not been of great importance to consider multidisciplinarity and unknown knowns, but with climate predictions it is, and our research systems are not geared up to dealing with this situation. Universities and research institutes have career paths which discourage it. Research funders, both publicly supported and philanthropic, haven't grasped the scope of the need or the barriers to its execution. The most high-profile academic journals want answers, not methods, conceptual understanding, or experimental designs. And researchers themselves hamper the efforts of funders and journals to engage with the deepest challenges, due to the limitations of the peer-review process. The drivers on individuals are directly counter to making progress on the integrated climate prediction problem.

Of course not everybody responds to those drivers. There are researchers around the globe who break with expectations and plough their own furrow. But they are both uncommon and—because of the drivers—restricted in how much they can achieve. Generally speaking, our research systems are holding back progress in the provision of reliable, relevant, robust, usable information to society about the details of the future under climate change.

The challenge is to create a research environment which is focused on the overlap between what matters for society and what is answerable with current—or future—science and technology. Where the two intersect may not be where either researchers or users expect it to be, so the two need to be brought together. Experimental design is a key challenge but we should not expect a one size fits all solution—we should not be aiming for a system that provides predictions of everything. Given the state of knowledge at the moment, the aim should be to build the foundations for future integrated climate predictions. These need to combine knowledge from multiple disciplines while focusing on those aspects that are relevant to particular users of climate predictions, and answerable within the constraints and uncertainties inherent in the problem. That's quite a mouthful and quite a task.

You may say: 'But what about the urgency? We need answers now! Don't let the perfect be the enemy of the good.' To which I reply: 'That misses the point'. These challenges are not about holding out for the perfect—they're about working out what we have today that might be good, or good enough, as opposed to simply what **looks** good. It's about knowing what information we have. The prevalence of models and massive computational datasets makes it easy to produce high-resolution pictures or videos which can be presented as representations of future climate or its societal consequences. But the important question is how reliable they are. Does the fine detail mean anything, or is it just very pretty noise? This is not a time in the history of climate science to be searching for more detailed predictions but rather for understanding what constitutes adequate and what constitutes robust.

We are not currently limited by technology—we are limited by our understanding of the constraints on, and the complexity of, the problem. Nevertheless, we have many many answers. Or at least many answers are within our grasp but they require bringing together knowledge and expertise: seeking out the unknown knowns. The challenges for the scientific community are to identify and use the reliable knowledge we already have in a way that is informative for society, while building the foundations for better predictions in the future. Current drivers encourage a focus that sits unhelpfully between the two: neither prioritizing the identification of what is robust in current knowledge, nor studying the conceptual barriers to better predictions in the future.

26.3 Drivers of models

The influence of contextual drivers of research is, of course, important well beyond the impact on individual researchers. Consider Global Climate and Earth System Models. They are created, maintained, and developed by a few large modelling centres around the world. There are hundreds of people working on them and, as discussed in Chapter 18, for each new release the models are subject to a process of tuning which adjusts them and ensures they produce results which are deemed credible. These centres are not working independently of each other, though, and nor are they independent of the wider issues and discussions about climate change. Inevitably any modelling centre whose next-generation model is perceived to be of low quality, that's to say, considered less realistic by comparison with either observations, current scientific opinion, or other models, would feel under pressure to 'correct it'. There is nothing nefarious or underhand about this. The modelling centres are aiming for models that represent scientific understanding as best they can, so this is, to some extent, a reasonable thing to do. Nevertheless, it's important to recognize the consequences of this pressure.

The difficulty for modern climate science is that the models are used as both a source of scientific insight and also a reflection of scientific understanding. There are conflicting interests in this dual role.

The framework within which model development sits—the contextual drivers— encourage the production of models that are in the middle of the pack, not too dissimilar to all the other models: not too extreme. Nobody wants a 'less realistic' model—nobody wants a model that represents the unlikely tail of the probability distribution for, say, future global warming, or indeed anything else. Yet to better understand the prediction process and better inform society, that is exactly what we need because it allows us to examine what is possible in a model and helps us consider what is credible in reality. It supports the study of the risks associated with climate change. As highlighted in Chapter 25, it can sometimes be that looking at the less-likely possibilities is the most societally relevant thing to do.

It is implausible, however, to think that a modelling centre could go to its funders and say: 'Hey look, we've just spent millions of your dollars/euros/pounds updating our model and it's now considered the least likely to simulate the actual outcomes of climate change. There's a roughly 98% probability (say) that the variable you're interested in will change by less (or more) than our model while other models give closer to what is thought to be the 50:50 value. Haven't we done well?'

Models and modelling centres are often national or regional efforts which are utilized to support studies of national climate impacts and to support regional adaptation planning, so the further they are from the consensus scientific view, the less useful they are perceived to be to society. By contrast, the greater the diversity in the collection of models, the more useful they are likely to be for scientific study. That's the conflict.

The drivers on modelling centres therefore mean we should expect models to cluster together. Not by design, just because a model perceived as less realistic is more likely to be re-examined and re-tuned, than one perceived as more realistic. It's an organizational and a human response to the context in which these models are built. It might come from considering what is most useful for the funders but it might equally arise simply from professional pride in wanting one's own model to be the 'best', or even just a familiarity with the academic literature leading to subconscious retuning to get something that closely represents what happened in the past without considering ways in which the past could have been different. It is not, however, a good thing when we are seeking to get a handle on uncertainty and asking ourselves whether our conclusions are robust. If we build models to reflect what we already believe to be the case, we shouldn't then use them as evidence for why we believe that to be the case. We need diversity in models, and this inevitably means we need some that are of apparent lower quality against which we can test our assumptions, and to help us understand potentially credible behaviour. But who in their right mind is going to invest millions in building a bad model?

We can see this process at work over the last twenty years by looking at climate sensitivity. In the three Intergovernmental Panel on Climate Change (IPCC) reports of 2002, 2007, and 2013, climate sensitivity was assessed to have a more than a two in three probability of being between either 1.5°C or 2°C, and 4.5°C. Of the major models developed in this period,[2] all but two had climate sensitivities in the 2°C to 4.5°C range, and the two which didn't, had values that were very close: 4.6°C and 4.7°C. All were in or very close to the central probability range; none were out in the tails of the climate sensitivity distribution despite scientific assessment putting up to one-third of the probability out there. If the models had reflected our uncertainty, then one-third of them should have been outside the central range. The models were therefore of limited value in exploring and quantifying uncertainty in the consequences of global warming, and hence in the scale of societal risks. At the same time, they were of limited value in helping us understand the processes involved in extreme global responses.

By 2021, when the IPCC released its sixth assessment report on the physical science of climate change, there were some signs of change. The new set of models used in this report had a much wider range of climate sensitivities: between 1.8°C and 5.6°C, with ten models, more than one-third, above 4.5°C.[e] They still don't cover the real tail of the probability distribution, beyond, say, 6°C, but it is a distinct improvement.

Nevertheless, the issue of contextual drivers in model development has not gone away and is closely linked to the issue of model tuning—of adjusting a model until it is by some definition acceptable. In practice the mere existence of tuning a model has sometimes been contentious. This was brought home to me at a conference sometime around 2004 at which I heard a modeller from one group vehemently deny that any modelling centre would ever try to tune their model to get a particular climate sensitivity. In another session a few hours later, I heard a modeller from another group discuss in detail how they had been frustrated by the high climate sensitivity of an earlier version of their model so they had made significant efforts to reduce it, only to find that it had increased further! This had led to renewed efforts to refine it until it was eventually brought into the acceptable band.

Nowadays, model tuning is beginning to get the acknowledgement it deserves. Surveys of modellers suggest that while some do take account of climate sensitivity when tuning, it isn't one of the main things that most of them say they focus on. Simulated twentieth-century warming, El Nino variability, Atlantic ocean circulation patterns that draw surface water down to the depths, and regional patterns of temperature and rainfall are among those aspects which are more likely to be actively tuned[f] according to model developers.

This move towards acknowledging the tuning process is a welcome, if tentative, step forward because it will allow us to see more clearly what controls there are on model behaviour. Nevertheless, unacknowledged and subconscious tuning in response to contextual drivers of the model development process will inevitably remain. These considerations imply a need to allow for sociological information

---

[2] The models of the third and fifth coupled model intercomparison projects: CMIP3 and CMIP5. There was never a CMIP4.

about modelling centres and model developers to be included in model interpretation, adding to the conceptual challenges of Chapters 18 and 19. This is yet another aspect of multidisciplinarity: one which is likely to be of importance to computer simulation modelling across many disciplines, wherever we are looking for new understanding or extrapolatory behaviour which observations or real-world experiments can't tell us about.

As an aside, it's worth emphasizing that changing the behaviour of a complex computer model to suit some target is at best difficult, and sometimes impossible. Developers have limited flexibility; they can't just turn a handle and get the answer they want. Nevertheless, if a model is too different from what is perceived to be desirable, it will likely be sent back for re-examination or adjustment, while if it isn't, it won't. This imbalance creates a nudge towards certain types of model behaviour which inevitably leads to biases in the results derived from multi-model ensembles.

Contextual drivers on model development are important and could have a potentially chilling influence on climate science unless they are appropriately accounted for. Consider, for instance, 'extreme' behaviour such as an eight-degree climate sensitivity. If this is seen as potentially unacceptable in a model, then a model with this behaviour is sent back for retuning, and the collection of models available for research is maintained in the central part of the uncertainty band. Over time researchers become increasingly used to seeing models consistently in this band, and this affects their opinion of what is possible, or at least likely. It is self-reinforcing. Expert opinion is influenced by the models, which are then influenced by expert opinion. Over time the idea of questioning whether extreme behaviour is likely or not becomes harder. Harder to get through funding peer review because of questions such as: why study something that all the models seem to show isn't possible? Harder to get through publication peer review because of objections such as: this paper studies the consequences of a high sensitivity but we know that is very unlikely because none of our models show it. In this way climate research risks being trapped in one of those scientific blind alleys, closing off certain lines of questioning. Even this would not be so bad if it were not for the urgency of the issue.

As we produce more and more detailed climate predictions, the potential for non-scientific, contextual drivers of model development to influence their behaviour has the risk of constraining the consensus opinion in the scientific community. This isn't just about climate sensitivity—it could influence many types of outcome, whether it be the likelihood of ice sheet collapse, more frequent El Nino conditions or extreme regional rainfall changes. Recognizing and considering the impact of these drivers is a fascinating, new and important challenge which needs to be addressed if we are to understand the reliability of climate predictions and create a research environment that can improve them.

Climate change is not, of course, just an academic subject though. It matters to us all and therefore how it is communicated also matters. Unfortunately the drivers of our communicators are also not well aligned to the subject.

26.4 Drivers of information

Climate change doesn't fit well with conventional journalism. It's not clear whether it's a science story, a political story, or a news story about the latest storm or heatwave. The consequence is that climate change communication often becomes a mash-up of all three, always accompanied by pictures of wildfires, droughts, floods, collapsing ice sheets, or something similar. The stories are often repetitive, bland descriptions of the issues, which are neither particularly informative about the state of the science, nor particularly helpful in elucidating what the issues might mean for individuals or for future generations and the societies they are going to inherit. Climate change reporting therefore has a tendency to be disempowering rather than empowering. It creates a sense of doom but without creating a structure within which we can debate the balances and compromises necessary in responding to the threat.

Often climate change is treated by the media as a science story, but only in a trivial way. The really interesting conceptual scientific issues regarding climate prediction and what we can and cannot know, are rarely, if ever, addressed. To me, much of the reporting comes across as not very interesting science and yet not very informative about the societal relevance either. It's neither one thing nor the other. The implications of climate predictions for how our societies work and for the economic consequences of climate change seem to be seen, at best, as secondary aspects. Climate change is presented as something external to society, something for scientists, politicians, and activists, not something that will directly impact you and about which you might wish to carefully consider your opinion and perspective.

Of course, climate change is an unfamiliar concept and the actions necessary to tackle it are also unfamiliar. So perhaps one goal of climate science communication should be to frame the conflicts and balances. Climate change communication could be about enabling individuals to make informed judgements that reflect their own values while avoiding, as far as possible, self-contradictory opinions. It could be about facilitating public and political debate. Climate predictions, along with their uncertainties and their character (distributions, tales, envelopes of possibility), are important for assessing what we as individuals think is an appropriate response, and therefore what we might be willing to accept from our governments, so communicating the state of the science very much matters but to do so we need to move beyond simplistic vanilla messages.

The problem with climate change communication is twofold. First, it fails to effectively separate confidence from uncertainty. Second, it fails to put the information in its social context and thus leaves viewers, readers, and listeners without the tools to integrate the information into their own worldview.

In climate change communication, both confidence and uncertainty are fundamental. The former is about the reality of the issues and the urgency of the problem. The latter is essential for us to examine our attitudes to risk in the context of climate change and, more practically, to build societies that are resilient to future climatic threats. Building resilient infrastructure, for instance, requires us to allow for the scientific uncertainties in climate predictions.

Confidence and uncertainty go hand in hand. We might be confident that under some scenario with little action to tackle climate change, global warming won't be less than 3°C by 2100, but we may also be uncertain about the actual outcome, with 5°C entirely plausible and significantly higher changes also very much credible. We are therefore confident that climate change is a huge threat, but just how huge is nevertheless open for discussion. The risk of 5°C or higher should be part of that discussion. Our uncertainty is part of what we know.

One area where so-called climate sceptics have had a particularly detrimental impact over the last thirty years is on the communication of climate science uncertainty. Discussing uncertainty in the context of climate change is often perceived by those trying to communicate the issue as likely to undermine the seriousness of the threat. Climate sceptics have often latched onto uncertainties in the science and re-interpreted them as an indication that we are confident about nothing. This, of course, is a fundamental misrepresentation of knowledge: part of our knowledge about many things is about uncertainty—you only need to think about dice to see this. Yet concerns about being misinterpreted regarding the seriousness of the threat provides a driver on communicators to pull back from serious discussions of uncertainty. This fundamentally undermines our ability to address the topic in an inclusive, comprehensive way. Responding to climate change has a component that is inherently about how we respond to risk in the context of a one-shot-bet. Uncertainty is a central element of the story and a reason for concern in itself. To not tell the uncertainty story is to not tell the climate change story.

It's not just about a reluctance to communicate uncertainty, though. The drivers on the media are to report what's new, and this often means the latest high-profile papers and reports. This leads to a constant stream of articles about new climate predictions or descriptions of how different aspects of the climate system are changing. The impression this gives is one of a steady process of building up a clearer and clearer picture of climate change and what it will look like in the future. It encourages a perspective that sees climate science as an applied science, one that can provide whatever predictions are necessary—if not now, then very soon. This misdirects politics and industry in their considerations of how to respond to the issue. It encourages them to treat climate science and climate models as simply sources of data rather than disciplines with which they need to engage closely to identify what relevant knowledge we have about particular vulnerabilities. It encourages separation rather than integration in the production of climate predictions and other types of climate information which is used by industry and communities to prepare and adapt.

The constant stream of reports about the latest results gives an impression of false confidence in what climate science and climate models can provide but at the same time, counterintuitively, it risks undermining communication of confidence in the existence and scale of the threat. It fails to separate what is solidly known basic science from the latest research results which might reasonably be open to question and debate. It gives an impression that climate science is like any other area of research science: it is muddling its way through to some sort of understanding. That's fair enough—that's what climate science research is like—but the knowledge of the threat of climate change is much more akin to school level science: science that we understand very, very well and it would be ridiculous to doubt. The basics of the physical aspects of human-induced climate change have been well-known since at least 1990. After the third IPCC report of 2002, one could argue that the central message of each subsequent report (2007, 2013, and 2021) has been: 'climate change is still a huge concern, but then we already knew that last time and our basic understanding hasn't changed; indeed we don't see any way it could'. The focus on the details in the latest paper, or the latest report can detract from the robustness of this key message even while failing to effectively communicate the uncertainties.

A lack of information about the uncertainties also goes hand in hand with a failure to communicate the exciting challenges and debates that exist in the subject. Climate science is not only about building a better future; it's also about pushing back the boundaries of human knowledge, understanding what we can know, and understanding how we might go about using fantastic new technology—computers, super computers, quantum computers and the like—to further our knowledge of the universe. Arguably physical climate science should be treated, at least partially, in the same way that science communication treats other profound or intensely complex issues: particle physics, cosmology, the understanding of consciousness. It's about working out how to understand the currently unknown. It's about being in awe of the complexity of the universe and wanting to understand it better. To do this, the next generation of researchers need to see and be inspired by the conceptual challenges we face. At the moment they rarely get mentioned, even in lengthy documentaries on the subject.

Finally, climate change is something that will affect almost all of us. Its communication needs to make a connection between the actions we can take and the outcomes we might see. We need help in interpreting the science and the economics in terms of what we personally see as worthwhile and important.

Whose job is this though? The drivers of academics don't encourage this. The drivers of environmental campaign groups perhaps do but they may be seen as partisan. The drivers of journalists and science communicators are to focus on the latest results, so they would struggle to invest the time and effort needed to bring together the different disciplines and concepts necessary to make the links. Making climate change real and personal falls through the cracks of climate change communication. And so we're left with inadequate communication of climate science and a public discourse which fails to grasp the risks which are understood, the risks which are not understood, and how our values influence our judgements and priorities.

Climate prediction creates challenges for science (physical and social), challenges for the communication of science, and challenges for how we react to and debate the issues it raises in the context of other more familiar social and political concerns. It doesn't fit neatly into the conventional boxes for any of them. There is a need, therefore, to reflect on the drivers which control how researchers approach the problem, how modelling centres develop and interpret their models, and how the interface between scientific knowledge and societal debate is curated. This involves studying the operation of complex organizational and funding systems. Expertise exists in the behaviour and optimization of complex organizations and in understanding how individuals and organizations respond to different types of information and the way it is presented.[3] The challenge is to engage and use that expertise. This is another example of the scale of the academic challenges related to climate predictions— challenges which may seem daunting in the context of the urgency of the issue but are also exciting and new in the context of human understanding.

---

[3] I'm thinking mostly about aspects of psychology and operations research.

27

Beyond comprehension? No, just new challenges for human intellect

The science of climate prediction is a science in its infancy, not its adulthood. It's a clever child. It's playing with new ideas and discovering how this or that works, often spending lengthy periods with one toy before shifting focus to another. It's intensely focused on what it's doing at the current moment but doesn't have an overarching view of what it's trying to achieve.

Having worked on climate predictions for more than twenty years, I find myself constantly disturbed by its Janus-like, conflicting characteristics. On the one hand, it's full of fascinating, conceptual, and practical challenges that are at least as profound and interesting as anything science has to offer. These challenges are found in a variety of diverse disciplines. If you're looking for challenging problems that question where the limits of human knowledge are and how we might approach them, then the science and social science of climate prediction are for you. And this is the case almost whatever your particular skills or subject speciality. I am regularly astounded by the fascinating insights of colleagues across a huge range of disciplines, and by the ability of diverse perspectives and approaches to illuminate different aspects of the climate change problem. It's an exciting subject to work on and one where I at least am constantly gaining new knowledge and learning new tools. It's fun.

Yet on the other hand, I find myself frustrated by the limited interest and imagination of so many specialists in the field: how little the reliability, relevance, and underlying assumptions implied by discipline-specific approaches are questioned. Experts can focus so intently on refining and improving an approach that they simply don't ask whether doing so is relevant to the ultimate problem they think they are addressing. As a consequence, assumptions get lost, and 'the robust' and 'the debatable' become annoyingly and unhelpfully intertwined. Today's climate change science undersells what we know with huge confidence and oversells what we know with little confidence. The fundamentals necessary to motivate many aspects of climate policy are not much beyond school-level science and are beyond serious debate.[1] Often, though, the details of what we might want to know—but don't necessarily need to know—to optimize policy and planning are at the very cutting

---

[1] That's not to say they should not be explained. Quite the opposite. The reasons for knowing they are robust should be regularly repeated and clarified so that public debates on policy don't go down the rabbit holes of misleading pseudo-science. Furthermore they are important for making the connections and identifying the dependencies across disciplines.

edge of conceptual understanding. Mixing the two misleads policymakers, business, industry, international negotiations, the public, and ourselves.

Assessment of the economic impacts, and judgements about the most appropriate policy responses, will vary according to the views and values of different societies and different individuals, but an effective global response relies on our agreeing about the underlying scientific and economic foundations. It is crucial therefore for researchers to better connect what we know and don't know about physical changes, with their potential economic and social consequences. We need to be able to separate what is 'beyond debate', from what is unknown or uncertain, from what is 'dependent on your values'. Of course highlighting what is robust intrinsically means acknowledging what is uncertain or open to question.

In an age with easy access to computer simulations, scientific research can become divorced from scientific understanding. Studying a model is often considered enough, without questioning its relationship to reality. This poses a real threat to the relevance of academic research results.

As with the 2008 financial crisis, a failure of the collective imagination of many bright people could be leading us to misrepresent the risks to the system as a whole. The physical and economic risks are, in my view, most likely to be underestimated because our model-based research doesn't reflect many of the uncertainties encapsulated by known unknowns, let alone the consequences of unknown knowns. There is also a tendency to rely on look-and-see information, either together with or separate from models, without questioning its relevance for the problem in hand.

These concerns mean we are missing risks. They are particularly worrying because it seems to be far easier to think of missing risk factors that would make the situation worse rather than better.

Independent of these concerns, the risks are not well presented or characterized. Their relevance is not made clear. We see pictures of floods, wildfires, collapsing ice sheets and degrading coral reefs, and we have concerns about extinctions and environmental degradation. These things matter. Of course they do. They matter a lot. We want to avoid them. Nevertheless, if we are talking about individual lives, societies, or national economies then these impacts are often to some degree distant. For many people they are inevitably secondary concerns because of the practical demands of living life and getting through the day: of earning an income, securing food and accommodation, maintaining health, providing education, looking after friends and family, and even having some fun. The first priority for most people is getting on with their daily lives. To make climate change real, even if distant in time, we need to describe what our daily lives will look like under future climates. How will you travel to work? What sort of work will be available in your nation or region? Will you be able to get the foods you want? Will you be more likely to be affected by conflict? If so, how? Will you be influenced by natural disasters and extremes? Will you be able to see relatives in far-away places? What will be considered 'far-away'? Will taxes go up or facilities go down because of the constant need to firefight the consequences of climate change?

In 2006 John Ashton, the then UK Climate Envoy, said: '*We need to treat climate change not as a long term threat to our environment, but as an immediate threat to our security and prosperity*'. I largely agree but if that's true what does that threat look like? How will it impact us? These are questions which have not been well addressed. The picture of climate change in this sense is still unclear.

Ultimately, like the financial crisis, the overarching issues are the risks to the system as a whole. Describing the risks to the system as a whole requires, to some extent, understanding how all the different components interact: how changes in them can cascade and potentially amplify each other. Understanding the behaviour of the whole requires understanding all the parts.

But understanding any individual part also often requires an understanding of all the parts. Getting a grip on the implications for the part of the system you care about is something that can only be done in the context of the system's other components. It can't be done in isolation. It is a 'whole system' problem. Focusing on one element can't give you a prediction for that element; you need an understanding of how all its interactions might change as well. The implications for South Korea or Italy can't be usefully assessed outside the global context; the implications for food availability can't be assessed without an understanding of the implications for supply chains. Climate prediction is in essence a back-and-forth between scales, disciplines, and concepts.

In 2008 Sir David King, the then Chief Scientist to the UK Government, said:

> '*The challenges of the 21st Century [. . .] require a re-think of priorities in science and technology and a redrawing of our society's inner attitudes towards science and technology.*'[a]

He went on to cause consternation among some physicists by saying:

> '*It's all very well to demonstrate that we can land a craft on Mars, it's all very well to discover whether or not there is a Higgs boson; but I would just suggest that we need to pull people towards perhaps the bigger challenges where the outcome for our civilisation is really crucial.*'

I suspect he was referring principally to the technological developments we might need in order to tackle climate change (renewable energy technologies, fusion, electric cars, carbon capture and storage, and the like) but the argument holds just as strongly for understanding the climate of the future and what it means for our societies and the way we will be able to live our lives. Gathering such knowledge is critical to being able to respond appropriately to climate change, and the challenges are as interesting and as complex as the search for an understanding of matter and fundamental forces and landing crafts on Mars.

Climate prediction represents a grand scientific challenge full of fascinating new questions about everything from how we study complex systems, to the role of computers in scientific research and how we identify robust knowledge across collections of disciplines in an era which has seen an explosion of publications, opinions, and data sources. Yet climate predictions are also important for society. They are critical in building policies that are robust to climate change, that respond to the threats of climate change, and which are supported by populations because their value is understood.

Is the future—that is the future conditioned on our actions—beyond comprehension? I don't believe so, but the challenge of understanding it requires new approaches, new methods, diverse collaborations, and a bucketload of imagination. The question is whether mankind's intellectual capacity can rise to the diverse demands and peculiar challenges of climate prediction.

Acknowledgements

I am grateful to many people for their help and support in both making this book a reality and in helping me develop my apparently unusual perspective on the issues it covers. Thank you to:

My family: Janet, Alan, Becca, Janet, Annie, Mark, Chris, Marc, Sue, Jane, Garip, Alex, Lydia, John, Gillian.

My illustrator, Annie Stainforth.

My readers, all of whom provided truly invaluable feedback and encouragement: Becca Heddle, Dan Taber, Erica Thompson, Chris Hughes, Raphael Calel, Chris Stainforth, Mark Stainforth, Neil Grandidge, Simon Dietz, Nicholas Watkins, Marina Baldissera Pacchetti, B.B. Cael.

Individuals who have encouraged and enabled me to be driven by interesting questions rather than career-progressing hoops: Leonard Smith, Myles Allen, Diana Liverman, Mark New, David Andrews, Peter Killworth.

The Grantham Foundation for the Protection of the Environment.

Academic colleagues and friends: I am particularly grateful to the following for their friendship, support and insight: Leonard Smith, Raphael Calel, Nicholas Watkins, Sandra Chapman, Erica Thompson, Roman Frigg, Joe Daron, Suraje Desai, Marina Baldissera Pacchetti, Michael Wehner, Tom Downing, Fernanda Zermoglio, Stephan Harrison, Seamus Bradley, Ana Lopez. Thanks also to the many other highly valued colleagues I have not named here for reasons of space but whose support has nevertheless been of great value to me and has been very much appreciated.

All those involved with climate*prediction*.net, including: Myles Allen, Dave Frame, Daithi Stone, Claudio Piani, Jim Hansen, Sylvia Knight, Carl Christensen, Tolu Aina, Neil Masey, Jamie Kettleborough, Nick Faull, Andrew Martin, Ben Booth, Ben Sanderson, Mat Collins, and many many others. I am also extremely grateful to all those who participated in the project by downloading and running the model.

OUP: My editors Dan Taber and Giulia Lipparini.

Keith Mansfield for initial discussions and John Grandidge for telling me 'of course you should write a book'.

My partner, Becca Heddle, for guidance on the accessibility of the content, helping me improve my writing skills, bouncing around ideas and concepts, and tolerating my frustrations with academic life.

David Stainforth, 2023

Glossary & Acronyms

AMOC: Atlantic Meridional Overturning Circulation.

Atlantic Meridional Overturning Circulation (AMOC): A large-scale circulation in the Atlantic ocean within which water flows northwards in the upper ocean and southwards in the deep ocean. It transports heat from the tropics to high northern latitudes and plays an important role in the circulation pattern of the global oceans. How it might change in response to increased atmospheric greenhouse gases could significantly affect the geographical pattern of global warming in the future.

Attractor: A collection of values for the variables of a stationary system—such as Lorenz '63 or perhaps the climate system without human-induced climate change or other external forcing—which the system approaches over time. Once a system is close to its attractor it stays close to it unless it is influenced by something external to the system. See Chapter 14. (Other definitions exist for more complex types of attractor.)

Average: A typical value. There are various definitions but in this book the term 'average' is usually used to describe what is more technically described as the 'mean', which is the result of adding together a set of numbers and dividing by the number of numbers.

BRD: Beyond Reasonable Doubt.

Climate: Please read Chapters 13–15.

Climate change: Please read Chapters 13–15.

Climate sensitivity: There are various definitions of climate sensitivity but in this book it refers to 'equilibrium climate sensitivity' which is the change in the global average surface temperature which ultimately results from a doubling of atmospheric carbon dioxide concentrations from pre-industrial conditions.

Climate system: A complicated multi-element system which includes the atmosphere, the oceans, sea ice, ice sheets, rainforests, ecosystems, and more.

Climate*predcition*.net: A public resource distributed computing project run out of Oxford University.

Deep uncertainty: The outcome of a situation may be uncertain but we might still know, or have a good idea of, the probabilities of different outcomes. When we do not know, or have a good idea of, the probabilities of different outcomes then the situation is described as having deep uncertainty.

Detection and Attribution (D&A): The process of identifying that some aspect of the climate system has changed and relating that change to some cause, usually the changes in atmospheric greenhouse gas concentrations resulting from human activities.

Economic system: The social system by which goods and services are produced, allocated, and distributed within a society or region.

Ecosystem: A collection of interacting organisms and the physical environment within which they reside.

ESM: Earth System Model.

Extrapolation: Drawing conclusions about the behaviour of a system in a state that is different to any state for which we have observations.

GCM: Global Climate Model.

GMT: Global Mean Temperature. This is sometimes referred to as Global Mean Surface Temperature (GMST). In this book it is usually referred to as global average temperature.

Greenhouse gases: Gases in the atmosphere that absorb and emit long wave radiation and thereby contribute to the greenhouse effect. The most significant ones are water vapour, carbon dioxide, methane and nitrous oxide.

IAM: Integrated Assessment Model.

ICE: Initial Condition Ensemble.

IPCC: Intergovernmental Panel on Climate Change.

Italic words: In this book words that refer to time within the world of a model are often italicized.

Land-use change: Changes in land cover that affect the exchange of greenhouse gases between the terrestrial biosphere and the atmosphere. One example is when forests are cleared for agricultural production or to build settlements.

Long-lived greenhouse gases: Greenhouse gases that stay in the atmosphere for long periods of time—typically decades to centuries. The term typically refers to carbon dioxide, methane, nitrous oxide and a collection of other gases that play a smaller role in the human-induced, enhanced, greenhouse effect. Methane is sometimes considered a short-lived greenhouse gas.

Mean: An average which is the result of adding together a set of numbers and dividing by the number of numbers.

Median: A value for which half of a set of numbers are higher, and half lower.

MIP: Model Intercomparison Project.

NAO: North Atlantic Oscillation.

Paleoclimate: The Earth's climate in periods before we had direct measurements of things like temperature or rainfall.

PPE: Perturbed Parameter Ensemble or Perturbed Physics Ensemble.

Pre-industrial: Before human societies began large-scale industrial activity—typically taken to be about 1750. For the purposes of estimating representative pre-industrial values for variables such as global average temperature, the period 1850–1900 is usually considered acceptable because between 1750 and 1900 human greenhouse gas emissions were relatively small.

RCM: Regional Climate Model.

Short-lived gases: Gases that interact with radiation and/or lead to changes in the atmosphere that affect how radiation passes through it, but that stay in the atmosphere for only short periods of time—typically hours to decades. The term typically refers to aerosols, ozone, black carbon, and sometimes methane. Methane is, however, often referred to as a long-lived gas.

SMILE: Single Model Initial condition Large Ensemble.

Social cost of carbon: The cost of the future damages—usually expressed in US dollars—related to the emission of one extra tonne of carbon at some point in time.

Stationary system: A system whose average and variability are not themselves changing over time.

UNFCCC: United Nations Framework Convention on Climate Change.

Weather: What you get. See Chapter 13.

Data Acknowledgements

I have used various datasets for some of the figures in this book and also for some of the numbers quoted in the text. I very gratefully acknowledge all the hard work done by many groups in generating these various datasets and in making them widely available. In the endnotes, I acknowledge them simply by name but the appropriate citations and acknowledgements are as follows:

E-Obs: I acknowledge the E-OBS data set from the EU-FP6 project UERRA (http://www.uerra.eu) and the data providers in the ECA&D project (https://www.ecad.eu).

Citation: Cornes, R., van der Schrier. G., van den Besselaar, E.J.M., & Jones, P.D. An Ensemble Version of the E-OBS Temperature and Precipitation Datasets, *J. Geophys. Res. Atmos.*, 123. doi:10.1029/2017JD028200 (2018).

CMIP6: I acknowledge the World Climate Research Programme, which, through its Working Group on Coupled Modelling, coordinated and promoted CMIP6. I thank the climate modelling groups for producing and making available their model output, the Earth System Grid Federation (ESGF) for archiving the data and providing access, and the multiple funding agencies who support CMIP6 and ESGF.

Citation: Eyring, V. et al. Overview of the Coupled Model Intercomparison Project Phase 6 (CMIP6) experimental design and organization. *Geosci. Model Dev.* 9, 1937–1958, doi:10.5194/gmd-9-1937-2016 (2016).

Output was used from the following models:

Model	Institution	Model	Institution
ACCESS-CM2	CSIRO-ARCCSS	GFDL-ESM4	NOAA-GFDL
ACCESS-ESM1-5	CSIRO	GISS-E2-1-G	NASA-GISS
AWI-CM-1-1-MR	AWI	HadGEM3-GC31-LL	MOHC NERC
BCC-CSM2-MR	BCC	HadGEM3-GC31-MM	MOHC
CAMS-CSM1-0	CAMS	INM-CM4-8	INM
CESM2	NCAR	INM-CM5-0	INM
CESM2-WACCM	NCAR	IPSL-CM6A-LR	IPSL
CIESM	THU	KACE-1-0-G	NIMS-KMA
CMCC-CM2-SR5	CMCC	MCM-UA-1-0	UA
CNRM-CM6-1	CNRM-CERFACS	MIROC-ES2L	MIROC
CNRM-CM6-1-HR	CNRM-CERFACS	MIROC6	MIROC
CNRM-ESM2-1	CNRM-CERFACS	MPI-ESM1-2-HR	MPI-M DWD DKRZ
CanESM5	CCCma	MPI-ESM1-2-LR	MPI-M AWI DKRZ DWD
CanESM5-CanOE	CCCma	MRI-ESM2-0	MRI

Model	Institution	Model	Institution
EC-Earth3	EC-Earth-Consortium	NESM3	
EC-Earth3-Veg	EC-Earth-Consortium	NorESM2-LM	NCC
FGOALS-f3-L	CAS	NorESM2-MM	NCC
FGOALS-g3	CAS	UKESM1-0-LL	MOHC NERC NIMS-KMA NIWA
FIO-ESM-2-0	FIO-QLNM		

HadCRUT5: HadCRUT.5 data were obtained from http://www.metoffice.gov.uk/hadobs/hadcrut5 in November 2022 and are © British Crown Copyright, Met Office 2021, provided under an Open Government License, http://www.nationalarchives.gov.uk/doc/open-government-licence/version/3/

Citation: Morice, C.P., et al. An Updated Assessment of Near-Surface Temperature Change From 1850: The HadCRUT5 Data Set. *Journal of Geophysical Research-Atmospheres* **126**, doi:10.1029/2019jd032361 (2021).

Notes

Chapter 1

a A discussion of issues related to the threats to butterflies in Europe—which references the Sierra Nevada Blue—can be found in: Warren, M.S., et al. The decline of butterflies in Europe: Problems, significance, and possible solutions. *Proc. Natl. Acad. Sci. U. S. A.* **118**, doi:10.1073/pnas.2002551117 (2021).
The Sierra Nevada Blue was highlighted in The Guardian's list of '10 Species Most At Risk from Climate Change', in 2017: https://www.theguardian.com/environment/2017/jan/19/critical-10-species-at-risk-climate-change-endangered-world
Specific discussion of the threats to the Sierra Nevada Blue can be found in: Munguira, M.L., et al. Ecology and recovery plans for the four Spanish endangered endemic butterfly species. *Journal of Insect Conservation* **21**, 423–437, doi:10.1007/s10841-016-9949-8 (2017).

Chapter 2

a For a discussion of the sources and history of this quote, see: https://quoteinvestigator.com/2013/10/20/no-predict/ (accessed on 1/1/2023).
Based on that website, it seems that the original source is: 1948, Farvel Og Tak: Minder Og Meninger by K.K. Steincke (Farvel Og tak: Ogsaa en Tilvaerelse IV (1935–1939)), page 227, Forlaget Fremad, *København* (Fremad, Copenhagen, Denmark).

b Guidance on carrying out the experiment is available from the Institute of Physics: https://spark.iop.org/boyles-law

c Date: 1/10/21.

Chapter 3

a See Craig, R., & Mindell, J., et al., Health Survey for England-2012, Chapter 10: https://digital.nhs.uk/data-and-information/publications/statistical/health-survey-for-england/health-survey-for-england-2012

b Based on E-Obs data.

c 2021.

Chapter 4

a Apparently George Stigler, who won the Nobel Memorial Prize in Economic Sciences in 1982, used to say: 'If you never miss the plane, you're spending too much time in airports.' See Jordan Ellenburg, J., 'Be more productive: miss some flights', *Wired Magazine*, August 2014: https://www.wired.co.uk/article/jordan-ellenburg

Alternatively see: Geiling, N., 'If You've Never Missed a Flight, You're Probably Wasting Your Time', Smithsonian Magazine, 23/6/2014: https://www.smithsonianmag.com/travel/case-missing-your-next-airline-flight-180951650/

b   Agriculture is the largest source of nitrous oxide emissions in the United States. A discussion of the sources of various greenhouse gases, including nitrous oxide, is available from the US Environmental Protection Agency website: https://www.epa.gov/ghgemissions/overview-greenhouse-gases

c   Source data is from the online data for Table 14.1 of the Australian Health Survey: First Results, 2011–2012. Available from: http://www.abs.gov.au/ausstats/abs@.nsf/Lookup/4338.0main+features212011-13 (accessed in December 2022).

d   Date: 2021.

e   IPCC, 2021: Summary for Policymakers. In: Climate Change 2021: The Physical Science Basis. Contribution of Working Group I to the Sixth Assessment Report of the Intergovernmental Panel on Climate Change. Masson-Delmotte, V., et al. (eds.)]. *Climate Change 2021: The Physical Science Basis.* Cambridge University Press, Cambridge, United Kingdom and New York, NY, USA, pp. 3–32, doi:10.1017/9781009157896.001.

f   See: Otto-Bliesner, B.L., et al. How warm was the last interglacial? New model-data comparisons. *Philos. Trans. R. Soc. A-Math. Phys. Eng. Sci.* **371**, doi:10.1098/rsta.2013.0097 (2013).

g   Dutton, A., & Lambeck, K. Ice volume and sea level during the last interglacial. *Science* **337**, 216–219, doi:10.1126/science.1205749 (2012).
Kopp, R.E., Simons, F.J., Mitrovica, J.X., Maloof, A.C., & Oppenheimer, M. Probabilistic assessment of sea level during the last interglacial stage. *Nature* **462**, 863–U851, doi:10.1038/nature08686 (2009).

h   These numbers, and Figure 4.2, arise from an extremely simplistic calculation where the daily summer temperatures for the period 1991–2020 are all shifted up by 1.1°C to get a new probability distribution and hence a change in the probability of exceeding certain values. I am in no way recommending this approach. There are better and more complex ways of examining what might actually happen in a particular location—global change is a bad starting point which underestimates what one should expect for most land areas. Furthermore, there are ways to consider how the whole shape of the probability distribution is changing and how it might change in the future. These crop up later in the book but at this point I am sticking with something simple, just to illustrate the point.

i   Data acknowledgement: Menne, Matthew J., Imke Durre, Bryant Korzeniewski, Shelley McNeill, Kristy Thomas, Xungang Yin, Steven Anthony, Ron Ray, Russell S. Vose, Byron E.Gleason, and Tamara G. Houston (2012): Global Historical Climatology Network–Daily (GHCN-Daily), Version 3. [Chicago Midway Airport]. NOAA National Climatic Data Center. doi:10.7289/V5D21VHZ [15/12/2022].
Citation: Matthew J. Menne, Imke Durre, Russell S. Vose, Byron E. Gleason, and Tamara G. Houston, 2012: An Overview of the Global Historical Climatology Network-Daily Database. *J. Atmos. Oceanic Technol.*, 29, 897–910. doi:10.1175/JTECH-D-11-00103.1.
Data available from: https://www.ncdc.noaa.gov/cdo-web/datasets

Chapter 5

a   See: Campbell, D., Crutchfield, J., Farmer, D., & Jen, E. Experimental mathematics: the role of computation in nonlinear science. *Communications of the ACM* **28**, 374–384, doi:10.1145/3341.3345 (1985).

And: Campbell, D. K. Nonlinear physics—Fresh breather. *Nature* **432**, 455–456, doi:10.1038/432455a (2004).

b   For a discussion of these issues, see 'The First Edge' in *What We Cannot Know* by Marcus du Sautoy. 4th Estate, an imprint of Harper Collins. Specifically the section: 'Knowing my dice'.

c   *A Very Short Introduction to Chaos*, Leonard Smith, Oxford University Press.

d   There is no agreed term for what is going on in this situation. It is different from what is usually described as 'chaos' but it includes many of the same characteristics. The term 'pandemonium' was used by Prof E. Spiegel in 1987 to describe a somewhat analogous situation but in a very different context. This was picked up on by myself and colleagues to describe the situation being discussed here, in a 2007 paper. These papers are:

Spiegel, E.A. Chaos—a mixed metaphor for turbulence. *Proceedings of the Royal Society of London Series a-Mathematical and Physical Sciences* **413**, 87–95, doi:10.1098/rspa.1987.0102 (1987).

Stainforth, D.A., Allen, M.R., Tredger, E.R., & Smith, L.A. Confidence, uncertainty and decision-support relevance in climate predictions. *Philos. Trans. R. Soc. A-Math. Phys. Eng. Sci.* **365**, 2145–2161, doi:10.1098/rsta.2007.2074 (2007).

e   See, for instance, Deser, C., Knutti, R., Solomon, S., & Phillips, A.S. Communication of the role of natural variability in future North American climate. *Nature Climate Change* **2**, 775–779, doi:10.1038/nclimate1562 (2012).

Chapter 6

a   'Measuring digital development: Facts and Figures 2022', International Telecommunication Union (ITU): https://www.itu.int/itu-d/reports/statistics/facts-figures-2022/

b   Manabe, S., & Wetherald, R.T. The effects of doubling the CO_2 concentration on the climate of a general circulation model. *Journal of the Atmospheric Sciences* **32**, 3–15, doi:10.1175/1520-0469 (1975).

c   See: 'The Evolution of Climate Science: A Personal View from Julia Slingo', *WMO Bulletin* **66** (1) (2017). (https://public.wmo.int/en/resources/bulletin/evolution-of-climate-science-personal-view-from-julia-slingo)
Also see: https://naturedocumentaries.org/13311/build-climate-laboratory-julia-slingo-royal-institution-2016/

d   Arrhenius, S. On the influence of carbonic acid in the air upon the temperature of the ground. *The London, Edinburgh and Dublin Philosophical Magazine and Journal of Science* **41**, 237–276 (1896).

e   See: Bauer, P., Thorpe, A., & Brunet, G. The quiet revolution of numerical weather prediction. *Nature* **525**, 47–55, doi:10.1038/nature14956 (2015).
And: Alley, R.B., Emanuel, K.A., & Zhang, F.Q. Advances in weather prediction. *Science* **363**, 342–344, doi:10.1126/science.aav7274 (2019).

Chapter 7

a   https://en.wikipedia.org/wiki/Standing_on_the_shoulders_of_giants

b   Besley, T, and Hennessy, P., 'The global financial crisis—why didn't anybody notice?'. *British Academy Review*, Issue 14 (November 2009): https://www.thebritishacademy.

ac.uk/publishing/review/14/british-academy-review-global-financial-crisis-why-didnt-anybody-notice/

Chapter 8

a https://en.m.wikipedia.org/wiki/List_of_ships_sunk_by_icebergs,   https://www.bbc.co.uk/news/magazine-17257653,   https://web.archive.org/web/20120315031947/http://www.icedata.ca/Pages/ShipCollisions/ShipCol_OnlineSearch.php

b https://quoteinvestigator.com/2020/03/17/own-facts/,   https://en.wikipedia.org/wiki/Daniel_Patrick_Moynihan

c https://www.ucsusa.org/climate/science,   https://www.reutersevents.com/sustainability/science-clear-about-climate-change-now-everyone-has-step,   https://www.gov.uk/government/speeches/the-science-is-clear-global-warming-is-real

d https://gml.noaa.gov/ccgg/trends/gl_data.html

e https://www.eea.europa.eu/data-and-maps/indicators/atmospheric-greenhouse-gas-concentrations-7/assessment

f Current global average, annual average temperature levels are already thought to exceed those seen over the last 12,000 years, so further warming will clearly take us beyond that experienced by human civilisation.
For an assessment of historic temperatures, see:
Bova, S., Rosenthal, Y., Liu, Z.Y., Godad, S.P., & Yan, M. Seasonal origin of the thermal maxima at the Holocene and the last interglacial. *Nature* **589**, 548–553, doi:10.1038/s41586-020-03155-x (2021).

g See: IPCC, 2022: *Climate Change 2022: Impacts, Adaptation, and Vulnerability*. Contribution of Working Group II to the Sixth Assessment Report of the Intergovernmental Panel on Climate Change. Pörtner, H.-O., et al. (eds.)]. Cambridge, UK and New York, NY, USA, Cambridge University Press, pp., 3056, doi:10.1017/9781009325844.
See also: Rising, J., Tedesco, M., Piontek, F., & Stainforth, D.A. The missing risks of climate change. *Nature* **610**, 643–651, doi:10.1038/s41586-022-05243-6 (2022).

h For information about climate change and sea turtles, see:
Blechschmidt, J., Wittmann, M.J., & Bluml, C. Climate change and green sea turtle sex ratio-preventing possible extinction. *Genes* **11**, doi:10.3390/genes11050588 (2020).
and: Esteban, N., et al. Optimism for mitigation of climate warming impacts for sea turtles through nest shading and relocation. *Scientific Reports* **8**, doi:10.1038/s41598-018-35821-6 (2018).
A list of seven animals affected by climate change—including sea turtles—is available at: https://www.gvi.co.uk/blog/6-animal-species-and-how-they-are-affected-by-climate-change/

Chapter 9

a 9/1/2020.

b Bank of England, Monetary Policy Report, November 2019: https://www.bankofengland.co.uk/monetary-policy-report/2019/november-2019

c Well actually this is true for forecasts up to nine months ahead—the one-year forecasts aren't quite so good.

See: Independent Evaluation Office, Bank of England, Evaluating forecast performance, November 2015: https://www.bankofengland.co.uk/independent-evaluation-office/forecasting-evaluation-november-2015

d See: https://data.worldbank.org/indicator/NY.GDP.MKTP.KD.ZG?locations=GB

e Data source: E-Obs

Chapter 10

a https://en.wikipedia.org/wiki/Black_swan_theory

b I should acknowledge that we don't fully understand all the laws of physics. How matter behaves in a black hole, inside a star or during the big bang may be different from what we currently understand and experience here on earth. There are fascinating questions around this. Nevertheless, the laws applicable to objects and materials familiar to us humans and to the earth's climate are well understood and it's only reasonable to expect that they will continue to apply here on earth, at least for many millennia. There is no evidence to suggest this isn't the case.

There is a parallel here with the applicability of Newton's laws of motion and gravitation. Einstein's theories of relativity explained how Newton's laws of motions were flawed; they aren't 100% correct for describing the universe in which we live. In terms of progress in physics and our understanding of the universe, this was a big leap forward. For the vast majority of practical situations which most humans experience, however, Newton's laws of motion remain more than adequate to describe what is going on. They had demonstrated their efficacy for more than 200 years before the flaw was found. This itself is an indication that, in most practical situations, they suffice.

The same is true regarding the constancy of the laws of physics and the accuracy of our current understanding of them. Future research may lead us to revise some of the details which are crucial for extreme situations such as in black holes or during the big bang, but that doesn't undermine their reliability to the here and now on earth.

c The middle plots have white noise added onto the underlying signal. The lower plots are based on a first-order auto-regressive process added on to the underlying signal.

d Source data: HadCRUT5.

e Source data: The Hadley Centre Central England Temperature (HadCET) dataset, www.metoffice.gov.uk/hadobs For details see: Parker, D.E., Legg, T.P., & Folland, C.K. A new daily Central England Temperature Series, 1772–1991. *Int. J. Clim.*, Vol 12, pp 317–342 (1992).

f These are two simulations with the MIROC model in the CMIP6 ensemble. Data acknowledgment: CMIP6.

g These two simulations use the Hasslemann model as utilised and described in Calel et al. 2021:

Hasselmann, K. Stochastic climate models Part I. Theory. *Tellus* **28**, 473–485, doi:https://doi.org/10.1111/j.2153-3490.1976.tb00696.x (1976).

Calel, R., Chapman, S.C., Stainforth, D.A., & Watkins, N.W. Temperature variability implies greater economic damages from climate change. *Nature Communications* **11**, doi:10.1038/s41467-020-18797-8 (2020).

h Take a look at Figure 5.2 of the IPCC Working Group 1 report from 2013 for a nice summary of what we know about historic atmospheric carbon dioxide concentrations going back about 65 million years.

See: Masson-Delmotte, V., et al., 2013: Information from paleoclimate archives. In: Climate Change 2013: The Physical Science Basis. Contribution of Working Group I to the Fifth Assessment Report of the Intergovernmental Panel on Climate Change [Stocker, T.F., et al. (eds.)]. Cambridge University Press, Cambridge, United Kingdom and New York, NY, USA.

i There are different ways to consider what constitutes the rise of human civilisation. The development of agriculture was about 12,000 years ago, while cities or proto-cities came a bit later.

See: https://en.wikipedia.org/wiki/Human_history

j Data from the Sea Ice Index from the National Snow and Ice Data Center.

Citation: Fetterer, F., Knowles, K., Meier, W.N., Savoie, M., and Windnagel, A.K. (2017). Sea Ice Index, Version 3 [Data Set]. Boulder, Colorado USA. National Snow and Ice Data Center. https://doi.org/10.7265/N5K072F8. Date Accessed 01-02-2023.

Source: https://nsidc.org/data/seaice_index/data-and-image-archive

k This is a quote from 1980s UK single 'Star Trekkin'. The original *Star Trek* TV series included something almost identical: 'I can't change the laws of physics. I've got to have thirty minutes.'

Chapter 11

a Although I discussed Boyle's law in terms of temperature, pressure, and volume, both it and the ideal gas equation can be expressed in terms of temperature, pressure, and either volume or density.

b Details of the resolution of various CMIP6 models is available at: https://wcrp-cmip.github.io/CMIP6_CVs/docs/CMIP6_source_id.html

For the slightly older CMIP5 models, there is a list of resolutions in Table 9.A.1 in Chapter 9 of the IPCC WG1 Fifth Assessment Report, 2013:

Flato, G., et al., 2013: Evaluation of climate models. In: Climate Change 2013: The Physical Science Basis. Contribution of Working Group I to the Fifth Assessment Report of the Intergovernmental Panel on Climate Change [Stocker, T.F., et al. (eds.)]. Cambridge University Press, Cambridge, United Kingdom and New York, NY, USA.

c For instance see: Boe, J., Somot, S., Corre, L., & Nabat, P. Large discrepancies in summer climate change over Europe as projected by global and regional climate models: causes and consequences. *Clim. Dyn.* **54**, 2981–3002, doi:10.1007/s00382-020-05153-1 (2020).

d It's difficult to be clear just how many GCMs contributed to the IPCC sixth assessment report as different sets were used in different experiments which contributed to different sections of the report via a huge number of academic papers. Furthermore, the model datasets have continued to expand since and, by some counts, are in excess of 100. The number I use in the text—forty—is likely a quite conservative estimate.

e Source: CMIP6.

f Source: HadCRUT5.

g For a list of Model InterComparison Projects (MIPS) see: https://www.wcrp-climate.org/modelling-wgcm-mip-catalogue/modelling-wgcm-cmip6-endorsed-mips. Citations for specific ones include:

Clouds: Webb, M.J., et al. The Cloud Feedback Model Intercomparison Project (CFMIP) contribution to CMIP6. *Geoscientific Model Development* **10**, 359–384, doi:10.5194/gmd-10-359-2017 (2017).

Ice: Nowicki, S.M.J., et al. Ice Sheet Model Intercomparison Project (ISMIP6) contribution to CMIP6. *Geoscientific Model Development* **9**, 4521–4545, doi:10.5194/gmd-9-4521-2016 (2016).

Geoengineering: Kravitz, B., et al. The Geoengineering Model Intercomparison Project Phase 6 (GeoMIP6): simulation design and preliminary results. *Geoscientific Model Development* **8**, 3379–3392, doi:10.5194/gmd-8-3379-2015 (2015).

h   AgMIP. Muller, C., et al. The Global Gridded Crop Model Intercomparison phase 1 simulation dataset. *Scientific Data* **6**, doi:10.1038/s41597-019-0023-8 (2019).

i   ISIMIP: Rosenzweig, C., et al. Assessing inter-sectoral climate change risks: the role of ISIMIP. *Environ. Res. Lett.* **12**, doi:10.1088/1748-9326/12/1/010301 (2017).

Chapter 12

a   Cook, J., et al. Consensus on consensus: a synthesis of consensus estimates on human-caused global warming. *Environ. Res. Lett.* **11**, doi:10.1088/1748-9326/11/4/048002 (2016).

b   Henrion, M., & Fischhoff, B. Assessing uncertainty in physical constants. *Am. J. Phys.* **54**, 791–798, doi:10.1119/1.14447 (1986).

c   A discussion of our tendency to underestimate uncertainty, specifically in the context of climate change, can be found in:

Morgan, M.G., & Keith, D.W. Improving the way we think about projecting future energy use and emissions of carbon dioxide. *Climatic Change* **90**, 189–215, doi:10.1007/s10584-008-9458-1 (2008).

d   This figure is constructed from data in Henrion and Fischhoff (1986)—see endnote [b]. They also discuss similar results for a number of other physical constants.

Chapter 13

a   1431 on 17/12/20.

b   IPCC. Annex VII: Glossary [Matthews, J.B.R., et al. (eds.)]. In: Masson-Delmotte, V., et al. (eds.). Climate Change 2021: The Physical Science Basis Contribution of Working Group I to the Sixth Assessment Report of the Intergovernmental Panel on Climate Change. Cambridge, United Kingdom and New York, NY, USA: Cambridge University Press, 2021, pp. 2215–2256.

c   https://www.metlink.org/resource/weather-glossary/ (downloaded on 18/11/2021).

d   American Meteorological Society Glossary of Meteorology 2002 and online https://glossary.ametsoc.org/wiki/Climate (accessed on 18/11/2021).

e   See, for instance,: https://en.wikipedia.org/wiki/Climate_of_Chicago (18/11/2021), https://weatherspark.com/y/14091/Average-Weather-in-Chicago-Illinois-United-States-Year-Round (18/11/2021).

f   The 1959 AMS definition in full is: *Climate*—'The synthesis of weather' (Durst, C.S.); the long-term manifestations of weather, however they may be expressed. More rigorously, the climate of a specified area is represented by the statistical collective of its weather conditions during a specified interval of time (usually several decades).

AMS. Glossary of Meteorology, 1st ed. Huschke RE (ed.). Boston, MA: American Meteorological Society, 1959.

g Thanks to Prof L. Smith for bringing the AMS 1959 and Kendrew 1938 quotes to my attention.

h Kendrew, W.G. *Climate: A Treatise on the Principles of Weather and Climate*, 2nd ed. Oxford: The Clarendon Press, 1938.

i Fryar, C.D., Carroll, M.D., Gu, Q., Afful, J., & Ogden, C.L. Anthropometric reference data for children and adults: United States, 2015–2018. National Center for Health Statistics. *Vital Health Stat* **3**(46) (2021). (https://www.cdc.gov/nchs/data/series/sr_03/sr03-046-508.pdf)

j Under some assumptions about the data collected on women's heights.

k In basic statistics, the standard error in the mean is equal to the standard deviation of the sample—roughly 7 cm for the height of American Women—divided by the square root of the sample size—the number of women whose height was measured. Reducing the sample size from 5000 to 50 would therefore increase the standard error in the mean by a factor of 10. In this particular study the authors used a more complex procedure to calculate the standard error in the mean to allow for the complexity of the sample design which covered different ages and races. The impact on the standard error of reducing the sample size could therefore be even greater than that deduced from the basic statistical formula.

l Data from Statistics Netherlands (CBS): https://opendata.cbs.nl/statline/#/CBS/en/dataset/81175ENG/table?fromstatweb

m Of course, comparing the heights of American and Dutch women in this way only works if the different studies are comparable. They may be done differently. They may be from different time periods—heights are changing over time. They may have different numbers of samples and different sampling strategies to allow for different ethnic groups etc., and as a result the uncertainties may be different. Some studies ask participants to measure their own height, while others get a third person to measure the heights: self-reported heights tend to be greater than those measured by an interviewer. All these things mean you shouldn't read too much into the specifics of the height example I give but the point still stands that averages are a good starting point for telling if distributions differ.

n Based on E-Obs data.

o See, for instance:

Hsiang, S., et al. Estimating economic damage from climate change in the United States. *Science*. **356**(6345), 1362–9 (2017).

Zivin, J.G., & Neidell, M.J. *Temperature and the Allocation of Time: Implications for Climate Change*. Cambridge, MA: National Bureau of Economic Research (2010).

Seppanen, O., Fisk, W.J., & Faulkner, D., in Annual Meeting of the American-Society-of-Heating-Refrigerating-and-Air-Conditioning-Engineers (ASHRAE), pp. 680–686 (2005).

p Kang, Y.H., Khan, S., & Ma, X.Y. Climate change impacts on crop yield, crop water productivity and food security—a review. *Progress in Natural Science-Materials International* **19**, 1665–1674, doi:10.1016/j.pnsc.2009.08.001 (2009).

Challinor, A.J., & Wheeler, T.R. Crop yield reduction in the tropics under climate change: Processes and uncertainties. *Agric. For. Meteorol.* **148**, 343–356, doi:10.1016/j.agrformet.2007.09.015 (2008).

Porter, J.R., & Semenov, M.A. Crop responses to climatic variation. *Philos. Trans. R. Soc. B-Biol. Sci.* **360**, 2021–2035, doi:10.1098/rstb.2005.1752 (2005).

Challinor, A.J., Ewert, F., Arnold, S., Simelton, E., & Fraser, E. Crops and climate change: progress, trends, and challenges in simulating impacts and informing adaptation. *Journal of Experimental Botany* **60**, 2775–2789, doi:10.1093/jxb/erp062 (2009).

Chapter 14

a See discussions in:
https://medium.com/choice-hacking/how-ikea-used-psychology-to-become-the-worlds-largest-furniture-retailer-7444a502daa1,
https://theconversation.com/the-ikea-effect-how-ingvar-kamprads-company-changed-the-way-we-shop-90896,
https://www.bbc.com/worklife/article/20180201-how-ikea-has-changed-the-way-weshop

b At time of writing, there are several examples of code to generate the Lorenz '63 paths on the its wikipedia page (https://en.wikipedia.org/wiki/Lorenz_system) and a number of websites providing interactive visualizations of them.

c A short, twelve-page, discussion of the history of chaos theory is:
Oestreicher, C. A history of chaos theory. *Dialogues Clin. Neurosci.* **9**, 279–289, doi:10.31887/DCNS.2007.9.3/coestreicher (2007).
This is available from: https://www.tandfonline.com/doi/full/10.31887/DCNS.2007.9.3/coestreicher

d Zhang, F.Q., et al. What is the predictability limit of midlatitude weather? *Journal of the Atmospheric Sciences* **76**, 1077–1091, doi:10.1175/jas-d-18-0269.1 (2019).

e Two of my favourites are *Chaos: A Very Short Introduction* by Leonard Smith, and *Does God Play Dice: The New Mathematics of Chaos* by Ian Stewart.

Chapter 15

a In Lorenz–Stommel, the fast variables actually represent a circulation pattern in the atmosphere known as the Hadley circulation, in which air rises near the equator, flows north or south high above the earth's surface, descends in the sub-tropics and then flows back towards the equator near the surface. The slow ones relate to a type of ocean circulation known as a thermohaline circulation which results from the changing density of seawater as a consequence of heating, cooling, rainfall and evaporation; there is a strong thermohaline circulation in the North Atlantic.

b Date: 6/1/2021.

Chapter 16

a The mathematicians haven't yet worked out quite how to describe the situation under climate change and the conventional definition of an attractor doesn't apply.

b Stainforth, D.A., Allen, M.R., Tredger, E.R., & Smith, L.A. Confidence, uncertainty and decision-support relevance in climate predictions. *Philos. Trans. R. Soc. A-Math. Phys. Eng. Sci.* **365**, 2145–2161, doi:10.1098/rsta.2007.2074 (2007).

c These results are from a 489-member micro-initial-condition ensemble from climate*prediction*.net. Apart from each simulation using the same model version, the experimental setup was the same as described in:

Stainforth, D.A. *et al.* Uncertainty in predictions of the climate response to rising levels of greenhouse gases. *Nature* 433, 403–406, doi:10.1038/nature03301 (2005).

d The actual experiment asked volunteers to simulate forty-five years, fifteen of which looked at the response to carbon dioxide doubling.

e Date: 2021.

f The National Center for Atmospheric Research Community Climate System Model version 3 (CCSM3).

g There are natural cycles and substantial variability in the circulation patterns of the ocean. These are sometimes studied by running simulations of the climate system for hundreds of years without climate change. This is kinda like producing observations of the complete state of the world's oceans from, say, the year 0 to 1000 CE, only it is all in the world of the model. A climate change simulation can then be initiated by picking a state of the oceans from any point in this simulation. This is a common approach. In this experiment though, we didn't take just one state but several, and for each one we repeated the first part of the experiment.

h Figure 16.2 shows results from the experiment presented in:

Hawkins, E., Smith, R., Gregory, J., & Stainforth, D.A. Irreducible uncertainty in near-term climate projections. *Clim. Dyn.* 1–13, doi:10.1007/s00382-015-2806-8 (2015).

i Figure 16.3 shows results from the same experiment as used in Figure 16.2.

Chapter 17

a Source data: E-Obs.

b Arias, P.A., et al. Technical summary. In: Masson-Delmotte, V., et al. (eds.). *Climate Change 2021: The Physical Science Basis Contribution of Working Group I to the Sixth Assessment Report of the Intergovernmental Panel on Climate Change.* Cambridge, United Kingdom and New York, NY, USA: Cambridge University Press, 2021, pp. 33–144.

c Source data: E-Obs. This plot uses the methodology in the paper below but is more up-to-date: it uses more recent data (up to 2020) and is based on ten samples where each one includes eleven years of data.

Stainforth, D.A., Chapman, S.C., & Watkins, N.W. Mapping climate change in European temperature distributions. *Environ. Res. Lett.* **8** (2013).

d The analysis in figure 17.3 is as in the paper below but using more recent data (as in figure 17.2) and looking at changes over a 51 year period.

Stainforth, D.A., Chapman, S.C., & Watkins, N.W. Mapping climate change in European temperature distributions *Environ. Res. Lett.* **8** (2013).

e See for instance: Hsiang, S.M. Temperatures and cyclones strongly associated with economic production in the Caribbean and Central America. *Proc. Natl. Acad. Sci. U. S. A.* **107**, 15367–15372, doi:10.1073/pnas.1009510107 (2010).

And: Zhu, J., Poulsen, C.J., & Otto-Bliesner, B.L. High climate sensitivity in CMIP6 model not supported by paleoclimate. *Nature Climate Change*, doi:10.1038/s41558-020-0764-6 (2020).

f See, for instance: Baccini, M., et al. Heat effects on mortality in 15 European cities. *Epidemiology* **19**, 711–719, doi:10.1097/EDE.0b013e318176bfcd (2008).

And: Gasparrini, A., et al. Mortality risk attributable to high and low ambient temperature: a multicountry observational study. *Lancet* **386**, 369–375, doi:10.1016/s0140-6736(14)62114-0 (2015).

g This figure uses the same methodology as in the paper below but is updated to use ten samples where each one includes eleven years of data, and presents results for changes over a fifty-one-year period.
Stainforth, D.A., Chapman, S.C., & Watkins, N.W. Mapping climate change in European temperature distributions *Environ. Res. Lett.* **8** (2013).

h Chapman, S.C., Murphy, E.J., Stainforth, D.A., & Watkins, N.W. Trends in winter warm spells in the central England temperature record. *JAMC* **59**, 1069–1076, doi:10.1175/jamc-d-19-0267.1 (2020)

Chapter 18

a Hourdin, F., et al. The art and science of climate model tuning. *Bull. Amer. Meteorol. Soc.* **98**, 589–602, doi:10.1175/bams-d-15-00135.1 (2017).

b See: Moon, W., Agarwal, S. & Wettlaufer, J. S. Intrinsic Pink-Noise Multidecadal Global Climate Dynamics Mode. *Physical Review Letters* **121**, 108701, doi:10.1103/PhysRevLett.121.108701 (2018), and references therein,
Weather, Macroweather, and the Climate by S. Lovejoy, ISBN: 9780190864217,
and
Franzke, C. L. E., et al. The Structure of Climate Variability Across Scales. *Reviews of Geophysics* **58**, e2019RG000657, doi:https://doi.org/10.1029/2019RG000657 (2020), and references therein.

c We don't observe global average temperature directly. Rather we observe local temperatures across the surface of the planet from day to day and we calculate the 'observed' global average temperature from those observations. Doing that calculation is not trivial, particularly for the pre-satellite era when local temperature observations far from covered the whole planet. Much more could be said about this, but it is not important for the issues discussed in the main text.

d Source data: CMIP6 and HadCRUT5. Note that each simulation has been individually 'rebased' to its 1850–1900 average i.e. the timeseries has had its 1850–1900 average subtracted from every value.

e Source data: CMIP6 and HadCRUT5.

f Source data: CMIP6.

Chapter 19

a Well, we don't know of any systematic way at the moment and we have no reason to expect that any such way exists.

b It's interesting to note that the lack of ten, twelve, and twenty-sided dice clearly limits which high numbers can be achieved, but the lack of the ten-sided dice also restricts the outcome on the low side because, unlike the other dice, a typical ten-sided dice has numbers 0 to 9 while all the others start with 1.

c There is actually good reason for it taking a shape somewhat like this because of the way it is related to physical processes—more on that in Chapter 23.

d PCC. Summary for Policymakers. In: Stocker, T.F., Qin, D., Plattner, G.-K., Tignor, M., Allen, S.K., Boschung, J., et al., Eds. *Climate Change 2013: The Physical Science Basis Contribution of Working Group I to the Fifth Assessment Report of the Intergovernmental Panel on Climate Change.* Cambridge, United Kingdom and New York, NY, USA: Cambridge University Press, 2013: pp. 1–30.

e Table 7.13 in: Forster, P., Storelvmo, T., Armour, K., Collins, W., Dufresne, J.L., Frame, D., et al. The earth's energy budget, climate feedbacks, and climate sensitivity. In: Masson-Delmotte, V., Zhai, P., Pirani, A., Connors, S.L., Péan, C., Berger, S., et al., Eds. *Climate Change 2021: The Physical Science Basis Contribution of Working Group I to the Sixth Assessment Report of the Intergovernmental Panel on Climate Change.* Cambridge, United Kingdom and New York, NY, USA: Cambridge University Press, 2021: pp. 923–1054.

f Climate sensitivity values taken from: Meehl, G.A., *et al.* Context for interpreting equilibrium climate sensitivity and transient climate response from the CMIP6 Earth system models. *Science Advances* **6**, doi:10.1126/sciadv.aba1981 (2020).

g Blastland, M., & Dilnot A. *The Tiger That Isn't: Seeing Through a World of Numbers.* London: Profile Books, 2007.

h Stainforth, D.A., *et al.* Uncertainty in predictions of the climate response to rising levels of greenhouse gases. *Nature* **433**, 403–406, doi:10.1038/nature03301 (2005).

i Murphy, J.M., et al. UKCP18 Land Projections: November 2018. Met Office Crown Copyright 2018.

j The numbers are not randomly distributed across the board but rather represent a tendency for high parameter values to be associated with lower numbers. This reflects the situation with models where we might expect some sort of relationship between parameter values and results. However, quite what that relationship looks like can be very difficult to know.

Chapter 20

a Frigg et al., 2015 (see below) reference an earlier paper—Murphy et al., 2004 (see below)—saying:

'Murphy et al. calculate the CPIs [Climate Prediction Indices] for all 53 model versions and find that it varies between 5 and 8. This means that the average of the model's retrodictions are 5 to 8 standard deviations away from the observations.'

Based on this, and assuming a Gaussian distribution, five or more standard deviations away from observations implies a less than 0.00003% probability that the model would have generated the observations. This is the origin of the figure in the text. The source papers are:

Frigg, R., Smith, L., & Stainforth, D. An assessment of the foundational assumptions in high-resolution climate projections: the case of UKCP09. *Synthese*, 1–30, doi:10.1007/s11229-015-0739-8 (2015).

Murphy, J.M., et al. Quantification of modelling uncertainties in a large ensemble of climate change simulations. *Nature* **430**, 768–772, doi:10.1038/nature02771 (2004).

b Palmer, T.N. A CERN for climate change. *Physics World* **24**, 14 (2011).

Palmer, T., & Stevens, B. The scientific challenge of understanding and estimating climate change. *Proc. Natl. Acad. Sci. U. S. A.* **116**, 24390–24395, doi:10.1073/pnas.1906691116 (2019).

Chapter 21

a Dictionnaire philosophique, Voltaire, 1770 ed, Arts Dramatique,
   See also: Ratcliffe, S. *Concise Oxford Dictionary of Quotations*. Oxford: Oxford University Press, 2011, p. 389., ISBN 978-0199567072. (https://www.oxfordreference.com/view/10.1093/acref/9780191866692.001.0001/q-oro-ed6-00011218)

b See, for instance, the IPCC Atlas:
   Gutiérrez, J.M., et al. Atlas. In: Masson-Delmotte, V., et al. (eds.). Climate Change 2021: The Physical Science Basis Contribution of Working Group I to the Sixth Assessment Report of the Intergovernmental Panel on Climate Change 2021.

c The values represent the difference between the thirty-year average at the end of the twenty-first century and the fifty-year average 1850–1900 under the RCP8.5 scenario. Where multiple simulations were available for an individual model, the result was further averaged across those simulations. As a result, some of the points in this plot represent the expected value for that model much better than others. Source data: CMIP6.

d Specifically *The Treachery of Images*.

e Specifically *Le Petit Prince*.

f Stainforth, D.A., Downing, T.E., Washington, R., Lopez, A., & New, M. Issues in the interpretation of climate model ensembles to inform decisions. *Philos. Trans. R Soc. A-Math. Phys. Eng. Sci.* **365**(1857), 2163–2177 (2007).

g Blastland, M., & Dilnot A. *The Tiger That Isn't: Seeing Through a World of Numbers*. London: Profile Books, 2007.

h Stainforth, D.A., et al. Uncertainty in predictions of the climate response to rising levels of greenhouse gases. *Nature*. **433**(7024), 403–406 (2005).

i For an accessible summary of the argument see: https://en.wikipedia.org/wiki/Private_language_argument
   This references the quote in section 258 of: Wittgenstein, L. (2001) [1953]. *Philosophical Investigations*. Blackwell Publishing. ISBN 0-631-23127-7.

j Stainforth, D.A., Allen, M.R., Tredger, E.R., & Smith, L.A. Confidence, uncertainty and decision-support relevance in climate predictions. *Philos. Trans. R Soc. A-Math. Phys. Eng. Sci.* 365(1857), 2145–61 (2007).

k Hazeleger, W., et al. Tales of future weather. *Nature Climate Change* **5** (2015).

l Dessai, S., et al. Building narratives to characterise uncertainty in regional climate change through expert elicitation. *Environ. Res. Lett.* **13**, doi:10.1088/1748-9326/aabcdd (2018).

m Hawkins, E., et al. Millions of historical monthly rainfall observations taken in the UK and Ireland rescued by citizen scientists. *Geoscience Data Journal* **10**, 246–261, doi:10.1002/gdj3.157 (2023).

n Source data: E-Obs.

o Dinku, T. Challenges with availability and quality of climate data in Africa, Chapter 7 in *Extreme Hydrology and Climate Variability*. IRI, Elsevier, 2019.

p Much more on these methods is presented in:
   Decision Making Under deep Uncertainty: From Theory to Practice, Springer. Editors: Vincent A.W.J. Marchau, Warren E. Walker, Pieter J.T.M. Bloemen, & Steven W. Popper. ISBN: 978-3-030-05251-5 Published: 15 April 2019 https://doi.org/10.1007/978-3-030-05252-2.

q Date: 2021.

Chapter 22

a Date: 9/3/22.

b United Nations Framework Convention On Climate Change, United Nations, 1992
   https://unfccc.int/process-and-meetings/what-is-the-united-nations-framework-
   convention-on-climate-change

c See discussion in: Naomi Oreskes, Erik M. Conway, *Merchants of Doubt*, 2010 Blooms-
   bury Press, ISBN 978-1-59691-610-4.

d Consider, for instance, the ClimateGate affair: https://en.wikipedia.org/wiki/Climatic_
   Research_Unit_email_controversy
   https://stories.uea.ac.uk/the-story-behind-the-trick/

e Medhaug, I., Stolpe, M.B., Fischer, E.M., & Knutti, R. Reconciling controversies about the
   'global warming hiatus'. *Nature* **545**, 41–47, doi:10.1038/nature22315 (2017).

f Date: Autumn 2021.

g John Gummer, quoted in an article by Andy McSmith: 'Climate change should be the top
   priority for governments of the industrialised world', Monday Interview, *The Independent*,
   3 April 2006.

h Michael Meacher, End of the world nigh—it's official, *The Guardian*, 14/2/2003: https://
   www.theguardian.com/politics/2003/feb/14/environment.highereducation

i Otto-Bliesner et al., 2013.

j https://en.wikipedia.org/wiki/Eemian
   Kopp et al., 2009.

k https://en.wikipedia.org/wiki/Kirkdale_Cave

l Tierney, J. E. et al. Glacial cooling and climate sensitivity revisited. *Nature* **584**, 569–573,
   doi:10.1038/s41586-020-2617-x (2020).

m Fairbanks, R.G. A 17,000-year glacio-eustatic sea-level record—influence of glacial melting rates on the younger dryas event and deep-ocean circulation. *Nature* **342**, 637–642,
   doi:10.1038/342637a0 (1989).
   For a brief discussion of sea level and climate, see: https://pubs.usgs.gov/fs/fs2-00/

n See: National Snow and Ice Data Centre: https://nsidc.org/learn/what-cryosphere

o The curves show the energy (actually power i.e. energy per second) emitted at different
   wavelengths by a black body at a temperature typical of the surface of the sun (orange)
   and of the surface of the earth (green)—see Figure 2.1 in *The Physics of Atmospheres*, John
   Houghton, Cambridge University Press.
   The distributions are drawn with equal areas since in a steady state the total energy
   received by the earth from the sun is the same as that it losses to space. They illustrate
   how the distribution of different types of radiation emitted, varies with temperature.

p This is the Stefan-Boltzmann law: the energy radiated per metre squared per second
   equals σT^4, where σ is the Stefan-Boltzmann constant and T is absolute temperature.

q It's known as the Iris effect.

r Carbon dioxide concentrations from Mauna Loa: https://scrippsco2.ucsd.edu/data/
   atmospheric_co2/primary_mlo_co2_record.html

s Data from the NOAA Global Monitoring Laboratory: https://gml.noaa.gov/ccgg/trends/
   gl_data.html

t Figures 6.6 and 6.7 in:

Ciais, P., et al. Carbon and other biogeochemical cycles. In: Stocker T.F., et al. (eds.). *Climate Change 2013: The Physical Science Basis Contribution of Working Group I to the Fifth Assessment Report of the Intergovernmental Panel on Climate Change*. Cambridge, United Kingdom and New York, NY, USA: Cambridge University Press, 2013: pp 465–570.

u Data acknowledgements to the Scripps CO_2 program at the Scripps Institution of Oceanography at UC San Diego. Data citation:

Keeling, C.D., Piper, S.C., Bacastow, R.B., Wahlen, M., Whorf, T.P., Heimann, M., & Meijer, H.A., Exchanges of atmospheric CO_2 and $^{13}CO_2$ with the terrestrial biosphere and oceans from 1978 to 2000. I. Global aspects, SIO Reference Series, No. 01-06, Scripps Institution of Oceanography, San Diego, 88 pages, 2001.

Data available from: https://scrippsco2.ucsd.edu/data/atmospheric_co2/primary_mlo_co2_record.html. Data made available under Creative Commons Attribution 4.0 International License (http://creativecommons.org/licenses/by/4.0/). No changes were made to the data.

v Figure 5.4 in:

Canadell, J.G., et al. Global carbon and other biogeochemical cycles and feedbacks. In: Masson-Delmotte, V., et al. (eds.). *Climate Change 2021: The Physical Science Basis Contribution of Working Group I to the Sixth Assessment Report of the Intergovernmental Panel on Climate Change*. Cambridge, United Kingdom and New York, NY, USA: Cambridge University Press, 2021: pp. 673–816.

w Schmidt, G.A., Ruedy, R.A., Miller, R.L., & Lacis, A.A. Attribution of the present-day total greenhouse effect. *Journal of Geophysical Research-Atmospheres* **115** (2010), https://doi.org/10.1029/2010JD014287.

x Schmidt et al. 2010.

y See: https://www.eea.europa.eu/data-and-maps/indicators/atmospheric-greenhouse-gas-concentrations-7/assessment

z Data comes from: Friedlingstein P, O'Sullivan M, Jones MW, Andrew RM, Hauck J, Olsen A, et al. Global Carbon Budget 2020. *Earth System Science Data* 2020;12(4):3269–3340.

For details of the uncertainties in the values quoted in the table, and of emissions due to land use change, and of the distribution of the remaining carbon between the ocean uptake and land uptake, see Table 6 of Friedlinstein et al., 2020.

aa Friedlingstein, P., et al. Global Carbon Budget 2020. *Earth System Science Data*. **12**(4), 3269–340 (2020).

bb See FAQ 8.1 on page 666 of:

Myhre, G., et al. Anthropogenic and natural radiative forcing. In: Stocker T.F. et al., (eds.). *Climate Change 2013: The Physical Science Basis Contribution of Working Group I to the Fifth Assessment Report of the Intergovernmental Panel on Climate Change*. Cambridge, United Kingdom and New York, NY, USA: Cambridge University Press, 2013: pp. 659–740.

cc Trenberth, K.E., Dai, A., Rasmussen, R.M., & Parsons, D.B. The changing character of precipitation. *Bull Amer Meteorol Soc*. **84**(9), 1205–1217 (2003).

Trenberth, K.E., & Shea, D.J. Relationships between precipitation and surface temperature. *Geophysical Research Letters*. **32**(14) (2005), https://doi.org/10.1029/2005GL022760.

dd Data source: HadCRUT5.

ee See Figure 2.31 in: Hartmann, D.L., et al. Observations: Atmosphere and surface. In: Stocker, T.F., et al. (eds.). *Climate Change 2013: The Physical Science Basis Contribution of Working Group I to the Fifth Assessment Report of the Intergovernmental Panel on Climate*

Change. Cambridge, United Kingdom and New York, NY, USA: Cambridge University Press, 2013: pp. 159–254.

Chapter 23

a See the IPCC sixth assessment report, technical summary of working group 1—specifically Figure TS.15—and also the fifth assessment report summary for policymakers from working group 1—specifically Figure SPM.5. The relevant citations are:

Arias P.A., et al. Technical summary. In: Masson-Delmotte, V., et al. (eds.). *Climate Change 2021: The Physical Science Basis Contribution of Working Group I to the Sixth Assessment Report of the Intergovernmental Panel on Climate Change*. Cambridge, United Kingdom and New York, NY, USA: Cambridge University Press; 2021: pp. 33–144. (https://oceanexplorer.noaa.gov/facts/ocean-depth.html)

IPCC. Summary for Policymakers. In: Stocker T.F., et al. (eds.). *Climate Change 2013: The Physical Science Basis Contribution of Working Group I to the Fifth Assessment Report of the Intergovernmental Panel on Climate Change*. Cambridge, United Kingdom and New York, NY, USA: Cambridge University Press, 2013: pp. 1–30.

b Friedlingstein, P., et al. Global Carbon Budget 2020. *Earth System Science Data* **12**, 3269–3340, doi:10.5194/essd-12-3269-2020 (2020).

c Some examples of work on this subject are:

Cox, P.M., Betts, R.A., Jones, C.D., Spall, S.A., & Totterdell, I.J. Acceleration of global warming due to carbon-cycle feedbacks in a coupled climate model. *Nature* **408**, 184–187, doi:10.1038/35041539 (2000).

Varney, R.M., et al. A spatial emergent constraint on the sensitivity of soil carbon turnover to global warming. *Nature Communications* **11**, doi:10.1038/s41467-020-19208-8 (2020).

Watson, A.J., et al. Revised estimates of ocean-atmosphere CO_2 flux are consistent with ocean carbon inventory. *Nature Communications* **11**, doi:10.1038/s41467-020-18203-3 (2020).

d Rahmstorf, S., et al. Exceptional twentieth-century slowdown in Atlantic Ocean overturning circulation. *Nature Climate Change* **5**, 475–480, doi:10.1038/nclimate2554 (2015).

e See Figure 9.26 in: Fox-Kemper, B., et al., in *Climate Change 2021: The Physical Science Basis. Contribution of Working Group I to the Sixth Assessment Report of the Intergovernmental Panel on Climate Change* (Masson-Delmotte, V., et al., eds). Cambridge: Cambridge University Press, 2021: pp. 1211–1362.

f The temperature feedback is also known as the Planck feedback. The number 1.2°C that I quote, is based on the radiative forcing due to carbon dioxide doubling being 3.75W/m^2 and the Planck feedback parameter being about -3.2 $W/m^2/°C$ which is a value consistent across climate models—see Figure 9.43 in:

Flato, G., et al. in *Climate Change 2013: The Physical Science Basis. Contribution of Working Group I to the Fifth Assessment Report of the Intergovernmental Panel on Climate Change* (Stocker, T.F., et al., eds.). Cambridge: Cambridge University Press, 2013: Ch. 9, pp. 741–866.

g There is a nice description of the role of clouds in FAQ7.1, p. 594 of:

Boucher, O., et al. in *Climate Change 2013: The Physical Science Basis. Contribution of Working Group I to the Fifth Assessment Report of the Intergovernmental Panel on Climate Change* (Stocker, T.F., et al. eds.). Cambridge University Press, 2013: Ch. 7, pp. 571–658.

h   See previous endnote.

i   Senior, C.A., & Mitchell, J.F.B. The time-dependence of climate sensitivity. *Geophysical Research Letters* **27**, 2685–2688, doi:10.1029/2000gl011373 (2000).
Bloch-Johnson, J., et al. Climate sensitivity increases under higher CO_2 levels due to feedback temperature dependence. *Geophysical Research Letters* **48**, doi:10.1029/2020gl089074 (2021).

j   The high emissions are RCP8.5, mid RCP4.5, and low RCP2.6. The values of the feedback parameter are: 1.23 (mid), 0.74 (low), and 1.85(high) in Watts/metres^2/Kelvin. The values of the effective heat capacity are: 0.8E9 (mid), 0.4E9 (low), and 1.5E9(high) in Joules/metres^2/Kelvin. When the parameters are changing, they change linearly from their mid value to their low value during the period 2040 to 2060.

k   Andrews, T., et al. Accounting for Changing Temperature Patterns Increases Historical Estimates of Climate Sensitivity. *Geophysical Research Letters* **45**, 8490–8499, doi:10.1029/2018gl078887 (2018).

l   Varney, R.M., et al. A spatial emergent constraint on the sensitivity of soil carbon turnover to global warming. *Nature Communications* **11**, doi:10.1038/s41467-020-19208-8 (2020)

m   https://www.climate.gov/news-features/understanding-climate/climate-change-global-temperature

n   https://unfccc.int/process-and-meetings/the-paris-agreement/the-paris-agreement

o   This figure is based on RCP2.6. The value of the effective heat capacity is 0.8E9. Values of the feedback parameter (roughly equivalent to climate sensitivity) are: 1.23 (3˚C), 0.74 (5˚C), 1.85(2˚C), 1.48(2.5˚C), 0.925(4˚C), 0.6167(6˚C), 0.5286(7˚C), 0.4625(8˚C).

Chapter 24

a   Date: June 2021.

b   This range covers the 'very likely' range (i.e. 90% probability range) for the medium–to–high and high reference scenarios used by the IPCC in 2021. See: IPCC, 2021: Summary for Policymakers. In: Climate Change 2021: The Physical Science Basis. Contribution of Working Group I to the Sixth Assessment Report of the Intergovernmental Panel on Climate Change. Masson-Delmotte, V., et al. (eds.). *Climate Change 2021: The Physical Science Basis.* Cambridge University Press, Cambridge, United Kingdom and New York, NY, USA, pp. 3–32, doi:10.1017/9781009157896.001.

c   Nordhaus, W.D. Revisiting the social cost of carbon. *Proc. Natl. Acad. Sci. U. S. A.* **114**, 1518–1523, doi:10.1073/pnas.1609244114 (2017).
Weitzman, M.L. GHG Targets as insurance against catastrophic climate damages. *Journal of Public Economic Theory* **14**, 221–244, doi:10.1111/j.1467-9779.2011.01539.x (2012).

d   See, for instance: Neumann, J.E., et al. Climate damage functions for estimating the economic impacts of climate change in the United States. *Review of Environmental Economics and Policy* **14**, 25–43, doi:10.1093/reep/rez021 (2020).

e   See, for instance: Burke, M., Hsiang, S.M., & Miguel, E. Global non-linear effect of temperature on economic production. *Nature* **527**, 235–239, doi:10.1038/nature15725 (2015).

Chapter 25

a See data from the European Environment Agency: https://www.eea.europa.eu/ims/atmospheric-greenhouse-gas-concentrations, checked on 1/4/22.

b The IPCC sixth assessment report gave 3°C as their best estimate for climate sensitivity.

c Rather than looking at the total value of the future, the comparison was done in terms of a representative size for the global economy for each scenario. To be more specific, an equivalent size of the global economy was calculated for each scenario, such that if it remained fixed at this size for all future times, then we would value it today at the same level as each scenario. The difference in the two scenarios was then expressed as the difference in the size of these two equivalent, fixed futures.

d Source data: HadCRUT5

e Three simulations from the CMIP6 ensemble. Source data: CMIP6.

f Hasselmann, K. Stochastic climate models Part I. Theory. *Tellus* **28**, 473–485, doi:10.1111/j.2153-3490.1976.tb00696.x (1976).

g The red lines are for RCP8.5, the blue for RCP2.6. For further details of the model and how it is solved, see:

Calel, R., Chapman, S.C., Stainforth, D.A., & Watkins, N.W. Temperature variability implies greater economic damages from climate change. *Nature Communications* **11**, doi:10.1038/s41467-020-18797-8 (2020).

Note, however, that in this plot the effective heat capacity takes a value of 0.8E9 whereas the Calel et al. paper used a value of 1.0E9. This makes this plot consistent with those in Chapter 23 (Figure 23.2 and Figure 23.1 top, left) but would lead to higher global damages than those presented in Figure 25.4 which represents results from Calel et al.

h Calel, R., Chapman, S.C., Stainforth, D.A., & Watkins, N. W. Temperature variability implies greater economic damages from climate change. *Nature Communications* **11**, doi:10.1038/s41467-020-18797-8 (2020).

i The scenarios presented are RCP2.6 (blue), RCP4.5 (yellow) and RCP8.5 (red). These results and further details are in: Calel, R., Chapman, S.C., Stainforth, D.A., & Watkins, N.W. Temperature variability implies greater economic damages from climate change. *Nature Communications* **11**, doi:10.1038/s41467-020-18797-8 (2020).

j Calel, R., Chapman, S.C., Stainforth, D.A., & Watkins, N.W. Temperature variability implies greater economic damages from climate change. *Nature Communications* **11**, doi:10.1038/s41467-020-18797-8 (2020).

k Calel, R., & Stainforth, D.A. On the physics of three integrated assessment models. *Bull. Amer. Meteorol. Soc.* **98**, 1199–1216, doi:10.1175/bams-d-16-0034.1 (2017)

Chapter 26

a See: https://en.wikipedia.org/wiki/There_are_known_knowns

b See: https://en.wikipedia.org/wiki/Johari_window

c Winston S. Churchill, 11th November 1947: https://winstonchurchill.org/resources/quotes/the-worst-form-of-government/

d Tennant, J.P., & Ross-Hellauer, T. The limitations to our understanding of peer review. *Research Integrity and Peer Review* **5**, 6, doi:10.1186/s41073-020-00092-1 (2020).

e Zelinka, M.D., et al. Causes of higher climate sensitivity in CMIP6 models. *Geophysical Research Letters* **47**, doi:10.1029/2019gl085782 (2020).

f Hourdin, F. et al. The art and science of climate model tuning. *Bull. Amer. Meteorol. Soc.* **98**, 589–602, doi:10.1175/bams-d-15-00135.1 (2017).

Chapter 27

a http://news.bbc.co.uk/1/hi/sci/tech/7603257.stm

Index